The glaciations of Wales and adjacent areas

The glaciations of Wales and adjacent areas

Edited by

Colin A. Lewis & Andrew E. Richards

Logaston Press

LOGASTON PRESS
Little Logaston, Logaston,
Woonton, Almeley, Herefordshire HR3 6QH

First published by Logaston Press 2005
Copyright © text by author(s) of each chapter 2005
Copyright © illustrations as acknowledged 2005

All rights reserved. No part of this publication
may be reproduced, stored in a retrieval system,
or transmitted, in any form or by any means,
electronic, mechanical, photocopying, recording
or otherwise, without the prior permission,
in writing, of the publisher

ISBN 1 904396 36 4

Set in Times by Logaston Press
and printed in Great Britain by
Arrowsmith, Bristol

Front cover illustration: Llyn Cwm Llwch from the summit of Pen y Fan, Brecon Beacons. The lake is enclosed by a moraine formed when a small cirque glacier occupied the shaded area below the backwall of Cwm Llwch and extended as far as the outer moraine, little more than 11,000 years ago. Glaciation has left many imprints on the landscapes of Wales and adjacent areas, from features associated with small cirque glaciers as at Cwm Llwch, to those caused by major ice-sheets, as in North Wales and the Cheshire-Shropshire Plain.
(Photo: Colin A. Lewis)

Contents

	Acknowledgements	*vii*
	Preface	*ix*
	List of Figures and Tables	*xi*
1	**Introduction**	1
	by Colin A. Lewis, *Department of Geography, Rhodes University, Grahamstown, South Africa*	
2.	**Stratigraphy**	17
	by Andrew E. Richards, *School of Applied Sciences, Geography and Archaeology, University College Worcester*	
3	**North-west Wales**	27
	by Danny McCarroll, *Department of Geography, University of Wales, Swansea*	
4	**North-east Wales**	41
	by Geoffrey S. P. Thomas, *Department of Geography, University of Liverpool*	
5	**The Cheshire—Shropshire Plain**	59
	by Peter Worsley, *School of Geography, University of Oxford*	
6	**The lower Severn valley**	73
	by Darrel Maddy, *Department of Geography, University of Newcastle* and Simon G. Lewis, *Department of Geography, Queen Mary College, University of London*	
7	**West Wales**	85
	by James L. Etienne (*current address: Department of Earth Sciences, ETH-Zurich*), Michael J. Hambrey, Neil F. Glasser *and* Krister N. Jansson (*current address: Department of Physical Geography and Quaternary Geology, Stockholm University*), *Centre for Glaciology, University of Wales, Aberystwyth*	
8	**The upper Wye and Usk regions**	101
	by Colin A. Lewis, *Department of Geography, Rhodes University, Grahamstown, South Africa and* Geoffrey S. P. Thomas, *Department of Geography, University of Liverpool*	
9	**Herefordshire**	129
	by Andrew E. Richards, *School of Applied Sciences, Geography and Archaeology, University College Worcester*	
10	**South Wales**	145
	by D. Q. Bowen, *School of Earth, Ocean and Planetary Sciences, Cardiff University*	
11	**South-west England**	165
	by Stephan Harrison, *School of Geography, University of Oxford* and David H. Keen, *Institute of Archaeology and Antiquity, University of Birmingham*	
12	**The Irish Sea basin**	177
	by Jasper Knight, *Department of Geography, Exeter University*	
	References	189
	Index	217

Acknowledgements

The respective authors thank Professor D.Q. Bowen for help in the preparation of Chapter 1; Professor G.J. Williams for allowing redrawn versions of his maps to be included in Chapter 8 and for his comments on drafts of that chapter; Alun Rogers, Peter Brabham, John Catt, John G. Evans, Fred Phillips and Gerald Sykes for help with Chapter 10, and Eric Brown, Tom Cronin, Rhodes Fairbridge, T. Neville George, Wyndham B. Evans, J. Cedric Griffiths, Gifford Miller, Gerry Richmond, Alec Skempton and John Wehmiller for the benefit of wide-ranging discussions on the Gower exposures with the author of that chapter over many years. As regards Chapter 7 James Etienne acknowledges funding from Natural Environment Research Council (CASE) studentship NER/S/A/2000/03690 in association with the British Geological Survey, and Neil Glasser and Michael Hambrey acknowledge funding through the British Geological Survey (Natural Environment Research Council) Research Agreement GA/98E/14; Neil Glasser also acknowledges partial funding from the Quaternary Research Association; Krister Jansson acknowledges funding from the Royal Physiographic Society in Lund; Drs. Richard Waters and Jeremy Davies (British Geological Survey) are thanked for their informative discussion and contribution of ideas.

The Editors thank Professor R. Hepburn for comments on Chapters 1 and 8, the anonymous referees for comments on these and other chapters, Dr. G.S.P. Thomas for consolidating and Professor Peter Worsley for checking the the References. The senior editor thanks the British Council and Rhodes University for financial support and his sons, Brychan and Hywel, for logistical support that facilitated his work on his book.

Figure 7.4 is reproduced with the permission of Wiley Interscience. Chapter 6 is a contribution to IGCP 449 (Global correlation of Late Cenozoic fluvial deposits).

Finally, Andy johnson of Logaston Press is thanked for his unfailing encouragement throughout the production process.

Preface

The landscapes of Wales and adjacent areas have been profoundly influenced by glaciation. The rugged mountains of Snowdonia, for example, owe much of their beauty to the effects of glacial erosion. The lowlands of Herefordshire, with their gentle landscapes and almost garden-like appearance, owe much to the effects of glacial deposition, as ice sheets from mid-Wales melted and deposited rich fertile sediments on the underlying bedrock. On the western side of the Irish Sea, in County Wexford, the hummocky kettled topography of the Screen Hills results from deposition due to melting of an ice sheet that occupied the Irish Sea basin during the Late Quaternary. Further north, in County Wicklow, impressive gorges cut through rocky uplands, such as The Scalp (near Enniskerry) and The Glen of the Downs (near Delgany), are the remains of glacial melt-water channels. Similar channels exist on the eastern side of the Irish Sea, as near Talybont, between Aberystwyth and Machynlleth.

Much attention has been paid to the origins of the Welsh landscape and, especially, to its geomorphology. In 1960, for example, Eric Brown published *The relief and drainage of Wales. A study in geomorphological development*. A century earlier, in 1860, Andrew Ramsay published a booklet on *The old glaciers of Switzerland and North Wales*, in which he remarked on the significance of glacial erosion: 'all glaciers must deepen their beds by erosion'.

In 1970 a group of eleven Geographers combined, under the leadership of a young Welshman then resident in Ireland, to publish *The glaciations of Wales and adjoining regions*. 'This book depicts the known glaciations not only of Wales but also of the Severn valley, South West England and the south and east coasts of Ireland', proclaimed the dust-jacket blurb. Three and a half decades have passed since that book appeared and there have been many advances in the knowledge of glacial events. Consequently a new group of fifteen scientists, under the leadership of the same but older Welshman, now resident in South Africa, and of a much younger geomorphologist from Herefordshire, have combined to produce the present book. Unlike the previous text, little attention is paid to the south coast of Ireland, since that area is discussed in *The Quaternary history of Ireland*, edited by Kevin J. Edwards and William P. Warren in 1985.

The authors thank all who have made this book possible: those who encouraged their interests in landscapes and, in particular, in glacial geomorphology; the funding agencies that have supported their research; the cartographers who prepared the final Figures; the Quaternary Research Association; and, of course, Logaston Press for publishing this text. The editors especially thank all the contributors for preparing their scripts and for their tolerance during the arduous editorial processes; the referees who read and commented on each chapter; and Mrs D. Brody, Miss B. Tweedie and Mrs J. Naidoo of Rhodes University for immense help with cartographic and secretarial matters.

The preface to the 1970 book stated that: 'Many details of the glaciations of Wales are still uncertain, whilst others are probably completely unknown.' That is still the case! In 1970 the editor wrote that:

'we have tried to present a synopsis of the knowledge available to us ... in the hope that it will encourage, and possibly guide, further research in the years ahead.' The aims of the present book are exactly the same.

Three of the authors of the 1970 book have also written for the present text (Bowen, Lewis, Worsley). They remember with affection and gratitude those who contributed to the 1970 book but who have passed to the life eternal, including Clifford Embleton, Edward Watson and Francis Synge. They also remember Frank [G. F.] Mitchell, Fred [F. W.] Shotton and many others with whom they had stimulating discussions about the Quaternary history of the English Midlands, Wales, South West England and the Irish Sea Basin. They remain fascinated with the study of the glacial history of Wales and its borderlands, the many remaining uncertainties of that history and the need to integrate the evolution of the area into a global overview. Like the other contributors to the present book, they hope that it stimulates a new generation of geoscientists and guides their researches in the years ahead.

Colin A. Lewis (Rhodes University) and Andrew E. Richards (University College, Worcester) (Editors), St David's Day, 2005.

List of Figures and Tables

Figures *page*

1.1 Milankovitch radiation curve for latitude 65° North — 5
1.2 The Global Conveyor or thermohaline circulation — 13
1.3 Heinrich Events and Dansgaard-Oeschger cycles — 14
2.1 Orbital and isotopic variations and climate stratigraphy — 18
2.2 Glacial limits in Wales and adjoining areas — 19
3.1 North-west Wales — 29
3.2 Ice movements in Snowdonia — 31
3.3 Ice directions in north-west Wales — 33
3.4 Glacial deposits at Dinas Dinlle — 34
3.5 Flow and uncoupling of Welsh and Irish Sea ice sheets in north west Wales — 35
3.6 Glacial deposits at Aberdaron — 37
3.7 Glacial sediments and landforms at Glanllynnau, near Cricceith — 39
4.1 Glacial features in north-east Wales — 42
4.2 The Alyn-Wheeler ice-marginal zone — 46
4.3 Landforms and sediments at Rhosesmor Quarry — 48
4.4 The Mold-Caergwyle ice-margin sandur system and the Wrexham fan-delta — 49
4.5 Sediment assemblages and ice margin conditions near Hope Quarry — 50
4.6 Cross-sections through the Wrexham fan-delta — 51
4.7 Sections in the Wrexham fan-delta — 53
4.8 Welsh and Irish Sea diamicts near Shrewsbury — 55
4.9 The Shrewsbury-Welshpool area — 56
4.10 Features associated with decoupled ice sheets between Welshpool and Shrewsbury — 57
5.1 Last Glacial Maximum ice limit in the Cheshire-Shropshire lowlands — 59
5.2 The Four Ashes Quarry seen in section as it was in 1968 — 61
5.3 Four Ashes Quarry, 1968 — 62
5.4 Oakwood Quarry, Chelford in 1978 — 63
5.5 Thrusted till in Wood Lane Quarry, Ellesmere — 65
5.6 The Shrewsbury Formation at Mousecroft Lane Quarry, 1969 — 67
5.7 The Stockport Formation overlain by the Shrewsbury Formation, Mousecroft Lane Quarry — 67
5.8 Ice retreat landforms around Newport, Shropshire, and adjacent Staffordshire — 68
5.9 Relationships between glacial and other deposits and the River Severn terraces — 70
5.10 Glacial drainage channels in the Bickerton Hills area — 71
6.1 Pre-Marine Oxygen Isotope Stage 12 drainage of the lower Severn river basin — 74
6.2 The lower Severn valley — 76
6.3 Longitudinal profiles of terraces in the Severn river valley — 77
6.4 Cold climate braided river sediments overlie temperate river sediments at Bushley Green — 78
6.5 Geological structure of the lower Severn river basin — 82
7.1 Glacial and fluvial landforms in west Wales — 86
7.2 Glacial erosional features in west Wales — 89
7.3 Glacial Lake Teifi and the Cippyn channel — 93
7.4 Valley-fill successions in the lower Teifi valley — rear cover
7.5 Deformed deposits at Traeth-y-Mwnt and at Gilfach-yr-Halen — 94
7.6 Ice-marginal deposits at Tonfannau, Gwynydd — 96
7.7 Periglacial phenomena in west Wales — 97
8.1 Terminal moraines in the Wye valley above Hereford — 102
8.2 Meltwater channels and other glacial features in the Talgarth-Hay region — 104
8.3 The remains of pingos near Llanidloes — 108
8.4 Sediment-landform assemblages in the middle and lower Usk valley — 109
8.5 Holocene river and Devensian fluvioglacial terraces near The Bryn, Usk valley — 112

8.6	Glacial and fluvioglacial features between Talybont and Abergavenny	113
8.7	Moraine ridges, sandur and river terraces near Gilwern	115
8.8	Moraine ridges, meltwater channels and sandur between Llanellen and Usk	117
8.9	Devensian maximum ice limit between Raglan and Llantilio Crossenny	119
8.10	Ice-marginal positions in the upper Usk and its right-bank tributary valleys	120
8.11	The Tyle-crwn meltwater channel	121
8.12	Ice-contact ridge south of Tregunter Farm, between Llanfillo and Bronllys	122
8.13	Kame moraine at Heol Senni	123
8.14	Cirque moraines at Llyn Cwm Llwch, Brecon Beacons	125
8.15	Reconstructed cirque glacier at Fan Hir, Black Mountain	126
9.1	River system development and glacial diversion in Herefordshire	130
9.2	Distribution of the Risbury Formation	132
9.3	Ice marginal environments associated with Glacial Lakes Bromyard and Humber	133
9.4	Glaciofluvial gravels at Franklands Gate	134
9.5	Deformed glaciolacustrine silts and clays, Burghill	135
9.6	Late Devensian morainic forms in the Hereford Basin	137
9.7	Fluvioglacial sediments at Stretford and at Lucton	140
9.8	The Late Devensian ice sheet in Herefordshire	141
9.9	Schematic representation of moraine formation in the Hereford Basin	143
10.1	South Wales: end-moraines and other locations	145
10.2	Glacial and glaciofluvial deposits in South Wales and lithostratigraphic Formations	146
10.3	Ice movement in South Wales	147
10.4	The Pennard Formation in south Gower	148
10.5	Head deposits near Gilman Point, Carmarthen Bay	149
10.6	The fossiliferous *Patella* beach and colluvial silts at Hunts Bay, Gower	150
10.7	Minchin Hole Cave, Gower, at low tide	150
10.8	D-alloisoleucine/L-isoleucine ratios in South Wales	152
10.9	Lacustrine clays of Lake Teifi, near Llechryd	154
10.10	The Pennard Formation from St Bride's Bay, Pembrokeshire, to east Gower	156
10.11	Signal of ice ages and interglacials from a site in the East Equatorial Pacific	157
10.12	Marine and terrestrial gastropod succession from Minchin Hole Cave, Gower	163
10.13	Terrestrial gastropod succession inside Minchin Hole Cave, Gower	164
11.1	South-west England and the Scilly Isles: location map	166
11.2	The Scilly Isles	171
11.3	The Punchbowl glacial cirque on Exmoor	173
12.1	Bathymetry of the Irish Sea Basin and location map	178
12.2	Glacial sediments at Dinas Dinlle, Lleyn peninsula, north-west Wales	186
12.3	Glacial sediments at Trwyn Maen Dylan, Lleyn peninsula, north-west Wales	187

Tables

2.1	Middle to Late Quaternary Cold and Warm Stages	20
2.2	Classification of till facies (types) and their potential for stratigraphic study	22
2.3	Glacial formations identified in each chapter of this book	24
4.1	Quaternary lithostratigraphic succession in north east Wales	43
6.1	Lithostratigraphic nomenclature for the lower Severn valley	75
7.1	Principal Quaternary surficial deposits in west Wales and their interpretation	90
9.1	Middle to Late Pleistocene stratigraphy of Herefordshire	131
10.1	Pleistocene Lithostratigraphical Formations in South Wales	146
10.2	Correlation of Pleistocene beds at Minchin Hole and Bacon Hole Caves, Gower	151
10.3	Correlation of the sequence of events in South Wales	158
12.1	Glacial marine and terrestrial ice models along the Irish Sea Basin coasts of Wales	182
12.2	Interpretations of Late Devensian sites along the Irish Sea coasts of Wales	185

1 Introduction

by Colin A. Lewis

The glacial theory
On the sixteenth of October, 1841, William Buckland wrote in the visitors' book of the Goat Hotel, Beddgelert:

> Notice to Geologists.- At Pont-aber-glass-llyn, 100 yards below the bridge, on the right bank of the river, and 20 feet above the road, see a good example of the furrows, flutings, and striae on rounded and polished surfaces of rock, which Agassiz refers to the action of glaciers. See many similar effects on the left, or south-west, side of the pass of Llanberis (Davies, 1969).

Buckland (1784–1856) was, in 1841, Rector of the living of Stoke Charity in Hampshire and a Canon of Christ Church Cathedral, Oxford, having resigned his Readership in Geology at Oxford University in 1825 (Hunt, 1886). In the late 1830s he came under the influence of a young Swiss scientist named Louis Agassiz (1807–73).

In 1836 Agassiz had accompanied Ignace Venetz (an engineer), and Jean de Charpentier (a graduate of the Freiberg Mining Academy and director of the salt mines at Bex, in Switzerland), on a tour of the area around Bex and the Valois to examine evidence for the former existence of larger glaciers than those that then existed in the region. Agassiz accepted the ideas of his companions: that the glaciers had formerly been larger; and thereby accepted the similar hypotheses of Bernard Kuhn (a Swiss cleric, 1787); James Hutton the Edinburgh geomorphologist (1795); John Playfair the geologist (1802); and Jean-Pierre Perraudin, a professional mountaineer from Lourtier in southern Switzerland (1815). Agassiz, however, extended the existing hypotheses by stating, in his paper to the *Societè Helvètique* in July 1837, '... that the northern hemisphere from the North Pole down to the latitude of the Mediterranean and Caspian Seas had until recently been shrouded beneath a thick mass of glacier ice'. In other words, Agassiz conceived of '... a full-scale ice-sheet glaciation of continental proportions' (Davies, 1969).

Buckland had been aware of Agassiz' non-glacial researches since at least the early 1830s, when the two men corresponded with each other. In 1834, during Agassiz' first visit to Britain, he and Buckland became friends. Thus it was not surprising that in 1838 Buckland and his wife travelled to Switzerland to stay with Agassiz and inspect the evidence for his glacial theory. The evidence: of erratics, and of polished and striated rocks analogous to those at the margins of existing glaciers; did not initially impress Buckland. [Erratics are sediments that differ from the underlying bedrock. Examples of striations in Wales are shown on Fig. 7.2E.] After taking their leave of Agassiz the couple toured the Oberland and looked at real glaciers, and Buckland became a convert to the glacial theory. Furthermore, Buckland realised that similar features to those accepted as evidence of former glaciation in Switzerland also existed in Britain. He therefore realised that Britain had formerly been glaciated, although it was Agassiz, in a paper at the 1840 meeting of the British Association for the Advancement of Science, who formally made this statement (Davies, 1969).

Initially British scientists were sceptical of the new theory. When Agassiz read a paper to the Geological Society of London in the autumn of 1840 he was asked whether he really thought that Lake Geneva (which is very deep) had formerly been occupied by ice some 3,000 feet thick? The depth of that lake had been explained as due to catastrophism, to some earthquake or other event that had cleft the rocks, forming a deep trench that the lake subsequently occupied. Catastrophism had theological implications. William Shakespeare, for example, had written in the seventeenth century about the perfect sphere, and it was understood in clerical circles that God had made the world without blemish: a perfect sphere. The sins of humans, or so it was believed, had caused the perfect sphere to lose its symmetry and to develop hills and mountains, valleys and chasms, even deep hollows such as that occupied by Lake Geneva.

Agassiz replied that 3,000 feet '... must be regarded as a minimum thickness for the ice in the area' (quoted from Davies, 1969. Quotes from this source are shown hereafter as Davies, 1969). His questioner responded that the glacial theory was the 'climax of absurdity in geological opinions'. Another sceptic asked whether, if scratches were explicable as the work of former glaciers, all scratches should be explained as caused similarly?

Buckland, like Agassiz, was not deterred by scepticism. In 1841 he undertook field work in Snowdonia, finding ample evidence of former glaciation. On the coastlands around Snowdonia, as at Dinas Dinlle near Caernarfon, Buckland noted deposits that had been 'disturbed'. In the following year Charles Darwin examined parts of North Wales and became thoroughly converted to the glacial theory. 'The valley[s] about here', he wrote, '... must once have been covered by at least 800 or 1,000 feet in thickness of solid ice!' (Davies, 1969).

Glacial submergence
During the 1840 meeting of the British Association Agassiz had suggested that the melting of glaciers that had formerly covered large areas of the globe was responsible for rises in sea levels. Consequently, large areas were inundated and '... marine currents had imported boulders and other debris and deposited them widely over the continental surfaces' (Davies, 1969). The concept of a late glacial submergence soon became part of the general understanding of the glacial theory and was used to explain the origin of sediments that we now consider glaciofluvial. Buckland believed that the 'disturbed' deposits at Dinas Dinlle (Chapter Three), and elsewhere, had been caused by ice-bergs floating in the glacial submergence that had thrust into superficial sediments (*drift*) as they grounded, thereby disturbing them.

In 1831 Trimmer had described '... the diluvial deposits of Caernarvonshire' and noted that sands with marine shells exist at an altitude of 396 m at the Alexandra Slate Quarry on Moel Tryfan on the Lleyn peninsula(520560; Davies,1969). Buckland (1841) explained these, and other shell-bearing deposits in the Vale of Clwyd, as due to deposition under the waters of the glacial submergence.

The concept of 'glacial submergence', the submergence of large areas by rising sea levels due to glacier melting, was long lasting. In the 1850s Sir Andrew Ramsay (1814–91) claimed, on the evidence of deposits on Carnedd Dafydd and Carnedd Llywelyn, that North Wales had been inundated to a depth of 2,300 feet. (Ramsay became Director General of the Geological Survey in 1871). In 1863, when Sir Charles Lyell (1779–1875) visited the Moel Tryfan deposits, he believed that 'These shells show that Snowdon and all the highest hills which are in the neighbourhood of Moel Tryfan were mere islands in the sea at a comparatively late period' (Davies,1969). Even in 1910 Edward Hull, a former Director of the Geological Survey of Ireland, believed that the Moel Tryfan deposits were evidence of marine transgression.

The deposits at Moel Tryfan are now interpreted (Chapter Three) as sediments dredged from the floor of the Irish Sea basin by ice moving southwards down that basin and impinging on the North Wales uplands. The shells have been radiocarbon dated to about 33,700 years BP, indicating 'that the Lleyn

Peninsula was glaciated, at least in part, by extraneous ice after this date' (Foster,1970). They have amino acid ratios that suggest that they predate the Last Glacial Maximum (Bowen *et al.*, 2002).

Chronology

By the 1850s a number of scientists had suggested chronologies of the British Pleistocene. Among the first of these was Ramsay's (1852) division of the glacial period in North Wales into three stages:

> i) the advance of valley glaciers from the mountains, forming striations and roches moutonnées on the valley floors.
>
> ii) marine submergence as glaciers melted back into the uplands and sea level rose, as evidenced by the Moel Tryfan and other deposits. During this submergence Ramsay postulated that 'marine currents swept a stream of ice-bergs south-westwards across Anglesey, and the frequent grounding of these ice-masses caused the island's rocks to become heavily striated in a north-east to south-west direction. Further to the east a few rogue icebergs left the main stream and floated into the Welsh valleys to deposit Scottish and other northern erratics' (Davies, 1969).
>
> iii) readvance of valley glaciers.

Other chronologies were produced for other areas of Britain. By the 1870s a tripartite stratigraphic division of the British drifts was generally accepted. At the base of the succession was the Lower Boulder Clay, supposedly deposited by an ice-sheet or by large valley glaciers; on top of this lay the Middle Sands and Gravels, which were thought to date to the supposed glacial submergence; finally, at the top of the sequence lay the Upper Boulder Clay, left by relatively small valley glaciers. Although it was believed that southern England had lain beyond the limits of glaciation nobody seems to have wondered how North Wales could be drowned by high sea levels of up to 2,300 feet while there was no such evidence of submergence in the former area!

The tripartite glacial chronology dominated thinking in the British Isles until the twentieth century, when new ideas resulting from the researches of Penck and Brückner (1909) in the Alps and surrounding areas suggested that the glacial sequence may have been more complicated.

Glacial erosion

Buckland, Darwin and many other scientists in the early days of the glacial theory concentrated on the recognition of striae and glaciated rock faces in order to establish that glaciation had formerly taken place, often overlooking the magnificent cirques, moraines, glacial breaches and other spectacular features that are the unmistakable evidence of the former work of ice. In 1859 Ramsay published an essay on 'The Old Glaciers of Switzerland and North Wales', which was published as a booklet in 1860. Ramsay stated '... that all glaciers must deepen their beds by erosion' and that they excavated rock basins. Over twenty years earlier an Irish engineer, Robert Mallet (1838), had claimed that '... the bed of a glacier is in continual process of degradation, or deepening by the resistless passage of these vast masses of ice and rocks over it'. It was Ramsay's booklet, however, that drew widespread attention to glacial erosion.

In 1860 Edward Hull suggested that many of the water-filled depressions in the English Lake District had been formed 'by the scooping action of glacier ice' and acknowledged his indebtedness to Ramsay for introducing him to the concept of glacial erosion. Nevertheless there was opposition to Ramsay's erosion concept.

In 1870 Sir Roderick Murchison (1792–1871), Director General of the Geological Survey and President of the Royal Geographical Society, wrote: '... where in any icy tract is there evidence that any glacier has by its advance excavated a single foot of solid rock? In their advance, glaciers striate and polish, but never excavate rocks'. Murchison's opposition was too late to elicit much support and, six years later when Judd questioned whether rock basins could have been caused by glacial erosion, many geologists wrote in support of the concept.

Subsequently a handful of geologists, led by Bonney and Garwood, argued that rivers have much greater erosive potential than glaciers and maintained that a covering of glacier ice actually protects the sub-glacial topography from experiencing as much erosion as it would have been subjected to if exposed to fluvial processes. Although this protectionist school was later scorned its proponents were not entirely incorrect, for *cold-based ice* (in which temperatures are below freezing from the surface to the base of the ice, Paterson, 2001) probably does have protective properties.

Wales played a major role in the development of the glacial erosion hypothesis, for it was research in the Principality, and especially in North Wales, that introduced the concept to the general scientific community and integrated it into the glacial theory.

Demise of glacial submergence

In 1862 Ramsay reconsidered his idea of a decade earlier: that ice-bergs had swept across Anglesey and, when grounding, striated the island's rocks. He noted that striations exist on the floors of deep hollows as well as elsewhere and now wrote that: 'An ice-berg that could float over the margin of a deep hollow would not touch the deeper recesses of the bottom'. In other words, he realised that a medium other than floating ice must have been responsible for the initiation of many striations. This medium, he now believed, had been '... sheets of true glacier-ice in motion, which moulded the whole surface of the country, and in favourable places scooped out depressions that subsequently became lakes.'

Ramsay was not the only, nor possibly the first, scientist to reject the theory of glacial submergence. Robert Jamieson, the Edinburgh scholar, argued in 1862 that 'striations, glacial polish, and roches moutonnées ... could only be the work of land-ice such as exists today in Greenland and Antarctica' (Davies, 1969). Geike (1863), Croll (1875) and other distinguished scientists also maintained that it was land-ice, and not glacial submergence, that had been responsible for the landforms and sediments that were so widespread in highland Britain and that were increasingly recognised as glacial. When Croll suggested in 1870 that shelly till in eastern Scotland, Orkney and Shetland was material dredged from the floor of the North Sea by glacier ice flowing westward from Scandinavia, the problem of shelly drifts was essentially resolved. Nevertheless Ramsay continued, into the 1880s, to believe that the shelly drifts of Moel Tryfan were due to glacial submergence. Others, however, were less blinkered and were increasingly aware of the findings that were beginning to be made in the Alps and in the upper reaches of the Danube valley and the valleys of its right bank tributaries.

Alpine glacial chronologies

In 1847 Collomb recognised evidence for two former glaciations of the Vosges Mountains, north of Switzerland (Bowen, 1978), then in 1856 Morlot discovered evidence of two former glaciations of Switzerland (Bowen, 1978). Thereafter increasing interest focussed on Switzerland and its surrounds as scientists sought to unravel the glacial chronology of that region. Their work peaked in 1909, when Penck and Brückner published *Die Alpen im Eiszeitalter*. In this book the authors presented evidence for what they argued were four different glaciations of the Alps, the main 'evidence' for which, they suggested, was a series of outwash terraces ('schotter') that could be traced upstream to end moraines in valleys of rivers south of Munich.

The basis for Penck and Brückner's argument came from mapping in Bavaria, especially in the valleys of the rivers Iller, Günz, Mindel and Lech, that Penck had first published in 1885. Subsequently mapping was extended to the valley of the River Würm and to other areas. This mapping showed that four outwash terraces could be traced up-valley to merge with end moraines. Each terrace and its associated end moraine, Penck and Brückner thought, evidenced one glaciation. The earliest glaciation, they reasoned, was evidenced by the outermost end moraine and the uppermost outwash terrace, which was associated with the outermost end moraine. Subsequent glaciations were evidenced by consecutively lower outwash terraces and by moraines. In other words their scheme was based mainly on morphology: the existence of end moraines and of terraces that could be traced back to those moraines.

Penck and Brückner's model of Alpine glaciations was essentially based on *morphostratigraphy*, which is the classification of bodies of sediment mainly on the basis of their surface morphology. They also examined the nature of the terrace deposits, concluding that the oldest (and highest) terraces were more weathered than the others. They maintained that the morphological evidence indicated that there had been four glaciations of the Alps, and they named these glaciations after river valleys in which they had studied the glacial and fluvioglacial landforms. The oldest glaciation, they thought, was the Günz, succeeded in turn by the Mindel, Riss and Würm. They also argued that the dissection of outwash plains down stream of end moraines, which transformed those outwash plains into outwash terraces, took place during interglacials. Their scheme therefore supposed that deposition occurred during glacials and erosion during interglacials and failed to consider that both processes could exist within glacials and/or interglacials.

The extent of weathering of terrace deposits led Penck and Brückner to suggest that the interglacial between the Mindel glaciation and that of the Riss was by far the longest of all interglacials that they recognised in Alpine Europe. They also suggested that the Riss glaciation lasted longer than the Würm because more sediments accumulated during the former glaciation. In some cases organic remains, that appear to be of interglacial character, are mixed with colluvial and scree deposits (such as the *Hottinger Breccia* near Innsbruck) and rest on terraces. These organic remains appeared to support the concept that outwash deposition was glacial and that erosion was interglacial.

In 1924 the German climatologist, Wladimir Köppen, and his son-in-law, Alfred Wegener (who introduced the concept of continental drift), published a book entitled *Die Klimate der Geologischen Vorzeit*. In it they included a graph by a great Serbian mathematician, Milutin Milankovitch, that showed how the intensity of summer sunlight (solar radiation) varied over the past 600,000 years (Fig. 1.1). Milankovitch correlated some of the low points on the graph with the four glaciations of the Alps as identified by Penck and Brückner. There was no geological proof for this correlation, but it became widely believed. Four years later, in 1930, Milankovitch published *Mathematical Climatology and the Astronomical Theory of Climatic Changes*. In this book he showed how the amount of solar radiation reaching the surface of the earth is influenced by the tilt of the earth on its axis, which takes place in a regular 41,000 year oscillation, and by the 22,000 year oscillation of the distance between the earth and

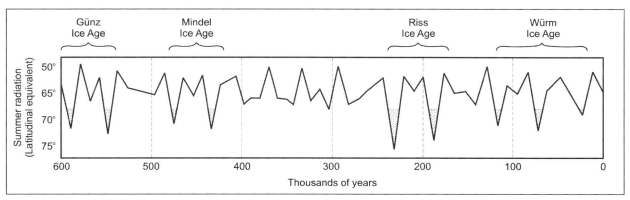

Fig. 1.1 Milankovitch's radiation curve for latitude 65°N was used by Köppen and Wegner to 'date' the classic glaciations of Alpine Europe. The radiation received 400,000 years ago at 65°N, for example, was equivalent to that now received at 67°N. Low points on the radiation curve were 'correlated' with glaciations. (Redrawn and simplified after Köppen and Wegener, 1924)

the sun, as described in Chapter Two. Milankovitch thereby provided an astronomical explanation for the existence of glacials and interglacials and is honoured scientifically by the almost universal recognition of *Milankovitch cycles*.

The Alpine model, as Penck and Brückner's classification came to be called, was extended by the work of Eberl (1930) and others, who believed that there had been at least one pre-Günz glaciation: the Donau. The Günz, they suggested, had been divided into substages, as had some of the other glaciations. Eberl (1930), like Penck and Brückner, mapped the morphology of landforms, but he also examined the *lithostratigraphy* of terrace gravels. *Lithostratigraphy* is the organisation of strata into units based on their lithological characteristics, as discussed in Chapter Two.

By the beginning of the Second World War it was widely believed that the four or five fold model of Alpine glaciations must have been repeated elsewhere. There was little, if any, appreciation that the model itself, which was by then an established classic, might be seriously flawed and might even be incorrect and misleading.

Bowen (1978) wrote that 'Pleistocene time, as conceived by Penck and Bruckner [*sic*] ... is represented in the type area by glacial and fluvioglacial (outwash) deposits that only represent part of that time. A good deal of Pleistocene time is represented by erosional breaks in the stratigraphic record. ... interglacial time is not represented by sediments, but merely by inferred erosion during which the outwash schotter were dissected into terraces. ... not only do the sediments not represent the full time-span ... but the unconformities between each successive terrace probably conceal events lost to the record ... As such ... because it is primarily a morphostratigraphical model, it is inherently deficient.'

Wales and the classic Alpine glacial chronology
By the 1880s it was widely accepted that large areas of Wales, and not just North Wales, had been glaciated. In 1883, for example, Edgeworth David published 'On the evidence of glacial action in south Brecknockshire and east Glamorganshire'. David showed, using erratic and other evidence, that ice from the Brecon Beacons/Fforest Fawr uplands had passed down the valleys of the adjacent South Wales coalfield to debouch onto the Vale of Glamorgan around Cardiff. Twenty years later, in 1903/4, Howard noted the distribution of erratic material and of striae and concluded, correctly, that ice had flowed from northerly sources (which were the plateaux of Pumlumon and its southerly extensions) down the valleys of the Wye and Usk. Howard and David had written in terms of one period of glaciation, but after the publication in 1909 of Penck and Brückner's findings in the Alps some scientists presented evidence that indicated, at least to them, that there had been four periods of glaciation in Wales (the same number as in the classic Alpine model).

In 1925 Pocock suggested that there had been two major glaciations of at least parts of Wales, as well as 'two limited and recent glaciations'. 'It seems probable that these last correspond with different stages of the Würm period, the more general glaciation with the Riss period and the maximum with the Mindel period'. He thought that these were evidenced by terraces and drift deposits in the Welsh borderlands. The basis for Pocock's dating of the various deposits was flimsy in the extreme: just because there were three or four terraces in the Wye or other valleys in Wales, and a similar number identified by Penck and Brückner in valleys leading north from the Alps, did not mean that they were of the same age. Pocock's 'dating' was therefore based on supposition rather than scientific proof.

The work of a number of Geological Survey Officers in Wales, including Jehu, Strahan, Cantrill, Thomas, Dixon and O. T. Jones, was of greater scientific value than that of those who strove to fit Welsh evidence into the (flawed) Alpine model. They mapped deposits and phenomena carefully throughout much of Wales. Another geological surveyor, albeit of the Geological Survey, Ireland, W. B. Wright (1914), formalised the division of the glacial deposits of these islands into an Older Drift and a Newer

INTRODUCTION

Drift. In other words, careful research in Wales and adjacent areas (including Ireland) had, by 1914, produced evidence for an older and a newer glaciation rather than for four glaciations as proposed by the Alpine model.

The delineation of glacial limits in and adjacent to Wales
A major advance in glacial studies came in 1929, when Charlesworth delineated 'The South Wales endmoraine', which he thought was the outermost limit of Welsh ice during Newer Drift times (*vide* Chapters Seven and Ten). Further delineations of the supposed limits of Welsh and Irish Sea ice were made subsequently, especially in the 1950s and thereafter (Bowen, 1981; Chapter Two). These were essentially based on a combination of morphological and lithological mapping. Mitchell (1960) suggested that there had been three glaciations of Wales and adjacent regions, although his map of the outermost limits of the most recent ice-sheet suggested that west Wales, from the uplands of the Lleyn peninsula south to Gower (apart from the fringes of Swansea Bay) lay beyond the ice-margin.

Detailed studies of glacial deposits and supposed glacial limits in various limited parts of Wales and its borderlands also appeared during the half-century following Charlesworth's seminal 1929 paper. Dwerryhouse and Miller (1930), for example, recognised a depositional feature in the northern area of the Hereford basin that they termed *The Kington-Orleton kettle moraine*. This feature is discussed in Chapter Nine. Elsewhere, as in Snowdonia, emphasis was placed on the identification of cirque moraines, as by Seddon (1957). Further south, in the upper Usk valley, moraines were linked to readvance and retreat stages of the last glacier to occupy that valley (Ellis-Gruffydd, 1977, *vide* Chapter Eight). Nevertheless new directions in Quaternary research were taking place in Wales, as elsewhere, and were to become increasingly influential.

Palynological studies
One of the most important forms of Quaternary research used in Wales from the late 1930s onwards has been *palynology*. This is the study of pollen and other spores, which had been pioneered in Scandinavia by Von Post in the early years of the twentieth century. Pollen grains from different plants have characteristics that enable scientists to identify the plants from which the grains are derived: oak (*Quercus*), elm (*Ulmus*) and so on. By identifying and counting the number of grains from each species, layer by layer stratigraphically, it is possible to gain an idea of what past vegetation used to be like. This enables scientists to reconstruct palaeoenvironments and to divide the sequential pollen record into zones of similarity, known as *pollen zones*.

In 1934 Knud Jessen, Professor of Botany at the University of Copenhagen, visited Ireland at the invitation of the Committee for Quaternary Research in that country in order to study pollen deposits. He continued to visit Ireland up to and including 1949 and worked there in collaboration with two Irish scientists: Anthony Farrington and Frank [G. F.] Mitchell (Lewis, 1984). The first paper resulting from these visits was published in 1938 and was written jointly by Jessen and Farrington, on 'The bogs at Ballybetagh, near Dublin, with remarks on late-glacial conditions in Ireland'. Further papers followed and that of 1949, followed by Mitchell's paper of 1956, established the pollen zonation of the Late Glacial and Post Glacial in Ireland.

Godwin and Mitchell pioneered pollen analysis in Wales with their 1938 paper on the stratigraphy and development of two raised bogs near Tregaron. Subsequently Godwin published two more important papers based on the analysis of pollen from various parts of Wales (1938b (with Newton), 1955). Professor Sir Harry Godwin (1901–85) was the founder of the Godwin Institute for Quaternary Research at the University of Cambridge, at which fundamental research into Quaternary history, and especially the dating of Quaternary events, has been undertaken.

After the end of the Second World War, Seddon (1957, 1962) published on the palynology of Late Glacial deposits in some of the cirque basins in Snowdonia, Bartley on a site in Radnorshire (1960), Trotman (1963) on evidence from parts of South Wales, P. D. Moore (*e.g.* 1966, 1968, 1970, 1972, 1978) on the vegetational history of mid-Wales, (in which he shed much light on Late Quaternary environments), J. J. Moore (1970) on the Late Glacial pollen sequence from Mynydd Illtyd, on the northern margins of the Brecon Beacons (a site that was later reworked by Walker, 1980, 1982), and Crabtree (1972) on Cors Geuallt in Snowdonia. These and other studies helped to show the complexity of Late Quaternary environmental changes and indicated the dynamic and ever-changing nature of climate. Some of them demonstrated that interstadial conditions (known as the Bølling–Allerød in northern mainland Europe; *e.g.* Bos *et al.*, 2001) existed after ice-sheets melted and before renewed climatic deterioration caused cirque glaciation to take place. Many palynological studies have been produced in recent decades, such as that by Robertson (1988) on the Brecon Beacons National Park region that is discussed in Chapter Eight.

The study of Coleoptera

Coleoptera are beetles and weevils. Beetles are very sensitive to climatic changes and different species of beetles exist at present under specific environmental conditions that may be related to temperatures and to precipitation. By identifying the remains of beetles in particular stratigraphic layers in Quaternary sediments, and then establishing the mutual climatic range of the beetles in individual layers, it is possible to establish the climatic conditions that existed when those animals lived. The *mutual climatic range* is that portion of the temperature range under which all the beetles in a particular stratum could have lived, and the portion of the precipitation range that is common to all the species in that stratum. Coleopteran remains have been discovered at a number of sites in North Wales and the Borderlands (*e.g.* Coope, 1972 (with Brophy), 1977: Chapter Three), Mathon in Herefordshire (Coope *et al.*, 2002; Chapter Nine), Llanilid in Glamorgan (Walker *et al.*, 2003; Chapter Ten) and from the Midlands (*e.g.* Coope and Sands, 1966; Morgan, 1973; Chapter Five).

Subdivision of the glacial deposits

The lithology of many of the superficial deposits of Wales and adjoining regions was mapped, in the late nineteenth and early twentieth centuries, by such Surveyors of the Geological Survey as Strahan, Cantrill, Thomas, Dixon, O. T. Jones and Jehu. In other words, they studied the physical and chemical characteristics of the rocks and finer sediments that are incorporated within superficial deposits. In the Wicklow Mountains, on the western side of the Irish Sea, glacial and associated deposits were also examined lithologically by a geologist of the Geological Survey of Ireland: Anthony Farrington (1893–1973), who was later employed by the Royal Irish Academy. In a series of papers Farrington (1934, 1942, 1944, 1949, 1957, 1966) established the lithostratigraphy of those deposits (Davies, 1963).

In 1973 a special report of the Geological Society of London attempted to correlate different regions with a standard classification based on inferred climate change (*climatostratigraphy*) in East Anglia (Mitchell *et al.*, 1973). The second edition of this report adopted a formal *lithostratigraphical* classification (Bowen, 1999) which, in Wales and the Borderland, closely followed an earlier systematic review of the stratigraphy of Pleistocene deposits (Bowen, 1973, 1974). Chapter Ten, with its subdivision of superficial deposits in south Wales into lithostratigraphic formations, bodies of sediment that are mappable and that may be subdivided into members and beds, exemplifies the use of a formal Quaternary lithostratigraphy.

Periglacial studies

Deposits of angular stony rubble held in a finer matrix, which were sometimes referred to as 'rubble drift' or *head* (de la Beche, 1839; Prestwich, 1892; George, 1933) have long been known in southern England and Wales. These deposits commonly infill valley bottoms and mantle the lower part of hill slopes and are semi-stratified. Spurrell (1886) thought that head accumulated by 'the intermittent flowing, under its own weight, of a soil undergoing thaw, that is, in a viscous state'. By the end of the nineteenth century head was considered, by most geologists, to be a periglacial deposit. *Periglacial* environments are those that are non-glacial but are dominated by cold climatic conditions in which frost is common but in which ground is seasonally snow free (Washburn, 1973). Head deposits have been mapped, particularly in coastal areas of Wales and in south-west England, by employees of the Geological Survey since the latter part of the nineteenth century, but little interest was otherwise shown in periglacial deposits and landforms in Wales until the second half of the twentieth century.

In 1961 Albert Pissart, a Belgian geomorphologist, visited Wales and identified the remains of pingos in the Llangurig area of mid-Wales. *Pingo* is 'An Eskimo [*sic.*, Inuit] word for a domed, perennial ice-cored mound of earth formed ... under permafrost conditions' (Whittow, 1984). *Permafrost* is a condition in which sediments below the ground surface are frozen for two or more consecutive years and commonly exists in periglacial environments.

Pingos melt as temperatures rise. This allows the inorganic sediments within the domed mounds, which were cemented by and incorporated within ice as the pingos formed, to sludge down the sides of the domes to form circular or sub-circular ramparts around the margins of the formerly active landforms. The remains of pingos are described in Chapters Seven and Eight.

Pissart published his findings in 1963, by which time he had interested Edward Watson, a geomorphologist in the Department of Geography at the University College of Wales in Aberystwyth, in periglacial studies. Pissart also identified fossil solifluction terraces in Wales (1963b). In the years that followed Watson provided firm foundations for periglacial studies in the Principality, writing on such periglacial sediments as *grèzes litées* (1965a), ice wedge casts (1981) and other periglacial structures (1965b), pingos (*e.g.* 1971), and nivation cirques with their associated landforms and sediments (1966, 1969). By the end of the 1960s periglacial studies were well established in Wales although it was not until the following decade that they were emphasised in Ireland (*e.g.* Mitchell, 1971; Lewis, 1979).

Sand and ice wedges form under specific environmental conditions and the casts of such wedges can be used to indicate former climatic conditions, as has been done for the Lower Severn Valley in Chapter Six. On a more extensive scale Huijzer and Vandenberghe (1998) have used ice wedge casts and other periglacial features to indicate climatic conditions in north-western and central Europe between ~72–13 ka, during the glacial stage known in the Netherlands, Scandinavia and the southern borderlands of the Baltic Sea, as the Weichselian.

The use of numerical dating
i) Radiocarbon dating
Nuclear research, especially during the Second World War, led to the appreciation that sediments could be dated by measuring the proportion of carbon atoms in an organic deposit. A radioactive form of carbon (radiocarbon) is produced in the atmosphere by cosmic rays. Radiocarbon is then absorbed into the bodies of living plants and animals but decays at a known and measurable rate once the plants/animals die.

Willard Libby, working at the University of Chicago, developed *radiocarbon dating* in the late 1940s (Imbrie and Imbrie, 1979). In the years that followed radiocarbon dating was applied to organic deposits in many parts of the world, including Wales and adjoining areas. As a result it was possible to construct a numerical timescale, in radiocarbon years, for geological events.

Radiocarbon dating was applicable to suitable deposits up to about 40,000 years old and now, with accelerator mass spectrometry (AMS), may be used to date suitable samples that are up to 50,000 years old. In 1970, when *The glaciations of Wales* was first published, some of the regional chapters did not contain a single date. Others, such as that on Pembrokeshire, contained 'radiocarbon age determinations' that some scientists regarded as questionable. The present book contains numerous radiocarbon dates, as in Chapters Three, Five and Eight. Many other numerical dating techniques now exist, such as uranium-series ages on stalagmites, thermoluminescence (TL), optically stimulated luminescence (OSL), electron spin resonance (ESR), amino acid dating of bivalves and gastropods and cosmogenic chlorine-36 rock exposure dating.

ii) Amino-acid dating
Aminostratigraphy is the correlation of stratigraphic units based on the ratios of particular amino acids preserved in the fossil protein of gastropods and bivalves. *Amino acid geochronology* relies on the age calibration of such ratios by, for example, radiocarbon, Uranium-series or other independent means. It has been applied to raised beach, terrestrial and shelly glacial deposits in Wales (Chapter Ten) and to raised beach deposits in south-west England (Chapter Eleven).

iii) Cosmogenic nuclide surface-exposure dates
Cosmogenic nuclide surface-exposure dates are quoted in Chapter Ten. Cosmogenic nuclides (^{3}He, ^{10}Be, ^{21}Ne, ^{26}Al, ^{36}Cl) are produced in rocks once they are exposed to cosmogenic ray bombardment at the surface of the Earth (*e.g.* Phillips *et al.*, 1994; Bowen *et al.*, 2002). Their accumulation is a measure of rock exposure. Arthur's Stone, on Cefn Bryn in Gower, is a glacial erratic with a Chlorine-36 age of about 23,000 years, indicating that deglaciation in that part of South Wales took place at about that time (Chapter Ten). At Cwm Idwal in Snowdonia, Chlorine-36 ages of boulders on the surface of the outer Late Glacial moraine (Phillips *et al.*, 1994) indicate that ice build-up at that site occurred during the interstadial preceding the Younger Dryas (the Bølling–Allerød), when precipitation took place, (a considerable proportion of which must have been of snow), derived from a relatively warm North Atlantic Ocean (Bowen, 1999).

Oxygen isotope stratigraphy ($\delta^{18}O$)
The sedimentary record of the continents is incomplete because of erosion, whereas sediments accumulated more or less continuously on the floor of the deep open (pelagic) ocean and contain benthic and planktonic micro-organisms (mainly the Foraminifera). The geochemistry of these organisms, such as their oxygen isotope, carbon isotope, cadmium-calcium and magnesium-calcium content, contain valuable information about the chemistry of the oceans from which they secreted chemicals for their shells. Oxygen isotope stratigraphy (Emiliani, 1966), expressed as $\delta^{18}O$, the ratio of ^{18}O to ^{16}O in fossil shells, has been used as a monitor of the changing isotopic composition of the oceans. During ice ages the evaporation of the lighter (^{16}O) isotope enriches continental ice-sheets, while residually enriching the isotopic composition of the ocean in the heavier (^{18}O) isotope. When deglaciation occurs, the ocean is enriched in the lighter isotope as the $\delta^{16}O$ enriched ice-sheets melt. An oxygen isotope stratigraphy from deep sea cores thus provides a record of the changing ice volume (glaciation) of the continents and, thereby, a record of Pleistocene glaciations, of which there were about 50 as well as a comparable number of interglacials back to some 2.5 million years ago (Bowen, 2004). These *oxygen isotope stages* are numbered backwards in time, with odd numbers denoting interglacials and even numbers denoting ice ages.

Correlation of oxygen isotope stages established from studies of ocean floor sediments with terrestrial events, such as the glaciations of Wales, is challenging but necessary if the global climatic system is to be understood. The deep sea isotopic record is tracked by astronomical calculations of the earth's orbital variations (the *Milankovitch cycles* that have already been discussed). Exactly how astronomical forcing of insolation changes is translated into climatic changes, and how millennial changes are superimposed on those of orbital origin, has yet to be established in causal terms. The estimation of sea surface temperatures is complicated and is usually done using a combination of oxygen isotopes, ecologic water masses characterised by combinations of plankton species and, recently, alkenone thermometry.

The ages of ice ages and interglacials, as established by oxygen isotope analyses, are calibrated by magnetic polarity through reference to the ages of major reversals of the Earth's magnetic field, as have been established on terrestrial lava flows (*e.g.* Shackleton and Opdyke, 1973).

Oxygen isotope stages are referred to in many of the following chapters, as in Chapters Three, Four, Six, Ten and Eleven, and their utility as palaeoclimatic indicators has helped scientists to realise that events in Wales and surrounding areas reflect developments in the global system. Such recognition has coincided with the development of a palaeoglaciological approach to glacial studies in and adjacent to the Principality.

Palaeoglaciology
In 1977, after publication of the first edition of this book, Boulton *et al.* attempted a reconstruction of the ice-sheet that covered much of Britain and Ireland during 'the last glaciation'. This reconstruction was based on 'flowlines derived from inspection of the glacial geology, to establish the ice surface topography' (Siegert, 2001) and proposed that the ice surface was some 1,200 m above modern sea level in North Wales, declining to under 400 m south of the Brecon Beacons, with large areas of south-west and south-east Wales being ice free. In 1985 Boulton *et al.* produced a revised reconstruction of the ice-sheet surface, indicating a thickness of little more than 250 m of ice in North Wales.

Lambeck (1993a, b) produced further models of the British and Irish ice-sheet, for 22,000 BP, 18,000 BP, 16,000 BP, 14,000 BP, 13,000 BP and 12,750 BP (BP means Before the Present, and the Present is regarded as 1950 AD). At 22,000 BP, according to the model, the ice-sheet had a surface of some 600 m in north central Wales but less elsewhere, thinning to 400 m by 18,000 BP. By 16,000 BP, or so the model indicated, only a small residual ice-cap existed in central Wales, remaining areas of the Principality being essentially ice free. Lambeck's model considered isostatic effects on the ice-sheet and concluded that by 18,000 years ago the Irish Sea was free of grounded ice, although geological and geomorphological evidence in north-east Ireland indicates otherwise (Clark *et al.*, 2004). Isostatic effects include the depression of the land due to the weight of the overlying ice, and the consequent rebound when the ice melts and that weight, or part thereof, is removed.

In contrast to the large-scale glaciological approach of ice-sheet modelling, reconstructions were published for central south Wales (Bowen, 1980), while Shakesby and Matthews (1993) published a reconstruction of a small glacier in the Mynydd Du uplands west of the Brecon Beacons in South Wales, showing the interpolated contours on the glacier surface (Chapter Eight). They also discussed the size of the most recent cirque glaciers in the Brecon Beacons region and computed the Equilibrium Line Altitudes (ELA) of some of them. The *Equilibrium Line Altitude* is the altitude at which accumulation on the higher portion of a glacier is exactly balanced by ablation (melting) on lower areas. Once the ELA is known it is possible to indicate former temperatures and/or precipitation.

Carr (2001) presented a more developed glaciological approach in his investigation of the most recent glaciers in the central area of the Brecon Beacons. From the geomorphological evidence Carr calculated ELAs, ablation gradients, glacier mass balance and ice discharge (mass flux), velocity, ice deformation, basal slippage/subglacial deformation and glacier flow through basal slippage. From these calculations he proposed that four of the five supposedly most recent glaciers were 'relatively slow moving...with a significant component of basal slip ... and compare well with modern [small glaciers] in southern Norway'. Carr concluded that 'the geomorphological evidence ... reflect[s] the activity of small glaciers'.

In 2004 Jansson and Glasser used satellite imagery to map glacial lineations in northern Wales, which enabled them to reconstruct former palaeo-ice-flow systems. They were then able to calculate former ice-sheet surface profiles and indicate the corresponding subglacial thermal regimes. They concluded that 'Landform creation and reorganization beneath the [last] Welsh Ice Cap appears to have been limited ... possibly because the ice cap was predominantly cold-based', which harks back to the protectionist ideas of Bonney and Garwood a century earlier. They also concluded that 'northern Wales was covered by glacier ice reaching an altitude of 1000–1200 m a. s. l.'

Glaciological studies are at an early stage of development in Wales and surrounding areas and it is important that they are consistent with the established facts of glacial geomorphology, lithostratigraphy and age estimates for the events they purport to model.

Wales and the global system
Introduction
Although there was general appreciation, when *The glaciations of Wales* appeared in 1970, that events in Wales and adjacent regions correlated with those elsewhere, there was no discussion of the role of global systems in causing those events. There was, for example, no discussion of the origins of ice ages, or of the role of atmospheric gasses in climate change, nor of the importance of ocean currents or even of glaciological events (such as the melting or surging of North American and Greenland ice-sheets and the consequent discharge of armadas of ice-bergs into the North Atlantic, with a concomitant lowering of sea temperatures with associated atmospheric consequences). That was mainly because research into those topics had scarcely commenced. Results from the first deep-polar ice core, in Greenland, were published in 1969 (Dansgaard *et al.*, 1969, 1971). They heralded a new scientific field of study in which millennial fluctuations of climate in both Greenland and Antarctica, and their interconnections, are investigated (*e.g.* EPICA, 2004).

i) The origin of ice ages
Changes in insolation caused by changes in the orbit of the Earth seem to be the primary pacemakers of climate change although the role of sub-orbital processes, especially at millennial timescales, is superimposed on these larger changes. Exactly how orbital variations are translated into climate change is still unknown, although it seems likely that variability in the irradiances of the Sun, as well as the circulations of the oceans and atmosphere, all interact with various feedback effects. The glaciations of Wales were influenced by modulation of the obliquity (41,000 years), by precessional (23,000 and 19,000 years) pacings and by eccentricity (100,000 years; Chapter Two). Variability on sub-orbital scales has also been important.

The role of plate tectonics in opening and closing seaways, as well as in the uplift of the Himalayas and Colorado Plateau (which influence global climates through the effects of high altitude areas on global atmospheric circulation), has also been important in preparing the global boundary conditions in which the glaciations of Wales have taken place (Raymo and Ruddiman, 1992; Ruddiman, 2000; Wilson *et al.*, 2000).

ii) Ocean Circulation
In 1987 Broecker suggested that there is a mass transfer of water globally as a result of salinity differences, resulting in the *Global Conveyor or thermohaline circulation*. Dense, saline water from the North Atlantic flows south and then east at depth into the Indian and southern Pacific Oceans. It then flows northward, in the Pacific, becoming warmer as it does so and becoming less saline. In the northern Pacific, around the latitude of Japan, the now warmed and less saline water starts to flow back near the surface, north of Australia, to the Indian Ocean and thence, around the Cape of Good Hope, into the Atlantic (Fig. 1.2). As the Global Conveyor flows northward through the Atlantic so it carries warmth into that Ocean. With progress northwards the Atlantic arm of the Global Conveyor becomes cooler and increases in salinity so that, between Scotland and Iceland, the current sinks below the surface and starts its return journey south and then east into the Pacific.

The Global Conveyor has many effects on climate, not only transferring heat around the globe but also, through releasing heat to the atmosphere in the North Atlantic, nurturing atmospheric water absorp-

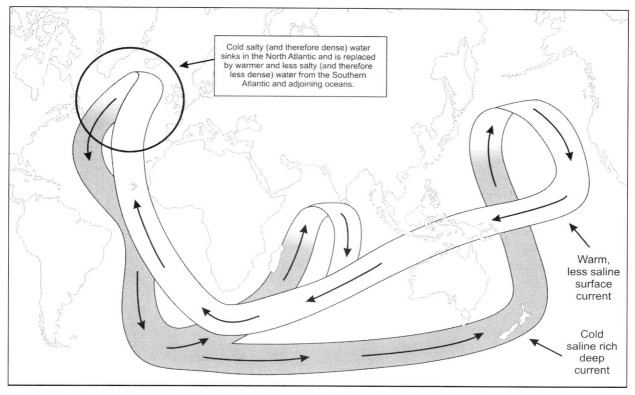

Fig. 1.2 The Global Conveyor, or thermohaline circulation, showing the main ocean currents that distribute heat from the Pacific and Indian Oceans northwards into the Atlantic Ocean. They return as cooler and more saline (and therefore denser and deeper) currents to flow south and then east into the Pacific Ocean where they flow northwards, warming and then returning to the Atlantic as less saline (and therefore less dense) surface currents

tion, cloud formation and, eventually, precipitation on adjacent continental areas influenced by the prevailing south-westerly winds that blow over that portion of the North Atlantic. The effectiveness of the Global Conveyor compared with atmospheric flux is debatable because, during transfer of heat poleward in the North Atlantic there is an atmospheric flux of 5 PW (one Peta Watt is 10^{15} watts/m^{-3}) compared with 1 PW through the ocean. There is also growing evidence that the Global Conveyor did not switch off during ice ages, as was previous thought (Raymo *et al.*, 2004). Atmospheric pulses of warmth derived from lower latitudes might have been the cause of at least some climatic changes that have been superimposed on larger changes of orbital origin (Cane and Clement, 1993; Bowen, 2000).

iii) Heinrich Events
There appear to have been a number of occasions in the past when the Global Conveyor was affected by the release of pulses of cold and fresh melt-water from decaying, or surging, ice-sheets in North America, Greenland and the islands of the arctic. Other cold freshwater pulses came from the (rapid) drainage of proglacial lakes as ice-barriers disappeared. Some pulses were associated with the release of ice-bergs into the Atlantic from the Laurentide (North American) and Greenland ice-sheets, forming *Heinrich Events* in which sediments from the melting ice-bergs were deposited over a broad swathe of the floor of the North Atlantic from Hudson Strait and Davis Strait to a zone west of the Celtic Sea/Bay of Biscay (Fig. 1.3). Another swathe of deposition from ice-bergs extended southwards from the vicinity of Spitsbergen towards Iceland and the Norwegian Sea.

Heinrich Events are named after Helmut Heinrich, the scientist who first reported their former existence (Heinrich, 1988; *vide* Andrews, 1998). The causes of Heinrich Events remain unclear, but may have included climatic amelioration, possibly associated with the release of pulses of warm water from the Caribbean and Gulf of Mexico, that caused ice-sheet decay. They may also have been the result of glaciological events, such as *surging*. Surging is when an ice-mass advances rapidly (MacAyeal, 1993; Clarke *et al*., 1999). During a surge large volumes of ice are transferred to the terminal part of the ice-mass, possibly leading to a reduction in altitude of the ice-mass. Surging is the result of instability within the glacier system, possibly due to a build-up of stress as ice thickens in the accumulation area, leading to a loss of cohesion within the ice or between it and its bed (Patterson, 2001). Surging may therefore be glaciologically controlled in at least some cases, rather than a reflection of climatic change.

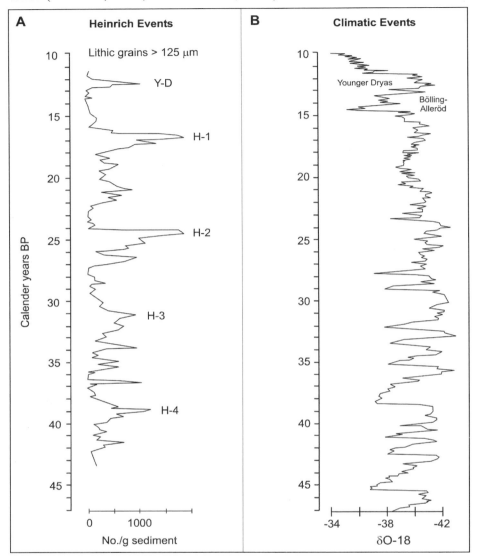

Fig. 1.3 a) *Heinrich Events, as evidenced by the number of lithic grains larger than 125 μm per gram of sediment recovered from a core taken from the floor of the Atlantic Ocean west of Scotland (Barra Fan core MD95-2006). The peak numbers of these grains were deposited as a result of major iceberg discharges (Heinrich Events) into the Atlantic Ocean. H-1 to H-4 were Heinrich Events. Y-D: Younger Dryas stadial, sometimes denoted as H-0*

b) *Climatic events during the last 47,000 years as evidenced by analysis of an ice core from a site in Greenland (GISP2). Fluctuations in δO-18 reflect temperature oscillations, many of which occurred rapidly and are grouped into Dansgaard-Oeschger cycles. (Source: Bowen et al., 2002)*

The freshwater pulses released into the Atlantic during Heinrich Events reduced the intensity of the Global Conveyor and, in some cases, caused it to turn southward considerably further south than it does at present in the North Atlantic. Temperatures consequently fell in the northern North Atlantic and adjacent areas, causing renewed glacial activity in some regions. Since the oceans form a coupled system there was an associated rise in sea temperatures in the tropical Atlantic, Southern Ocean and in the tropical Pacific, albeit after a time-lag of several hundred years (Clarke *et al.*, 2004). This tropical warming may have then influenced higher-latitude deglaciation.

Heinrich Events are referred to in Chapters Ten and Twelve. Based on cosmogenic rock exposure ages on glacial boulders it has been suggested that Heinrich Event 2 corresponded with the Last Glacial Maximum in Ireland and Wales (Bowen *et al.*, 2002). A ^{36}Cl rock exposure age on an erratic deposited at the ice-margin in Gower (Chapter Ten) apparently confirms this suggestion.

iv) Dansgaard-Oeschger and Bond cycles
During the 1960s the first cores were extracted from the world's major ice-sheets: from Camp Century in Greenland in 1966 and from Byrd Station in the Antarctic in 1968. Cores have since been recovered from ice-masses in many parts of the world, including Dye 3 (1981), Renland (1988), GRIP (1992) and GISP (1993) in Greenland; Devon Island in arctic Canada; Dome C (1979) and Vostok (1985) in the Antarctic, and from much smaller ice-masses including the Quelccaya ice-cap in the Peruvian Andes; Lewis Glacier on Mount Kenya and Kilimanjaro, both in Africa; the Inilchek Glacier in Kyrgystan; and many other locations (NOAA, 2004).

In 1969 Dansgaard *et al.*, from their study of evidence from the Camp Century core from north-west Greenland, showed that there have been rapid temperature changes over the past 100,000 years. Further work showed that these changes form cycles, now known as *Dansgaard–Oeschger cycles*, that have pronounced 1,450–1,500 year pacings. The warmer events are *interstadials* and the colder are *stadials*. There appear to have been rapid changes of temperature between stadials and interstadials, sometimes of as much as 10°C in less than a decade (Fig. 1.3). In some cases, particularly between 20,000 and 80,000 years ago, Dansgaard–Oeschger events were grouped into cooling cycles of some 3-6,000 years. These are know as *Bond Cycles* (Bond and Lotti, 1995). Chapter Twelve suggests that ice in the Irish Sea basin retreated from still-stand positions after the Last Glacial Maximum with the same pacing as Heinrich Events, Bond Cycles and Dansgaard-Oeschger cycles, indicating that glaciation in Wales and adjacent areas was related to much larger, probably global, systems.

v) Greenhouse gasses
Bubbles of air within ice-sheets provide information on greenhouse gas concentrations in the atmosphere at the time when those bubbles were sealed: lower concentrations of greenhouse gasses (carbon dioxide and methane) existed during glacial ages and higher concentrations during interglacials. Greenland and Antarctic ice cores show that greenhouse gas variability pulsed on orbital and millennial time scales. Whether greenhouse gasses changed climate, or whether they amplified changes in temperature caused by insolation variability, is as yet uncertain. Air temperature variability in Antarctica has now been estimated for the last 740,000 (EPICA, 2004) and 800,000 years (Jouzel *et al.*, 2004), with the analysis of greenhouse gas variability to follow soon (EPICA, 2004). Such variables, at orbital and millennial timescales, will have influenced climatic and other events in Wales (Bowen, 2000, 2004, 2005a, b), as is hinted at in Chapter Twelve.

Conclusion
A thorough understanding of glacial events, of glacials and interglacials, stadials and interstadials, requires interdisciplinary collaboration, an appreciation of the evolution of the global climate system and of its modelling.

The aim of the present book is to provide a compendium of existing knowledge of the glaciations of Wales and surrounding districts so as to provide a benchmark for further studies. The book consists essentially of a series of regional chapters that describe glacial and associated landforms and deposits in the different regions of Wales and surrounding areas.

Chapter Two provides an introduction to glacial deposits, Quaternary stratigraphy and Welsh climate history (Bowen, 1999). The Chapter also lists the major lithostratigraphical formations of glacial origin in and adjacent to the Principality. The latter part of Chapter Eight pays particular attention to events towards the end of the most recent (Devensian) glacial stage, when an interstadial began before 14,000 BP, following melting of the last regional ice-sheet and before reversion to cold, stadial conditions at about 12,450 BP, when cold conditions returned during the Younger Dryas Stadial. A small ice-sheet developed in the Western Highlands of Scotland at approximately the same time, reaching its maximum extent at the southern end of Loch Lomond, after which it has been named the *Loch Lomond Glaciation* (Sissons, 1976). By contrast, Chapters Six, Nine, Ten and Eleven draw attention to glacial deposits that may be more than 400,000 years old, possibly correlating with Oxygen Isotope Stage 16, 'the first major glaciation of the hundred thousand eccentricity world' (Bowen, this volume) or with Oxygen Isotope Stage 12. No evidence of earlier glacial deposits has yet been found in Wales and adjacent areas so that events during the earlier stages of the Quaternary in the region remain unknown.

Much of the initial work in Britain that was related to the establishment of the glacial theory was undertaken in Wales, and especially in North Wales. Recent research in South Wales has concentrated on the establishment of a geochronology that should enable scientists to correlate events in Wales and adjoining districts with the evolution of the global climate system. Wales therefore holds an honourable place in the continuing development of ice age investigations.

2 Stratigraphy

by Andrew Richards

Glacial deposits and Quaternary stratigraphy

Stratigraphy provides a framework within which the development of the landscape can be understood. The record of Quaternary landscape evolution is recorded in deposits formed at different times and places and is far from complete. Sedimentary strata provide evidence for geological, climatic, and biogenic changes that have occurred throughout Quaternary time. The basic role of stratigraphy is to define the characteristics of sediments and to determine the order in which the sediments were formed. Deposits are grouped into units that have similar characteristics and their position is related to the temporal and spatial distribution of other distinct units. This process of defining groups of rocks and sediments by their characteristics forms the basis of our understanding of events in geological history. The role of stratigraphy in the Quaternary Era is particularly important as the modem landscape, fauna and flora, as well as humankind, evolved during this period.

The Quaternary Era is marked by massive climatic fluctuations and the alteration between Cold and Warm Stage environments, driven by variations in the amount of solar radiation received by the Earth's surface. There have been over 100 such oscillations since the start of the Quaternary Era, approximately 2.5 million years before present (Shackleton *et al.*, 1990). The flux in climate during this period is believed to be largely associated with changes in the shape of the Earth's orbit and variation in the orientation of the earth's axis. These astronomical variations are known as Milankovitch cycles after the Serbian astronomer who is generally credited with calculating their magnitude. There are three principal cycles (see Fig. 2. 1a) which, taken in unison, impact on the seasonality and location of solar energy around the Earth, thus causing contrasts between the seasons. These times of increased or decreased solar radiation directly influence the Earth's climate system, thus influencing the advance and retreat of Earth's glaciers.

Scientists have pieced together the record of Quaternary climate fluctuation by recovering and analysing sediments from the ocean floor. This sediment contains calcium carbonate shells from organisms that have lived in the ocean. One of the most prevalent organisms, foraminifera (or forams), extract minerals from sea-water in order to secrete a $CaCO_3$-rich shell. Up to nine shells may be secreted, shed and deposited at the ocean floor during the lifetime of a foram. At any point in time the shell of a foram will reflect the chemistry of the sea-water in which it has lived. Oxygen is found in two isotopic forms. Atoms of 0^{16} contain 8 protons and 8 neutrons, while a small fraction (1 in a 1000) of oxygen atoms contain 8 protons and 10 neutrons. This isotope of oxygen, 0^{18}, is heavier and therefore less susceptible to evaporation (Fig. 2.1b). As a consequence, during colder conditions more 0^{16} becomes tied up within the world's glaciers and ice-sheets, and the ratio of $0^{16}:0^{18}$ in foram fossils within ocean sediments reflects the amount of ice present on Earth during the lives of these forams. The fluctuating record of $0^{16}:0^{18}$ within $CaCO_3$ rich ocean sediments can therefore be considered as a record of global temperature. Scientists have subdivided this record into Oxygen Isotope Stages (OIS); with even numbered stages representing Cold Stages and odd numbered stages representing the intervening Warm Stages (Fig. 2.1c).

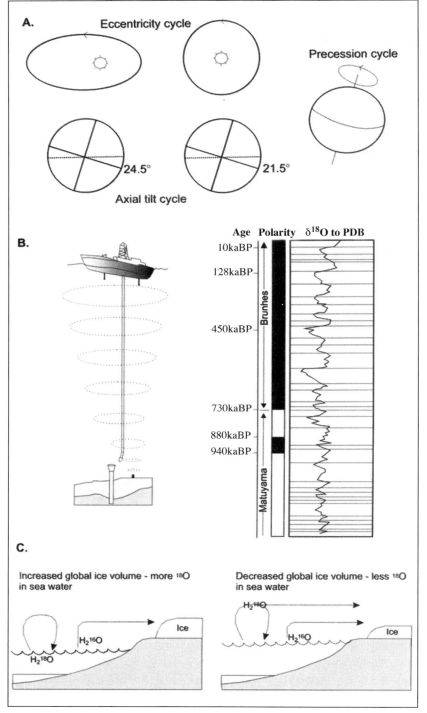

Fig. 2.1a
i) Eccentricity cycle: the orbit of the Earth around the Sun is elliptical rather than circular. When Earth is nearer the Sun it is warmer than when it is further away.
ii) Axial tilt cycle: the tilt of the Earth on its axis varies from 24.5° to 21.5°
iii) Precession cycle: Earth wobbles as it spins on its axis. These Milankovitch cycles combine to influence the flux of incoming solar radiation and the temporal and spatial distribution of that energy, thereby affecting climate and global ice volumes

Fig. 2.1b
The Ocean Drilling programme has retrieved cores of calcium carbonate rich sediment from the ocean floor. This record of isotopic variations provides a proxy record of Quaternary climatic changes that can be dated by reference to changes in polarity of the Earth's magnetic field, which is recorded in core sediments

Fig. 2.1c
The fluctuation in the ratio of ^{18}O to ^{16}O reflect the growth and decay of global ice masses, and therefore record periods when the Earth was much colder and, probably, underwent glaciation. These fluctuations in isotopic ratio form the basis of climate stratigraphy and the subdivision of Quaternary time into Warm and Cold Stages

During Cold Stages, processes active on the surface of Wales and neighbouring areas would have been markedly different from those that are present in the modem landscape. Sea levels were much lower (in excess of 100 m lower during some Cold Stages) and river systems became more aggressive, sometimes changing from relatively inert, single-thread meandering systems into braided patterns that have a

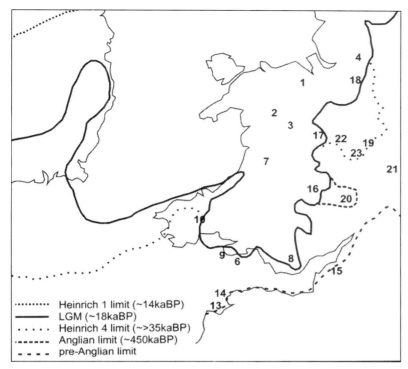

Fig. 2.2 The limits of glaciations in Wales during the cold stages of the Quaternary. Location numbers relate to the glacial formations noted in Table 2.3. Many of the limits drawn are approximations only, particularly the limits associated with earlier Cold Stages

greater capacity for moulding the landscape. During the most intense phases of Cold Stages, which are known as Stadials, there was also a marked increase in the volume of ice at, and beneath, the Earth's surface. As well as shaping the landscape, extensive ice cover during these Stadials has left Wales and its surrounds with a partial record of Cold Stage environments prior to the Late Devensian glaciation (Fig. 2.2). There is also a very poor record of intervening temperate episodes.

Stratigraphy provides a framework within which the sedimentary and geomorphological record of Quaternary Cold and Warm Stages can be studied. There are many facets to stratigraphy and a number of methodologies have contributed to the framework currently used in Britain and Ireland. This chapter initially discusses strategies developed in order to construct the glaciostratigraphic framework. The chapter then provides an overview of current understanding of Cold Stage environments that have affected Wales and adjacent areas, highlighting the spatial and temporal occurrence of glaciers and ice-sheets during each stage. Finally, there is a short discussion of how the stratigraphic and geomorphological record of glaciation is likely to be refined and developed by future work.

The Quaternary Stratigraphic Framework of Britain and Ireland

The Quaternary sedimentary sequence of Britain and Ireland was first formally classified into Warm and Cold Stages by Mitchell *et al.* (1973). This framework (Table 2.1) was based on the recognition of *stratotypes*. These stratotypes represent the original or designated representative of a specific stratum that best record the environments associated with each stage. This framework, explicitly or implicitly, informed the work of Quaternary scientists in the following decades. The methodology that was employed in this classification was driven largely by the recognition and differentiation of Warm Stage deposits. While other fossils were often used for palaeoenvironmental reconstruction, the Warm Stages were mainly defined by pollen spectra. Underpinning this rationale was the belief that each Warm Stage was marked by a particular type of vegetation dynamic; driven by climate, geology, soil processes, migration, extinction and competition. Therefore where a characteristic pollen spectra had been recorded from a unit, its stratigraphic classification permitted its use as a chronological marker horizon for units underlying or overlying the Warm Stage unit.

Recent discoveries and the advent of new techniques have resulted in reassessment of the classification proposed by Mitchell *et al.* (1973). Advances in the accuracy and utility of absolute dating techniques, detailed studies of the fauna and flora represented in organic rich sediments and comparison with high resolution records of palaeoclimate from marine and ice-core archives revealed that the terrestrial record of Quaternary environmental change is much more complex than was previously thought. The reclassification proposed by Bowen (1999) records major changes in our understanding of the terrestrial record of landscape development when compared to the sequence proposed by Mitchell *et al.* (1973). Firstly, the *stratotype* that Mitchell *et al.* (1973) used to define the characteristics of the Cold Stage (Wolstonian) that they believed to precede the Devensian, has proved not to be valid. The sequence of glacigenic deposits recorded in the Avon Valley at Wolston, the location after which the Cold Stage had been named, were found to be unrelated to any interglacial marker horizon (Bowen, 1978). They correlated with Anglian Stage glacial deposits of the Lowestoft Formation (Perrin *et al.*, 1979; Maddy *et al.*, 1991; Sumbler, 1983) and have recently been found to underlie Hoxnian Stage organic deposits (Keen *et al.*, 1997). While the stratigraphic issues relating to deposits at Wolston are detailed and beyond the remit of this book, the fundamental problem was that our understanding of the penultimate Quaternary Cold stage had been influenced by sediments that were, in fact, the product of cold stage environments that existed over 200 ka earlier!

The classification provided by Mitchell *et al.* (1973) has also been expanded to incorporate the evidence for additional stages. Consequently, the original Wolstonian Cold Stage now includes two Cold Stages separated by an additional pre-Ipswichian/post-Hoxnian Warm Stage. This reclassification is widely accepted and is based on a wealth of evidence, including relative and absolute dating from a large number of locations, using a wide variety of techniques. Important evidence substantiating the validity of this reclassification is given in Chapters 5 and 6.

The classification of Warm and Cold stages proposed by Bowen (1999) is shown in Table 2.1. Like the earlier framework, this classification accepts the underlying role of Quaternary climate change in defining discrete stages and incorporates all of the events so far documented by Quaternary deposits in the British Isles. However, the terrestrial record of palaeoenvironmental change is far from complete and, as a consequence, will certainly be subject to revision.

Original framework (Mitchell et al., 1973)	*Revised Framework (Bowen, 1999)*	*Oxygen Isotope Stage*
Holocene (Warm)	**Holocene** (Warm)	1
		2
Devensian (Cold)	**Devensian** (Cold)	3
		4
		5a-d
Ipswichian (Warm)	**Trafalgar Square** (Warm)	5e
Wolstonian (Cold)	?	6
Hoxnian (Warm)	**Strensham** (Warm)	7
Anglian (Cold)	?	8
Cromerian (Warm)	**Purfleet** (Warm)	9
	?	10
	Swanscombe (Warm)	11
	Anglian? (Cold)	12
	Cromerian (Warm)	13

Table 2.1 *The classification of Warm and Cold Stages of the Quaternary (after Bowen, 1999)*

The complexity of Climate change and relationships with glaciers and ice-sheets

When the classification of British Warm and Cold Stages is compared with the climatic stratigraphy indicated by ocean sediments, it is clear that accurate correlation between terrestrial and ocean records is fraught with problems. Terrestrial and marine records of palaeoenvironment have advantages and disadvantages. The marine record is of high resolution and involves a very small time lag (Ruddiman, 1987). However, such records give us little information about the geomorphological processes that affected our landscape. Climatic fluctuation results in a dynamic landscape and therefore more recent events tend to remove much of the evidence for preceding episodes in the evolution of the landscape. Thus, the record of Devensian Stage glaciation is widespread, but evidence for previous glaciation is in short supply. This leads to a further problem. It is clear from the marine record that Quaternary climate change is very complex. While cold stages were dominated by temperatures much colder than those associated with Warm Stages, temperatures were by no means stable, and a number of Cold Stages experienced more than one period when conditions were sufficient to maintain considerable increases in global ice volume. Many of these intensely cold periods were separated by phases where temperatures improved, almost to Warm Stage conditions. Such changes were often very rapid, occurring over a few thousand years, or often shorter intervals. As our studies of the terrestrial sequence evolve, we will acquire a more accurate understanding of just how terrestrial geomorphological processes are related to the short-term climatic changes that are documented by marine sediments.

Recent advances in the understanding of the behaviour of glaciers and ice-sheets have led to an increased knowledge of glacial history. Ice-sheets are very sensitive to climatic changes, particularly in temperate regimes, or in areas in proximity to the ocean. While the general global volume of ice results from climatic events associated with Milankovitch cycles, short-term ice-sheet and glacier oscillations may be due to other factors. Heinrich Events are episodes of massive discharge of ice-rafted debris in the North Atlantic, and are thought to be caused by abrupt changes in ice-sheets, usually accompanied by melting and partial collapse (McCabe and Clark, 1998) as ice-sheets and glaciers surged into the ocean releasing large volumes of ice-bergs. These ice-bergs and associated glacial meltwaters had a huge effect on the way heat was transported by oceanic currents. During these cold phases the North Atlantic current, that is responsible for the generally warm, moist conditions associated with north-west Europe today, was displaced south of 40°N into what are now tropical waters. Such events are documented in high-resolution palaeoclimate records from marine sediments by a sudden increase in silt and sand grains.

The recognition of such events has important implications for ice-sheet, and therefore landscape, evolution. Firstly, these events record climate variability over much shorter time-scales (*i.e.* millennia) than those predicted by orbital forcing (*i.e.* Milankovitch cycles). Secondly, it appears that global climate and the behaviour of ice-sheets are even more closely associated than previously thought. Ice-sheets actively participate in climate change; amplifying or driving abrupt changes by cooling the atmosphere, changing salinity patterns in the world's oceans, displacing jet-streams and reorganising or strengthening mid-latitude low pressure centres.

Glacial stratigraphy

Stratigraphy involves the study of layers of rock and sediment in order to build up a chronologically sound record of past environments. There are differing strands of stratigraphy. *Lithostratigraphy* involves the study of the *physical* characteristics of the rock units, while *biostratigraphy* concerns biological characteristics. *Chronostratigraphy* provides the time scale that constrains these events. Despite the many changes in Quaternary science since the first version of this book was published in 1970, the need for a sound lithostratigraphic framework is still of fundamental importance. Dates may be inaccurate, methodologies flawed and new information gathered. However, providing the basic building block of stratigraphy, the lithostratigraphic unit, has been accurately described, and its temporal and spatial relationships with other units is understood, the framework will stand the test of time, yielding a robust archive for the reference of future work. The basic use of glacial deposits in stratigraphy is no different to the methodologies employed for the study of rocks and sediments derived from other geological eras. However, the recent nature of these glacial deposits often means that glacial stratigraphy must also include the incorporation of geomorphology within the scheme. Thus, in order to accurately represent the nature, timing

Glacier dynamics	Till facies (type)	Basic process	Sediment characteristics	Stratigraphic reliability
Active ice	Lodgement till	'plastering on' of debris from glacier sole	Homogenous, compacted diamicton	High • composition generally reflects load of depositing glacier • normally persist regionally as till sheets
	Deformation till	Shearing and reorganisation of subglacial materials	Varies from deformed rock masses to deformed soft sediment	Variable • exist as extensive sheets • composition of till is heavily influenced by variability in substrate
Passive ice	Melt out till	Deposition of debris from melting, stagnant, glacial ice masses	Variable textural characteristics dependent on nature of glacial debris and speed of melt-out process	Low • composition reflects load of depositing glacier • exists as localised lenses rather than persistent till sheets
	Sublimation till	Deposition of glacial material by stagnant ice mass as ice changes phase rapidly to water vapour		
	Flow till	Remobilisation of formerly deposited till, in contact with glacial ice	Variable textural characteristics dependent on primary process of deposition and speed/nature of subsequent remobilisation	Low • composition reflects load of depositing glacier • often dominated by supraglacial material • exists as localised lenses rather than persistent till sheets

Table 2.2 A simplified classification of till facies (types) and their potential for use as a part of a stratigraphic study

and extent of glaciation, we must first establish the relative order of glacial landforms and sediments and then interpret these features in terms of the environments in which they formed.

In Wales and adjoining regions, glacial deposits are widespread. During glaciation, an ice-sheet or glacier may deposit a variety of poorly sorted sediments derived directly from glacial ice, known as till. Outwash from the melting ice may also deposit gravels which may be associated with these tills. The recognition of glacial till provides the scientist with the means to accurately record the former presence of an ice-sheet in a locality. However, this is not without problems. Glacial till is a diamicton, which is a non-sorted/poorly-sorted sediment containing a wide variety of particle sizes. Diamictons may also be formed in other environments; commonly as a result of hillslope processes. However, there are a number of diagnostic criteria that can be used to distinguish tills from diamictons derived from other environments; such as the presence of far-travelled pebbles, stone shape and roundness characteristics, stone surface textures, stone fabrics and glaciotectonic structures. There are also a number of varieties of glacial till, each type (or facies) having different characteristics (Table 2.2). While the identification of these features is of particular importance for reconstructing glacial environments, till variability, especially in terms of lithological variability, may have implications when defining glacial stratigraphic units. Tills derived from the same ice-mass, during the same phase of glaciation, may show lithological variability that reflects differing positions of entrainment, transport and deposition. The Quaternary scientist must be aware of this when constructing a stratigraphy.

Stratigraphic subdivision of glacial deposits

The basic principle that guides the use of glacial deposits in stratigraphy is that the lithological composition of glacigenic materials reflects the characteristic load of the depositing ice-mass. This principle determines how glacial deposits are subdivided into the following stratigraphic building blocks:

1. *Formation*: a mappable unit, the fundamental unit in lithostratigraphical classification. This is defined by the lithological similarity of glacigenic materials within an area, reflecting the characteristic load of a glacier or ice-sheet.
2. *Members*: units within a Formation with characteristics that allow them to be distinguished from other parts of that Formation. In the case of glacial deposits, a Member of a Formation may represent a distinct phase in the dynamics of an ice-sheet or glacier, a period of re-advance for example.
3. *Beds*: parts of a Member of a Formation. In Quaternary sequences, beds are often defined on the basis of their biological characteristics or for dating purposes. As a consequence, they rarely incorporate glacial sediments.

Due to the complexity of glacial processes, and inherent variability of both forms and sediments, the aim of glacial stratigraphy is to identify, as a Formation, the deposits and landforms that are associated with discrete glaciations within a region. The Formation then forms the mappable unit that can be correlated or distinguished, allowing a stratigraphic framework to be defined based on the temporal and spatial distribution of the Formations. The detail of each Formation may then be characterised by subdivision and the definition of component Members and Beds. However, the subdivision of glacial deposits beyond Formation level must avoid the proliferation of many components in order to preserve the utility of the framework (Salvador, 1994). Each Formation of glacial deposits is related to a type section, region or locality, where the characteristics of the suite of deposits are best described. Subsequently, palaeoenvironmental inferences may be made. Formations may be dated by absolute or relative means, and a robust record of glacial history, based on the availability of exposures and accuracy of techniques, can be defined.

The Glacial stratigraphy of Wales and adjoining regions

The glacial Formations that have been recognised by contributors to this volume can be seen in Table 2.3. The proposed extent of each unit is illustrated in Fig. 2.2. Some of the units, in particular those derived from earlier stages of the Quaternary, cannot be shown as mappable units. By far the most widespread Formations are those that are thought to be derived from the last phase of glaciation. The Last Glacial Maximum is widely accepted to have occurred at approximately 18 ka BP during the Dimlington Stadial of the Late Devensian. The Dimlington Stadial is named after the type-site at Dimlington in eastern Yorkshire. Here moss remains (from a species that colonises proglacial meltwater channels), dated at around 18,000 B.P., underlie Late Devensian glacigenic sediments. This suggests a maximum age for the Late Devensian ice-sheet in this area. More locally, bones found in a cave in Tremeirchion, North Wales, have also been dated to 18,000 B.P. Tills derived from ice moving from the Irish Sea seal the entrance to this cave. Near Dimlington, organics above the till have been dated at 13,000 B.P. and this gives a minimum age for deglaciation and the onset of the Late Glacial interstadial. This latter date is confirmed from many sites around the British Isles (Table 2.3).

The Last Glacial Maximum event has been correlated with the Heinrich Event 2 ice-berg rafting events recorded by glacial debris in marine cores from the North Atlantic. As yet, there is no evidence in Wales, or the rest of the mainland of the United Kingdom for glaciation during Heinrich Event 1, at approximately 14 ka BP. A Devensian phase of glacial advance that precedes the Last Glacial Maximum

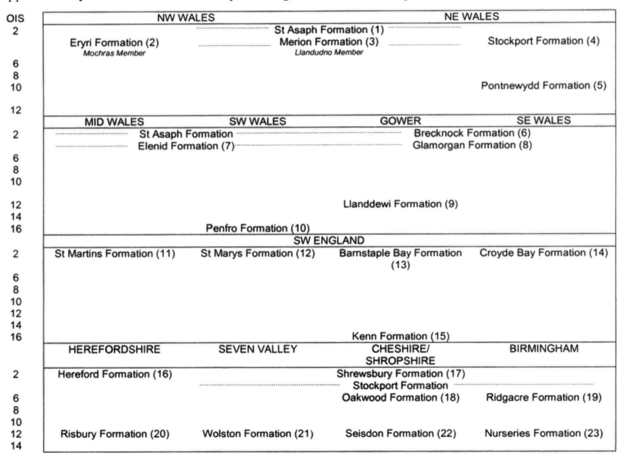

Table 2.3 The glacial formations identified by the authors' in each chapter of this book. For locations, refer to Fig. 2.2

was originally recognised in the English Midlands by Wills (1937; 1952), Boulton and Worsley (1965), Shotton (1968) and Worsley (1970). The limit of this glaciation became known as the 'Wolverhampton Line' and was believed to mark the southerly extent of an ice-sheet that emanated from the Irish Sea. In recent decades, the concept of an earlier Devensian phase of glaciation has largely fallen out of vogue (*i.e.* Jones and Keen, 1993). However, recent work by Bowen *et al.* (2002) suggests that the original proposals for more than one phase of glaciation during the Devensian Cold Stage may be correct. The advance of Irish Sea ice at approximately 38 ka BP may be represented by deposits in the English Midlands, the Cheshire-Shropshire Basin and Pembrokeshire. This event is thought to be associated with Heinrich Event 4 (Bowen *et al.*, 2002).

Due to the erosional capacity of Devensian ice-sheets, evidence for pre-Devensian glaciation in Wales and adjoining regions is scarce. However, there is appreciable stratigraphic evidence that the Cold Stage immediately preceding the Ipswichian Warm Stage, correlating with Oxygen Isotope Stage 6, is marked by the occurrence of glacial deposits in the English Midlands. Worsley (1991) suggests that tills and other glacigenic deposits of the Oakwood Formation are derived from this period. In addition, dating of glacial boulders associated with the Ridgeacre Formation of the West Midlands and Severn Valley by Maddy *et al.* (1995) suggests that an ice-sheet advanced into the region during the same phase, at approximately 160 ka BP.

Quaternary scientists generally accept that the Anglian Stage marks the most extensive glaciation to have affected the British Isles. The limit of this glaciation extends east from the Severn Estuary to areas in north London. While this phase of glaciation is marked by extensive glacial deposits in East Anglia, Anglian glacial deposits are relatively rare in Wales and adjoining areas, largely due to the extent of subsequent ice-sheets and marked river system development throughout the area. Anglian stage glacigenic deposits have been described in the Cheshire-Shropshire lowland (Seisdon Formation; Worsley, 1991), the West Midlands and Severn Valley (Nurseries Formation, Wolston Formation and correlatives of the Seisdon Formation; Horton, 1974; 1989; Kelly, 1964; Worsley, 1991; Chapter 5). While these deposits give little impression of the extent of the Anglian ice-sheet in this area, they are locally overlain with organic sediments derived from the Hoxnian Warm Stage. The lack of dateable horizons means that many of these deposits cannot be firmly assigned to the Anglian Cold Stage (Maddy, *In* Bowen, 1999). The Risbury Formation occurs as small remnants of ice-marginal glaciofluvial, glaciolacustrine and ice-contact deposits beyond the Last Glacial Maximum in eastern Herefordshire. In localities immediately west of the Malvern Hills, the Formation is underlain by organic silts of Cromerian age and overlain by organic deposits of Hoxnian age (Chapter 9).

There is little evidence for Anglian Stage glaciation within Wales, or for an earlier, more extensive glaciation before the Anglian Cold Stage. The Pontnewydd Formation is represented by the occurrence of pebbles of volcanic rocks and flint in Pontnewydd Cave in north-west Wales. These gravels are believed to be glacial in origin, representing the incursion of an ice-sheet into the surrounding landscape during or before Oxygen Isotope Stage 8 (Green *et al.*, 1981; 1984). There are also some mysterious deposits that occur beyond the Devensian ice-limit on Gower, in Pembrokeshire and in south-west England that may be of even greater antiquity. The Llanddewi Formation occurs on the Namurian Shales of the Gower Peninsula. It includes till and glaciofluvial sands and gravels and stratigraphic evidence suggests that it was deposited during a Cold Stage before Oxygen Isotope Stage 7. A further pre-Devensian glacial deposit occurs in Pembrokeshire. The dissected morphology of the Penfro Formation suggests that it may be older than the Llanddewi Formation, and it has been ascribed to Oxygen Isotope Stage 16 (Bowen, 1994; Campbell and Bowen, 1989). The Kenn Formation, found as remnants on high ground in Avon and North Somerset, consists of till, glaciofluvial and glaciomarine deposits that are overlain by Cromerian organic deposits (Gilbertson and Hawkins, 1978ab; Andrews *et al.*, 1984) and therefore must have been deposited during, or before, Oxygen Isotope Stage 16. The Kenn Formation has been correlated with the

Penfro Formation of Pembrokeshire (Bowen, 1994; Campbell and Bowen, 1989) and with the latter may represent the oldest Quaternary glacial deposits in the British Isles. The only other evidence for pre-Anglian glaciation is derived from the extensive sequences of Quaternary deposits in East Anglia and gravels of the Thames terraces. However in these areas, inferences are based on erratic suites and surface textures, and there are no firm records of glacial till (Hey, 1991).

Future prospects in glacial stratigraphy
A robust record of the glacial history of Wales and adjoining regions has been formulated since the publication of the predecessor of this book in 1970. This stratigraphic framework is based on the geometry of unequivocal lithostratigraphic units and their spatial relationships (Campbell and Bowen, 1989; Bowen, 1999).

Throughout the 1970s and 1980s, most emphasis was placed on the characterisation of floral/faunal content and relative/absolute dating of intervening Warm Stage deposits. However, it has been recognised that the floral and faunal response to climatic change, especially if it is very rapid, is difficult to predict and even more difficult to describe from the geological record. As a result, Quaternary scientists have recognised that a number of discrete Warm Stages may be represented by very similar floral and faunal characteristics. Conversely, temperate deposits of the same age may have widely diverging characteristics, as a consequence of geographical and geological influences, that may preclude correlation (*cf.* Bowen, 1978; Campbell and Bowen, 1989).

It appears that a strict lithostratigraphic scheme coupled with new advances in absolute dating techniques may offer Quaternary scientists the most suitable avenue for research into Quaternary landscape evolution. Such a scheme offers a number of advantages for those attempting to unravel the landscape history of a region. Scientists are beginning to understand the complexity of the relationship between ice-sheets and climatic change. The relationships of orbital forcing and Milankovitch cycles with Quaternary climatic change, and therefore glaciation, have long been established. However, it is only in the last few decades that Quaternary scientists have learnt how quickly glaciers can react to very rapid changes in temperature and how ice-sheets also play a role as forcing agents in climate change over short time-scales.

Bowen *et al.* (2002) have recently published work that details dates obtained from glacial deposits in the British Isles, and it appears that age groupings cluster around Heinrich Events. Attempts are being made to relate these short-term events to the terrestrial record of environmental change. Previously, most Quaternary scientists in Britain accepted one phase of widespread glaciation in the last cold stage, and set about defining similar records in previous cold stages.

The record of glacial deposition has always been complex, but the recognition that Heinrich Events are related to short term behaviour in ice-sheets allows us to reinvestigate Welsh glacial history from a new angle. The work of Bowen *et al.* (2002) and some localised studies in Ireland (*i.e.* Evans and O'Cofaigh, 2003; O'Cofaigh and Evans, 2001; Richards, 2002; Richards, Waller and Bloetjes, 2002) suggest that Ireland, including the Irish Sea basin, was subject to up to three phases of glaciation during the Devensian Cold Stage. Absolute dating of glacial materials, in association with other techniques, has also been used recently to resolve problems concerning the glacial history of the English Midlands and the suitability of the Wolstonian Stage Type Site (Maddy *et al.*, 1995).

The use of new dating methods in conjunction with strictly defined glaciostratigraphic units is crucial for defining a framework within which glacial history may be understood. The presence of ice-sheets and glaciers in the landscape represents intensely cold conditions. Such conditions are readily exhibited in the climate stratigraphy derived from high-resolution, accurately dated marine records. As ice-masses have been the main agents for driving landscape change in the Quaternary, such a strategy is appropriate in order to relate terrestrial and marine records of palaeoenvironmental change and as a result, fully appreciate the nature, timing and extent of important events in Quaternary landscape evolution.

3 North-west Wales
by Danny McCarroll

Introduction

The north-west hosts the highest and most spectacular scenery in Wales, much of its rugged beauty resulting directly from erosion by glaciers during successive glaciations, and also some of the finest lowland and coastal scenery. For the glacial geomorphologist it offers a rich and varied selection of landforms and deposits, rivalling those of any other part of these islands. The high ground ensures glacier formation even in short-lived cold phases, such as the Loch Lomond Stadial, whilst in major glaciations ice moving down the Irish Sea basin would cross the low ground of Anglesey and western Lleyn, meeting local Welsh ice in Arfon and eastern Lleyn. Proximity to the sea offers the great advantage that many of the sediments are exposed in large and extensive coastal sections, kept fresh by the action of winter storms. It is unsurprising, therefore, that the landforms and deposits of north-west Wales have featured strongly in debates about the Quaternary history of Britain and Ireland since the concept of glaciations was first conceived.

When Whittow and Ball (1970) reviewed the evidence from this area more than 30 years ago, it was approaching a time of transition in Quaternary geomorphology. Up to that time glacial events were often separated on the basis of contrasts in the 'freshness' of glacial landforms, by the number of 'tills' exposed and by the presence of 'weathered horizons' within drift sequences (Mitchell, 1960, 1972; Synge, 1964). It was on the basis of such evidence that they erected their sequence of events, which included three glacial 'stages'. Shortly after the work was published, and even in the same volume, each of these criteria was questioned. Bowen (1970, 1973), for example, argued against division on the basis of morphology and number of tills ('counting from the top'), advocating a more geological approach, using stratigraphic marker horizons to correlate between sections. A little later Boulton (1977), using Glanllynnau on Lleyn as the example, demonstrated that multiple till sequences can be produced in a single glaciation. This eventually proved a death-blow to the concept of a widespread tripartite division of drift deposits into lower and upper tills, representing different glacial events, separated by interglacial or interstadial sands and gravels. The concept of weathered horizons, once so critical in debates about glacial limits, but for which there was never very clear evidence, took longer to fade away and has only recently been challenged directly (Walden and Addison, 1995). The review by Whittow and Ball (1970), therefore, though the first to cover the region, rapidly became outdated.

Since that time the area has continued to provide critical evidence in debates over the extent, timing and environmental conditions of glaciations and retreats. The calcareous Irish Sea drifts of Lleyn, for example, with their abundant microfauna of marine foraminifera, have once again been invoked as evidence for deposition in the sea, with the sand and gravel terraces of Lleyn being interpreted as marine Gilbert-type deltas (Eyles and McCabe, 1989). The ensuing debate between the 'glacimarine' and 'terrestrial' schools is strangely reminiscent of the debate between the 'diluvialists' and 'glacialists' more than a century earlier (McCarroll, 2001). The excellent exposures of the region have continued to be used as examples of glacial sedimentology and deformation structures, and to provide the venues for many university field classes. The glaciated mountains are even more popular venues for field classes at all levels, although some of the common interpretations with regard to ice thickness, direction of flow and the origin of the major glacial troughs, have been challenged (Gemmel *et al.*, 1986; McCarroll and Ballantyne, 2000). It is timely, therefore, to once again review the evidence for glaciations in north-west Wales, in the vain hope that advances in Quaternary science will not render this version outdated as quickly as that in the previous edition.

Pre-Devensian events

Although at a local scale the landforms of north-west Wales are dominated by the direct effects of glaciations, and the periglacial and paraglacial conditions that bracket them, the larger-scale topography of the region retains much evidence of the pre-Quaternary landscape. The most striking feature is the general accordance of summits in Snowdonia and the contrast with the much lower ground of Anglesey, Arfon and Lleyn. Brown (1960) placed the high ground in his upland plateau and the lowlands in the coastal plateau. Anglesey, with its broad plain and isolated hills, for example, was interpreted as the product of two episodes of Tertiary marine planation (Brown, 1960; Embleton, 1964). Battiau-Queney (1984) has reinterpreted the landscapes of both Anglesy and Snowdonia in terms of deep tropical weathering followed by removal of the saprolite to reveal an uneven 'etchplain', with resistant rocks protruding as inselbergs. Deep weathering may have begun as early as the Triassic, since Anglesey and probably Lleyn must have emerged by that time to supply sediment to the area of the Mochras borehole, near Harlech (Harrison, 1971; Allen and Jackson, 1985).

The steep scarp between Snowdonia and Anglesey has long been regarded as an ancient feature, fossilized by Mesozoic rocks and then exhumed by Cenozoic denudation (Jones, 1952). However, as long ago as 1938, Greenly suggested that it must be much younger, since Eocene dykes, of distinctive lithology, crop out at a low elevation on Anglesey but at 750–780 m in Snowdonia. Had the topographic difference existed, the dykes would have become extrusive lava flows over Anglesey. The fact that the dykes are intrusive demonstrates that there could have been little if any difference in the level of the Anglesey and Snowdonian surfaces at that time. The available evidence thus suggests that the high plateau of Snowdonia is equivalent to the coastal plateau of Anglesey, that both pre-date the Tertiary, and that the uplift of Snowdonia is relatively recent (post Eocene).

Study of the larger scale landforms of Britain has been out of fashion for some decades, and the origin of the pre-Quaternary landscape of north-west Wales remains obscure. However, the accordance of summits and the presence of distinct surfaces in the landscape have not become any less obvious. A clear explanation of the large-scale pre-Quaternary landscape features of north-west Wales, supported by dating evidence, remains an outstanding challenge for British geomorphology.

Evidence for pre-Devensian glaciations of north-west Wales is restricted to limestone caves in the Vale of Clwyd. The most famous site is Pontnewydd Cave in the Elwy Valley, near Rhyl (Green, 1984). The deposits in this cave are not in primary position, having been re-worked by debris flows, but there is a range of dating evidence suggesting that they extend back to at least Oxygen Isotope Stage 8. The cave sediments have yielded the oldest evidence of human occupation of Wales, from deposits attributed to the OI Stage 7 interglacial. The finds include seven human fragments, both teeth and bone, as well as many artefacts, including 32 handaxes. The presence of Irish Sea erratics demonstrates that there must have been an earlier Irish Sea glaciation of the region, probably in OI Stage 8.

The Tremeirchion Caves in the Vale of Clwyd also contain pre-Devensian sediments, and have yielded human remains and artefacts indicating occupation somewhere in the range 40 to 20 ka, most probably in OI Stage 3 (Campbell and Bowen, 1989). The cave is sealed by Irish Sea till and a radiocarbon date of 18,000 (+1400, -1200 BP) on a mammoth carpal has been used as a maximum age for the last glaciation (Rowlands, 1971). The Cefn Caves of the Elwy Valley have also yielded human remains and artefacts, probably from the Upper Palaeolithic and Neolithic, as well as both temperate and interglacial faunas (Campbell and Bowen, 1989). The records are poor, so the context of the finds is uncertain.

Beyond the caves, sediments pre-dating the Devensian are sparse. Cemented gravels in the Irish Sea drift at Porth Oer, in western Lleyn were interpreted as a raised beach by Jehu (1909) and Synge (1964) and assigned by Bowen (1973) to the Ipswichian (5e) and used as a stratigraphic marker. Gibbons and McCarroll (1993) have demonstrated that such cemented gravels are common in western Lleyn, occurring throughout the highly calcareous drift sequences, and that they are not raised beaches. The other raised beach used as a stratigraphic marker in north-west Wales was Red Wharf Bay on Anglesey, where rounded limestone pebbles occur within the local head deposits. There are no shells present, in contrast to the Ipswichian (limestone) beaches of Gower, and the 'marine platform' is clearly striated, so the interpretation of the deposits as beach must be treated with caution. Gravels at Llanddona on Anglesey have also been interpreted as raised beach and assigned to OI Stage 5e (Campbell *et al.*, 1995).

Outside of caves, the only post-Ipswichian pre-glacial sediments in north-west Wales that have received more than a passing mention are those exposed in the brickworks at Pen-y-bryn, near Caernarfon. Here there are organic horizons within muds, sands and gravels overlying locally-derived gravel and underlying both Welsh and Irish Sea tills. Pollen analysis suggests relatively warm conditions and they are tentatively correlated with Oxygen Isotope substages 5a and 5c (Addison and Edge, 1992; Chambers *et al.*, 1995; Bowen, 1999).

Other sediments that post-date the Ipswichian interglacial but underlie Devensian glacial sediments are widespread around the coast of north-west Wales but remain poorly described. Generally described as 'head', comprising gravitationally transported locally derived material, they occur at many of the well-known Quaternary sections of southern Lleyn, where they were protected from glacial erosion in the lee of pre-glacial cliffs. They occur at Aberdaron, Porth Ceiriad and Porth Neigwl, but are perhaps best displayed at Porth Ysgo, to the east of Aberdaron in western Lleyn, where 10 m of angular local material is incised by a small stream. The overlying Irish Sea till at this site includes huge boulders of ultrabasic rock from the local area (Gibbons and McCarroll, 1993). These 'head' deposits have generally been ascribed to periglacial conditions in advance of the last glaciation. However, more than one hundred thousand years separates the warmest phase of the last interglacial and the Last Glacial Maximum, and during that time climatic fluctuations led to quite marked changes in environment, with associated changes in flora and particularly fauna (Currant and Jacobi, 2001). Apart from some limestone cave deposits, generally poorly excavated, little is known of this period in north-west Wales, so there may be a lot more to be gleaned from these widespread but rather neglected deposits.

Fig. 3.1. Location map of the study area. Shading indicates high ground. The three 'sarns' (Welsh for ford) extending into Cardigan Bay are submarine boulder ridges interpreted as lateral moraines of glaciers flowing out of the major valleys. The box indicates the area covered by Fig. 3.2. After McCarroll and Ballantyne (2000)

Glaciation of the Mountains

The mountains of north-west Wales are the highest in Britain, south of the Scottish highlands (Fig. 3.1), reaching 1085 m on Yr Wyddfa (Snowdon). In Snowdonia the high mountains are concentrated in three massifs (Fig. 3.2), aligned roughly on a south-west/north-east axis, dissected by the north-west/south-east orientated glacial troughs of Llanberis Pass and the Ogwen Valley (Nant Ffrancon and Nant-y-benglog). The Snowdon range includes seven peaks above 800 m forming a 'horseshoe' around the stepped cirques of Glaslyn and Llydaw. Steep cliffs form the backwalls of cirques on all sides. Across the Llanberis Pass, which cuts through the mountains at an altitude of only 356 m, the central Glyderau massif forms a broad curving ridge rising to 999 m at Glyder Fawr. To the north, beyond the famous cirque of Cwm Idwal (Addison, 1983, 1988), another glacial trough cuts through the mountains at low altitude, reaching 303 m east of the junction of Nant-y-benglog and Nant Francon, where the Ogwen river turns to the north. Beyond this the Carneddau rise abruptly at Pen yr Ole Wen (979 m) and continue to the north-east, reaching 1064 m on Carnedd Llewelyn before descending gently to the north and north-west. To the west of Snowdon Moel Hebog (782 m) and Garnedd Goch (700 m) form the eastern margin of Lleyn. To the south-east of the main massifs, Nantygwryd separates the Glyders from the north-west/south-east orientated ridge of Moel Siabod (872 m), which merges to the south-west with the Moelwyn range, with Moelwyn Mawr reaching 770 m.

To the south of Snowdonia the mountains continue (Fig. 3.1). Along the west coast, between Porthmadog and Barmouth there is just a narrow lowland strip rising steeply to the north/south ridge of the Rhinogs, rising above 750 m at its southern end. South-east of Snowdonia lie the high moors of the Migneint, with the Arenigs to the south. South of Barmouth and Dolgellau, Cadair Idris, one of the most spectacular massifs in Wales rises to 892 m, with the high and remote Arans to the east (Aran Fawddwy 905 m). The Arans and Arenigs are separated by the fault-guided valley occupied by Llyn Tegid (Bala Lake). This fault continues on the south-east side of Cadair Idris, forming the Tal-y-llyn Valley. To the west of these mountains, the margins of glaciers leaving the main valleys are marked by the 'sarns', which are lines of boulders (medial moraines) leading out into Cardigan Bay, visible at low tide (Fig. 3.1). The eastern end of Snowdonia is marked by the major north/south Conwy Valley, with the high ground of the Denbigh Moors and Clwydian hills stretching on towards the Cheshire–Shropshire lowlands. The east/west orientated Dee Valley, passing through Llangollen, separates this area from the Berwyns to the south.

Whittow and Ball (1970) argued for two 'stages' in the glaciation of the region, with contrasting effects on the mountains (Fig. 3.3). In the first stage they suggested that ice moved radially from a centre near Llyn Tegid (Bala) and that it was able to force its way through Snowdonia, carving the major troughs of Llanberis Pass and the Ogwen Valley. In the second stage they envisaged thinner ice in the mountains, with periglacial features, including tors and blockfields, developing on nunataks. Both stages were regarded as post-dating the last full interglacial. For decades, field classes have stood on the roche moutonnée beneath Cwm Idwal, close to the former watershed between Nant Francon and Nant-y-benglog and been told of the great ice-sheet that tore its way through from the south-east. In fact there is no field evidence that ice has ever over-ridden the mountains of Snowdonia and the breaches can be explained quite satisfactorily by the action of local ice (McCarroll and Ballantyne, 2000).

The many cold stages in the marine oxygen isotope records suggest that the mountains of Wales have probably been glaciated many times. Indeed, since most of the Pleistocene is characterised by conditions much cooler than the present interglacial, cirque and small valley glaciation is probably the 'average' condition for north-west Wales. Most of the glacial erosional landforms, therefore, including the cirques and valleys, can be regarded as the product of multiple glaciations. Under such 'average' conditions ice would radiate away from the highest massifs, including Snowdonia. Only in unusually large, or 'major' glaciations would there be sufficient ice in north-west Wales to impede and disrupt the radial drainage, leading to the breaching of former ice divides.

The surface altitude of the last major glaciation of the mountains was reconstructed by McCarroll and Ballantyne (2000) by mapping the 'periglacial trimline' distinguishing rock showing evidence of glacial erosion from the periglacial landscapes of former nunataks (Fig. 3.2). Evidence for glacial erosion

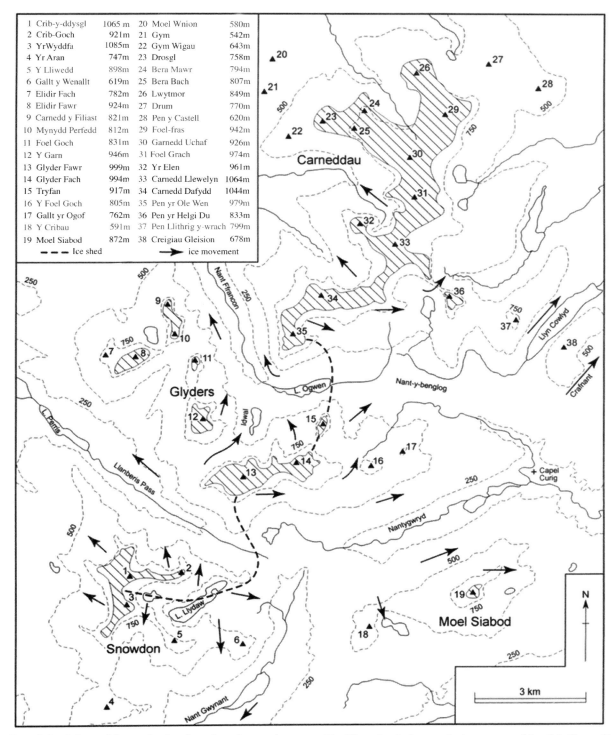

Fig. 3.2. Map of Snowdonia showing the major massifs. The shaded area is interpreted by McCarroll and Ballantyne (2000) to have lain above the last ice-sheet, forming nunantaks. Arrows indicate the general direction of ice movement at the maximum extent of the last glaciation. Note that ice radiated away from an ice-shed aligned north-east/south-west rather than crossing the area from the south-east. After McCarroll and Ballantyne (2000)

includes striae, grooves, perched boulders and ice-scoured surfaces, whereas former nunataks are dominated by frost-weathered debris, *in situ* blockfields and remnant tors. Remnant soils on former nunataks often contain gibbsite, an end product of the weathering of silicate minerals. This methodology has been used successfully elsewhere (Ballantyne *et al.*, 1998), but the trimlines in Snowdonia are amongst the clearest in Britain. In the Glyders, the contrast between the ice-scoured terrain and the former nunataks is particularly striking. At the eastern end of the ridge ice moulding, with roches moutonnées, extends up to 820 m and at 860 m on the eastern spur of Glyder Fach there are rounded boulders perched on bedrock slabs. On the summit of Glyder Fach, and on Castell y Gwynt, however, there is a most spectacular blockfield, with great cantilevered slabs. At the opposite end of the ridge, the col (at 715 m) above Cwm Idwal hosts clearly ice-moulded slabs, showing ice spilling out of Lanberis pass into the head of the cirque, whilst the summit of Elidir Fawr (923 m) is thickly mantled in large angular blocks, with no evidence of glacial erosion.

Across the Ogwen Valley in the Carneddau, the summits are again swathed in blockfield but there are also many examples of remnant tors, the finest of which rises above blockfield on Yr Aryg (865 m). The whole of the main Carneddau ridge stood above the last ice-sheet as a nunatak and it presents one of the most impressive periglacial landscapes in Britain, with blockfield, tors, erected slabs, gelifluction lobes, large sorted circles and sorted stripes.

McCarroll and Ballantyne (2000) placed the surface of the last ice-sheet at a maximum altitude of about 850 m over the three main massifs of Snowdonia (Fig. 3.2). On Moel Siabod, to the south-east, the ice surface was lower (800–830 m) and erosional evidence shows that ice moved around Moel Siabod from the west-north-west, from Nantygwryd, crossing the ridge high on the mountain. There is no evidence of ice moving from the south-east towards Snowdonia.

On Cadair Idris there is a clear trimline at 740–750 m, above which there are tors and blockfields, with remnant soils also yielding gibbsite (Ballantyne, 2001). Ice crossed this area from the east, but on Aran Fawddwy, 19 km to the east, the ice crossed the 900 m ridge from the west, leaving perched boulders (McCarroll and Ballantyne, 2002). This places the ice-shed higher than 900 m somewhere between the two mountains, consistent with the conclusions of Foster (1970), who mapped striations up to the summit of Rhinog Fawr (720 m) in the Rhinog mountains to the north of Cadair Idris. A high ice-shed is also supported by the evidence on Arenig Fawr (854 m), which McCarroll and Ballantyne (2000) suggest was also over-run by ice from the west, though this is disputed (Hughes, 2002a, 2002b).

The picture that emerges for the last glaciation of north-west Wales is of a major centre of dispersal, probably with a north/south axis, located in Merioneth north of Dolgellau (Fig. 3.3). The highest point must have been above 900 m and ice moved to the west and the east, covering much of the Rhinogs and the Arans, Arenigs and Berwyns but leaving a trimline on Cadair Idris at 740–750 m. Snowdonia was an independent ice centre, initially radiating ice in all directions, but eventually impeded to the south by the Merioneth ice. Where the ice from the two centres met it was forced west, along the Vale of Ffestiniog and out into Tremadog Bay, where a lobe of ice reached as far west as St Tudwal's Peninsula on Lleyn (Nicholas, 1915; Young *et al.*, 2002). Sarn Badrig is a medial moraine formed between this lobe and another leaving the mountains near Barmouth. To the east of Snowdon the ice was pushed east, crossing the ridge beyond Capel Curig to join a major outlet glacier in the Conwy Valley. Merioneth ice moved down the valley of the Dee, past Llangollen, and south of the Berwyns a lobe reached as far east as Shrewsbury (Thomas, 1989).

The age of the last major glaciation of the mountains remains very poorly constrained. The only date from the mountains is a ^{36}Cl exposure age from a site above Cwm Idwal. However, the sampling site has been described as both 'the plateau above Cwm Idwal' (which was a nunatak during the last glaciation) and as 'a col above the cirque headwall' (which was ice-scoured) and the date has changed from 32.9 ± 2 ka to 38.5 ± 5.4 ka (Bowen, 1999; Bowen *et al.*, 2002). Bowen *et al.* (2002) have argued that the last major glacia-

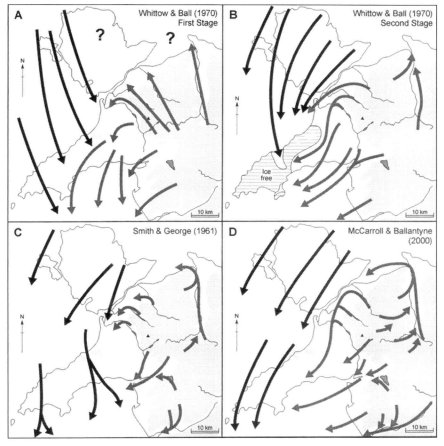

Fig. 3.3. Ice directions in north-west Wales as suggested by Whittow and Ball (A and B), the Geological Survey (C) and by McCarroll and Ballantyne. (D) After McCarroll and Ballantyne (2000)

tion of the mountains pre-dates the Late Devensian 'Last Glacial Maximum' and may correspond with Heinrich event 4 at around 40,000 BP. Clearly, more dates are required to test this hypothesis.

Anglesey and Arfon

The glacial deposits of Anglesey, including drumlins and erratic trains, were mapped by Edward Greenly (1919) and the Menai Straits and the Arfon lowlands, north-west of Snowdonia were studied by Clifford Embleton (1964). All of the evidence suggests that Anglesey was glaciated from the north-east, by ice moving down the Irish Sea. This probably happened several times, but any evidence pre-dating the last glaciation has been erased. During each major glaciation Irish Sea ice on Anglesey must have met ice radiating northwards out of Snowdonia and also from the major outlet glacier in the Conwy Valley. Ice would have been pushed to the south-west and at some stage, though not necessarily the last glaciation, meltwater would have carved out the Menai Straits. Greenly (1919) noted the odd distribution of mainland Welsh erratics, which are absent from the Anglesey shores of the Menai Straits but occur along a zone aligned with the Straits a little inland. Whittow and Ball (1970) attributed these to a very early glaciation, but a much simpler explanation is that they represent the flow line of the ice which moved out of the Conwy valley to join the Irish Sea ice as it crossed Anglesey towards Lleyn.

The glacial deposits of Anglesey contrast markedly with those on Lleyn to the south in that they are dominated by local materials. Only along the east coast are erratic-rich tills common. This led Whittow and Ball to propose a later 'Liverpool Bay phase', with ice just impinging on the east coast of Anglesey, with a lobe extending from Pentraeth to Newborough. They did, however, raise the possibility that the erratic-rich tills and the underlying locally-sourced tills could belong to the same event. The latter explanation now seems the most likely. Harris (1991), for example, obtained access to excavated and cleaned drift sections at Wylfa Head, northern Anglesey, and was able to show the locally-derived material interdigitates with the erratic-rich till, suggesting that they were formed in a single event. He suggests that as the Irish Sea glacier, which rested on soft sediment, met Anglesey, subglacial variations in meltwater pressures caused localised zones of sediment deposition and bedrock erosion. The landscape of Anglesey represents a subglacial landsystem, with widespread erosion and transport of local bedrock and formation of drumlins.

Perhaps the best known and certainly the most spectacular drift section along the Arfon coast is that exposed by coastal erosion of a small hill at Dinas Dinlle, south of Caernarfon. Here there are both Welsh and Irish Sea tills exposed, but the most remarkable feature is the steep northward dip of the beds. The site has been interpreted in a variety of ways, but a multidisciplinary study by Harris *et al.* (1997) has demonstrated that the whole section has been compressionally deformed (Fig. 3.4). An original sequence of two diamicton units separated and overlain by sands and gravels has been compressed by about 54%, with the shear taking place along discrete planes, with striated and slickensided surfaces, between thrust blocks, or nappes. The feature is best described as a thrust-block moraine. On the south side of the hill Late Glacial organic deposits (Coope, 1977) overlying vertically deformed sands and gravels represent a kettle hole formed by the melting of an ice block.

Perhaps the most enigmatic glacial deposits in this area are the 'shelly drifts' that occur at 400 m on the flanks of Moel Tryfan, in the foothills of Snowdonia, south of Caernarfon. The once extensive deposits were exposed by slate quarrying and became a focal point in the debate between the 'diluvialists' and 'glacialists' of the nineteenth century. The history of research at the site is reviewed by Campbell and Bowen (1989), who also describe the few remaining and generally degraded sections. Although there are records of Irish Sea till at the site, these have not been confirmed and the shell-bearing sediments are mainly gravel. Most of the shells are broken and polished, but there were layers of complete shells, from which 56 different species have been identified, indicating a mixed assemblage. The Irish Sea sediments are overlain by a till of Welsh origin. Although Moel Tryfan is the most famous site, shelly drifts have been widely reported from other locations in the Snowdonian foothills (Campbell and Bowen, 1989).

The age of the high level 'shelly drifts' of Moel Tryfan has been much debated and remains unresolved. Foster (1968, *in* Campbell and Bowen, 1989) obtained a radiocarbon date of 33,740 BP (+2100,

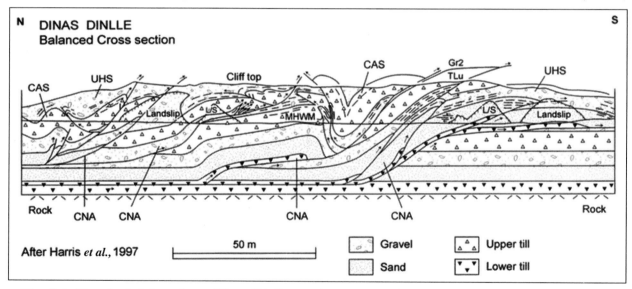

Fig. 3.4. The glacial deposits at Dinas Dinlle, on the coast south of Caernarfon with the sub-surface geology reconstructed using seismic reflection techniques. An originally simple sequence of two diamicts (tills) separated by sands and gravels has been compressed to form a thrust-block moraine. Codes used to describe the deformation structures as produced by pure shear (P), simple shear (S), Compressional (C), Vertical (V) or Undeformed (U) are: CAS anticlinal and synclinal folding, UHS horizontal bedding, CNA nappes. MHWM is mean high water mark. After Harris et al. *(1997) and McCarroll and Rijsdijk (2003)*

-1800 BP) on bulk shell material, and suggested that since the shells represent a mixture of environments, and must be derived, the glaciation must post-date *c.* 34,000 BP. However, radiocarbon dates on bulked shell material are generally regarded as unreliable. Campbell and Thompson (1991) examined quartz grain surface textures, confirming that the sediments have been dredged from the floor of the Irish Sea and then deposited in a fluvioglacial environment. They noted only slight post-depositional weathering, which might support the interpretation of the sediments as Late Devensian.

Most recently, Bowen *et al.* (2002) have re-interpreted the high level shelly drifts of Moel Tryfan and adjacent mountains as part of the evidence for a very extensive pre-Late Devenisan glaciation of Britain and Ireland. The argument is based on amino-acid ratios and the radiocarbon date on bulked shell, neither technique being particularly reliable (McCarroll, 2002). Whatever the age, the sediments demonstrate that Irish Sea ice at some time was able to push high into the foothills of Snowdonia and that the sediments it deposited were later overrun by a glacier moving out of the Welsh mountains. Whether this occurred in an extensive pre-Late Devensian glaciation or during advance of the Irish Sea glacier during the Late Devensian remains moot.

Lleyn

For the glacial geomorphologist, the Lleyn Peninsula represents one of the most interesting areas in the whole of Wales (Fig. 3.5). It was here that the glaciers from the Irish Sea basin and the Welsh mountains met and eventually uncoupled, and marine erosion ensures that the wide range of sediments are well exposed. The topography of the peninsula is important in explaining the glacial sediments, and particularly the marked difference between Lleyn and Anglesey. Whereas Anglesey is relatively flat, Lleyn has a series of hills stretching along the north coast, reaching 564 m at Yr Eifl. On the south coast St Tudwal's Peninsula is an area of relatively high ground forming a natural barrier bounded on the west by the large, drift-filled embayment of Porth Neigwl (Hell's Mouth). A series of smaller drift-filled embayments lie to the west of St Tudwal's, including Aberdaron Bay.

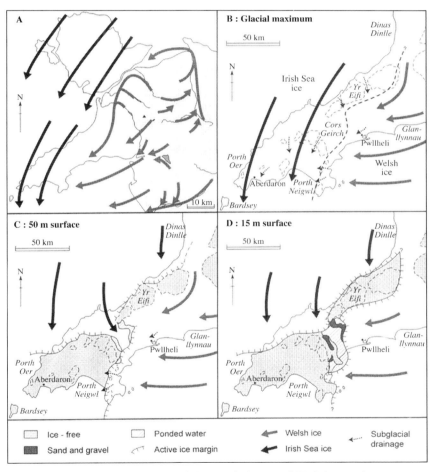

Fig. 3.5. Uncoupling of the Irish Sea and Welsh ice-sheets on the Lleyn Peninsula and formation of the sand and gravel terraces at 50 m and 15 m above sea level around Cors Geirch. After McCarroll (2001) and Young et al. (2002)

The hills of north Lleyn formed a barrier to ice from the north and are cut by a series of impressive meltwater channels, now hosting diminutive streams. The largest, Nant Saethon, has walls reaching 70 m high and cuts straight towards the Porth Neigwl embayment, ending at Nanhoron. To the west, Nant Llaniestyn is wider, and just as impressive, but more difficult to access. There are also channels cutting generally east to west through a bedrock ridge north of Pwllheli, one of which carries the A499, and on St Tudwal's Peninsula. The most impressive is that which now carries the Afon Soch. This route is unusual in that the river flows across the drift of the Porth Neigwl embayment to within 1 km of the coast, and at less than 5 m above sea level it turns abruptly north-east to flow through the channel to Abersoch.

Close to the northern hills there is an abundance of sand and gravel banked against the hills and extending through the gaps, and this has become known as the Clynnog Fawr moraine. South of the hills in central Lleyn, on either side of a low wetland (Cors Geirch), extensive sand and gravel terraces occur at about 50 m, dropping abruptly to about 16 m nearer the south coast. The highest remnants of sand and gravel occur at 78 m. In eastern Lleyn there are also extensive sand and gravel deposits, providing a valuable resource (Crimes *et al.*, 1992; Thomas *et al.*, 1998). Glacial diamicts occur extensively in southern and western Lleyn but there is a clear divide marked by the high ground of St Tudwal's Peninsula. To the west the material came from the Irish Sea, whereas to the east it came from the mountains of Snowdonia.

Whittow and Ball (1970) regarded the Irish Sea drift of Lleyn as older than that on Anglesey and suggested that 'all of the striae and stone orientations of Bardsey Island and south-western Lleyn show ice movements from north-north-west to south-south-east' (p.25). In fact the striae are aligned north-east/south-west, veering to the south around the hills of the west coast, and the clast fabrics of *in situ* tills, such as the lower diamict at Aberdaron, also show clasts aligned towards the north-east, parallel to the ice movement over Anglesey. It seems more reasonable to assume that Anglesey and Lleyn were glaciated in the same event, and that the Clynnog Fawr moraine marks a topographically controlled still-stand during retreat, rather than a re-advance (McCarroll, 1991).

The glacial deposits of Lleyn have attracted attention for more than a century. Those of western Lleyn are rich in northern erratics and they have all of the characteristics of marine sediments scraped from the floor of the Irish Sea, including an abundance of broken marine shell fragments and well preserved foraminifera. Such characteristics were originally seen as evidence for the Biblical Flood, but for almost a century they have been regarded as terrestrial tills with fluvioglacial outwash. Eyles and McCabe (1989) questioned that axiom, preferring to interpret the 'Irish Sea tills' of Lleyn and elsewhere as evidence for high relative sea levels during retreat of the last Irish Sea glacier. Their argument was based mainly on sedimentology, but they used the foraminifera, amino acid age estimates and the presence of supposed 'Gilbert-type deltas' as supporting evidence. They used the sediments at Aberdaron and Porth Neigwl as examples of glacimarine sediments and argued that the terraces around Cors Geirch represent glacimarine deltas, formed as the ice-margin was temporarily pinned at the north Lleyn coast. All of the evidence that they used has been disputed (McCarroll, 2001), and Lleyn has provided key sites at which to test the competing 'glacimarine' and 'terrestrial' hypotheses.

Aberdaron, which Eyles and McCabe describe as a glacimarine 'valley infill complex' has been interpreted by McCarroll and Harris (1992), Gibbons and McCarroll (1993) and McCarroll and Rijsdijk (2003) as entirely terrestrial (Fig. 3.6). There are two diamict units, with similar sedimentary characteristics but very different deformation structures. The lower diamict shows intense deformation by simple shear, suggesting subglacial deformation. The upper diamict is relatively undeformed, apart from some gentle flexures and small scale normal faulting, and includes channels of sand and gravel. It is interpreted in terms of paraglacial remobilisation of Irish Sea tills into the Aberdaron embayment, with sediment flows stacking or disaggregating as they entered ephemeral streams. The foraminifera are not *in situ*, since their abundance does not change up the section, in what would be increasingly distal glacimarine sediments. They behave like small clasts, with abundance determined by sediment grain-size and

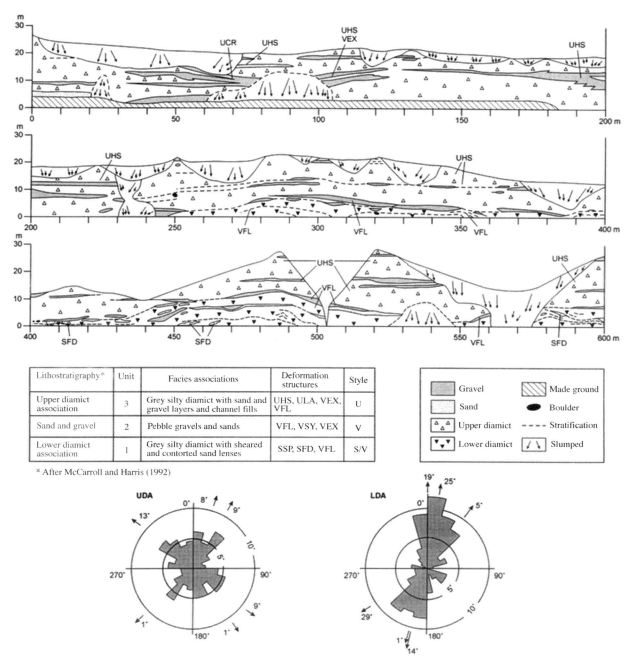

Figure 3.6. The glacial deposits at Aberdaron, the western end of the Lleyn Peninsula. The fabric diagrams show the long axis orientations of 180 clasts from within the 'upper diamict association' and 'lower diamict association'. Arrows around the circumference indicate the preferred orientation and dip of six separate samples. Codes used to describe the deformation structures as produced by pure shear (P), simple shear (S), Compressional (C), Vertical (V) or Undeformed (U) are: UHS horizontal bedding, ULA lamination, VEX extensional fractures, VFL flexures, VSY synforms, SSP simple shear profiles, SFD shear thrust fractures and dykes. After McCarroll and Harris (1992) and McCarroll and Rijsdijk (2003)

sorting (Austin and McCarroll, 1992). The Irish Sea glacial sequence at Porth Neigwl was also interpeted by Eyles and McCabe in terms of glacimarine sedimentation, but here they suggested that the sequence was capped by a 'regional mud drape', representing distal rainout of fine sediment. Young et al. (2002) have argued that the waterlain sediments are restricted to the eastern end of the section, filling a small local pond.

Eyles and McCabe (1989) used the Cors Geirch terraces as an example of 'Gilbert-type' glacimarine deltas on the basis that there is no source for the sand and water other than the Irish Sea glacier and no topographic barrier against which fresh water could be ponded. They also argued that 'south of these deltaic complexes there is no geomorphic evidence of ice-margins that may have dammed lakes against the peninsula' (p. 331). In fact there is abundant and rather obvious evidence for an ice-margin south of the terraces because all of the drift in southeastern Lleyn is of Welsh rather than Irish Sea origin, deposited by a glacier moving west out of the Vale of Ffestiniog (Nicholas, 1915; Young et al., 2002). Interpreting the terraces as glacimarine requires relative sea level to drop from 86 m to 16 m whilst the glacier remained stationary (McCarroll, 1995). It is much simpler to interpret the sand and gravel terraces in terms of ponding of fresh water between the two glaciers as they uncoupled, with the drops in height representing a change in the lake level as the ice thinned and retreated (Fig. 3.5). A similar explanation can be used for sand and gravel deposits in eastern Lleyn (Crimes et al., 1992; Thomas et al., 1998), which Eyles and McCabe (1989) also interpreted as glacimarine deltas.

The glacier moving down the Vale of Ffestiniog reached as far as St Tudwal's Peninsula, where it carved meltwater channels, and may have extended over into Porth Neigwl after the Irish Sea ice had stagnated and decayed (Young et al., 2002). When the Welsh ice decayed and melted it left a landscape dotted with kettle holes, formed by the melting of buried ice blocks. Where this kettled topography meets the coast, at Glanllynnau near Criccieth, it provides one of the best known Quaternary sites in Wales.

The section at Glanllynnau is more than 360 m long and reaches 10 m high. A ridge of clay-rich dense grey (lower) diamict is overlain by sand and gravel and by a cryoturbated gravelly (upper) diamict. The surface of the lower diamict is incised by wedge-shaped fractures, which are stained brown, and contorted laminated clays separate the lower diamict from the overlying sand and gravel. The sequence has been interpreted as the result of two separate advances, with the wedge-shaped fractures and contorted clays representing intermediate periglacial conditions (ice wedges and weathering). Saunders (1968) suggested that the lower diamict was deposited by ice flowing from the north-east and the upper diamict by ice from the east, implying two advances separated by a cold interstadial. This interpretation was followed by Whittow and Ball (1970), who ascribed the lower diamict to the 'Criccieth Advance' and the upper diamict to the 'Arvon Advance'. The subdivision was retained by Bowen et al. (1986) who proposed a Gwynedd re-advance of both the Welsh and Irish Sea glaciers, later raising the possibility that the lower diamict may be Early Devensian, the zone of weathering and ice wedges Middle Devensian and the upper diamict Late Devensian (Campbell and Bowen, 1989). Boulton (1977) interpreted the whole sequence in terms of a single glaciation, with the upper diamict representing flow tills.

The glacial sediments at Glanllynnau were examined in detail by Harris and McCarroll (1990) and Young et al., (2002), who followed Boulton (1977) in interpreting the whole sequence in terms of a single glaciation (Fig. 3.7). Buried debris-rich glacier ice formed ridges feeding flow tills into intervening fluvial channels, accumulating sand and gravel. When the buried ice melted the topography was reversed, with the gravel forming ridges and buried ice kettle holes. The inclined wedges are regarded as tension fractures on the lee side of a drumlin-like subglacial bedform rather than as ice wedges, so there is no evidence for a period of weathering under periglacial conditions (Harris and McCarroll, 1990; Walden and Addison, 1995). The 'laminated clays' may represent a shear zone at the base of the glacier. There seems to be no need to invoke a 'Gwynedd re-advance' of either the Irish Sea or the Welsh glaciers (McCarroll, 1991).

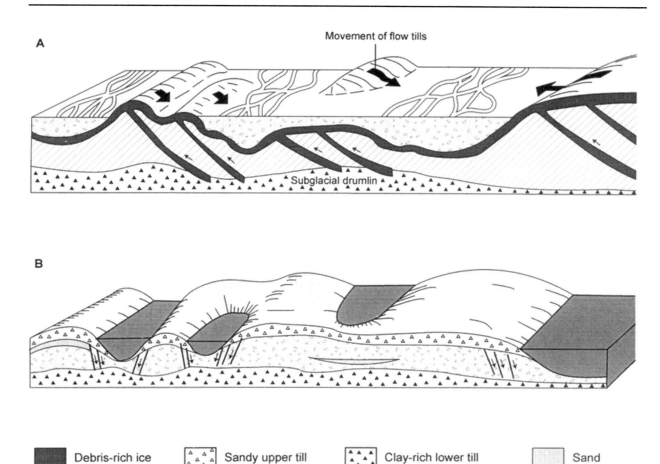

Fig. 3.7. Sketch of the origin of the sediments and landforms at Glanllynnau near Criccieth. Ridges of stagnant ice feed flow tills into braided channels, but when the ice melts it forms the kettle holes, resulting in a reversal of topography. After Harris and McCarroll (1990) and Young et al. (2003)

At the western end of Glanllynnau, kettle holes have been exposed at the coast, though the best sections have now been eroded away. Sediments at the base of these depressions accumulated immediately after the ice retreated and record conditions during the Devensian Late Glacial. A radiocarbon date from the lowest organic material yielded a date of 14,628 ± 300BP (Coope and Brophy, 1972), and deglaciation before 14,000 BP is supported by dates on lake sediments at Llyn Gwernan, near Cadair Idris (Lowe and Lowe, 2001). Deglaciation may have been triggered by moisture starvation rather than warming, since cold conditions continued until about 13,000 BP. Glanllynnau was one of the key sites used to demonstrate the value of *coleoptera* (beetles) in recording the rapidity of warming at the start of the Windermere insterstadial (Coope and Brophy, 1972).

The Late Glacial
During the Loch Lomond Stadial of the Late Devensian Late Glacial, glaciers reformed in many of the mountain cirques in north-west Wales, the best examples being in Snowdonia and on Cadair Idris. The cirque moraines of Snowdonia were mapped by Gray (1982), who suggested a rise in the equilibrium line

from 450 m in the south-west to 700 m in the north-east, implying a July temperature at sea level of about 8.5°C. Perhaps the best example of the effects of a Loch Lomond Stadial glacier can be seen on the flanks of Snowdon, where lowering of the water level in Llyn Llydaw has revealed beautifully striated bedrock surfaces. There are two sets of striae here, with an earlier set indicating ice movement during the glacial maximum, and a second set marking the more constrained flow of the Loch Lomond ice (Gray and Lowe, 1982; Sharp et al., 1989).

Gray (1982) also mapped periglacial features formed during the Loch Lomond Stadial, including protalus ramparts formed by debris moving over the surface of semi-permanent snowpatches. The largest and most controversial example occurs on the eastern side of Nant Ffrancon. Other interpretations for this impressive feature include a landslide and a remnant of a lateral moraine (Curry et al., 2001). Smaller examples of pro-talus ramparts occur at the base of talus slopes on the north side of Cadair Idris. On Moelwyn Mawr, south of Snowdonia, Rose (2001) has described a substantial fossil rock glacier, also attributed to the Loch Lomond Stadial. The spectacular periglacial landscapes of the summits of the Carneddau and Glyders formed on nunataks during the glacial maximum, though they may have been reactivated during the Loch Lomond Stadial. Impressive areas of patterned ground occur in areas that were undoubtedly glaciated during the last glaciation, including the Rhinogs and Moelwyns and these probably date to the Loch Lomond Stadial, though some were certainly active during the Little Ice Age of the seventeenth and eighteenth centuries and may still be active today (Rose, 2001).

Conclusion

There has been a series of advances in our understanding of the glaciations of north-west Wales since the review by Whittow and Ball (1970), but controversies remain and we still have much to learn. There seems to be little evidence to support the three phases of glaciation that Whittow and Ball proposed, and it is possible to interpret all of the glacial deposits as formed during a single event of the Late Devensian and during the Loch Lomond Stadial. However, we know from marine and ice cores that there were many climatic fluctuations during the Devensian, and it remains possible that more than one event is represented. The new techniques of cosmogenic isotope surface exposure dating may help to place the glaciations of north-west Wales into a more secure framework, but only if they are applied with great care and due regard to the many potential sources of error. A less rigorous approach will produce dubious dates, leading to confusion rather that enlightenment.

4 North-east Wales

by Geoffrey S.P. Thomas

Introduction

During successive glacial stages, north-east Wales was affected by two major ice-streams. Welsh ice radiated north, north-east and east towards Liverpool Bay and the borders from major ice source areas in Snowdonia, the Arenigs and the Berwyns (Fig. 4.1). The direction of ice movement is recorded by striae and erratics and the orientation of drumlin fields and ice-moulded landforms and the dominant flows were north down the Vale of Conway, east down the Dee and the Vyrnwy and north-east down the Severn. Erratic distributions show that Snowdonian volcanic rocks were carried as far as the Great Orme and Colwyn Bay by the Conway ice-stream (Hall, 1870; Read, 1885) and Arenig volcanics reached the Clwydian Hills and Halkyn Mountain (Mackintosh, 1874, 1879). The occurrence of these far-travelled erratics suggests that even the highest parts of north-east Wales were inundated by Welsh ice at some stage and that ice thicknesses reached a maximum of at least 600 m.

Irish Sea ice, originating from source areas in western Scotland, the Southern Uplands of Scotland and the English Lake District, flowed south across the floor of the Irish Sea, pressed against the North Wales coast and passed south-east across the northern margin of the Clwydian range into the Cheshire-Shropshire lowlands (Fig. 4.1). In the lower Conway valley Irish Sea ice penetrated only a limited distance south of Llandudno Junction. In the Vale of Clwyd it passed some 18 km inland from the present coast and Scottish and Lake District erratics occur as far south as Denbigh. To the east the ice limit skirted around the northern margin of the Clywdian Hills to a height of 185 m, rode up the flanks of Halkyn Mountain to 295 m (Strahan, 1886), penetrated the Wheeler valley, ran along the foot of Hope Mountain, rose to 350 m against Ruabon Mountain, rose against Gloppa Hill, west of Oswestry, and then ran south-eastwards towards Shrewsbury (Fig. 4.1).

Stratigraphy and chronology

A revised Quaternary stratigraphic succession in north-east Wales is outlined in Bowen (1999) and the oldest sediments occur in caves in the Elwy Valley, west of Denbigh. In Pontnewydd Cave (SH014710) a complex stratigraphic sequence of cave earths, silts, breccias, debris flows, travertine, stalagmite and sands and gravels, together with human teeth and bones and a rich mammalian fauna, provides the most extensive record in Wales as well as the oldest evidence for human occupation (Green and Walker, 1991; Cambell and Bowen, 1989). These sediments are classified as the Pontnewydd Formation (Table 4.1) and range from the Middle Pleistocene (*c.* 250 ka) to the Late Holocene (*c.* 5 ka). The earliest recorded event is deposition of sands and gravels containing flint and volcanic erratics indicative of an extensive Irish Sea glaciation, probably during Marine Oxygen Isotope Stage 8 (*c.* 250 ka). This event is succeeded by debris flows containing a temperate mammalian fauna, artifacts and human remains, including teeth, dated to *c.* 200 ka (Stage 7). Overlying stalagmite has yielded dates of 225-160 ka and 95-80 ka, indicating that no clastic sedimentation occurred in the extended interval covering much of Stages 6 and 5, probably due to sealing of the cave. Irish Sea diamict of the Llandudno Formation,

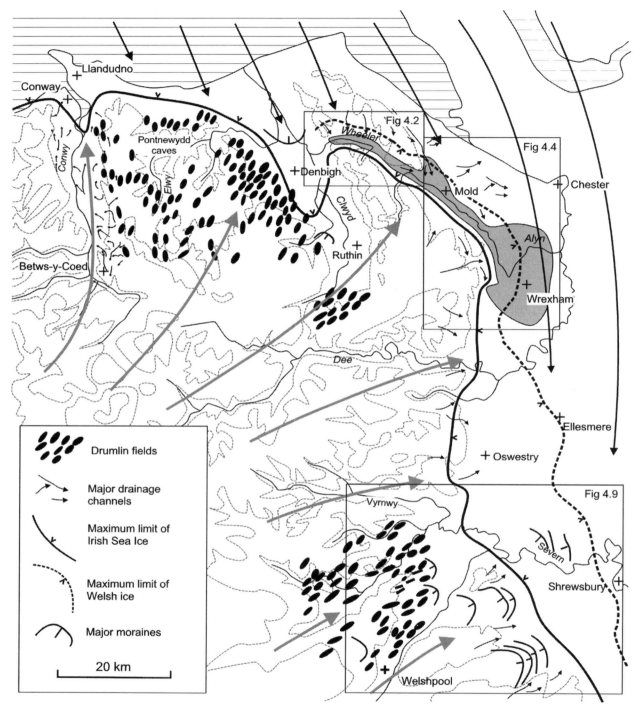

Fig. 4.1 Map of north-east Wales showing general direction of ice-flow, distribution of moraines and drumlin fields, the location of the Alyn-Wheeler ice-marginal drainage system (shown shaded) and the maximum limits of Welsh and Irish Sea ice. Area of geomorphological maps in Figs. 4.2, 4.4 and 4.9 also shown

<table>
<tr><th rowspan="3">Isotope Stage</th><th colspan="4">Province</th></tr>
<tr><th colspan="2">Irish Sea</th><th colspan="2" rowspan="2">Welsh</th></tr>
<tr><th>(North Coast)</th><th>(Eastern Border)</th></tr>
<tr><td>1</td><td colspan="3">Caerwys Fm
Tywi Fm
Kenfig Fm
Gwynllwg Fm
Tregaron Fm</td><td rowspan="8">Pontnewydd Fm</td></tr>
<tr><td>2</td><td>St Asaph Fm</td><td>Stockport Fm
Llai Mb
Wheeler Mb
Singret Mb
Dee Mb
Ruabon Mb</td><td>Meirion Fm

Ruabon Mb</td></tr>
<tr><td>3</td><td></td><td></td><td></td></tr>
<tr><td>4</td><td>Hiraethog Fm</td><td></td><td>Hiraethog Fm</td></tr>
<tr><td>5</td><td></td><td></td><td></td></tr>
<tr><td>5e</td><td>Deganwy Fm</td><td></td><td></td></tr>
<tr><td>6</td><td>Llandudno Fm</td><td></td><td></td></tr>
<tr><td>7/8</td><td></td><td></td><td></td></tr>
</table>

Table 4.1 The Quaternary lithostratigraphic succession in north-east Wales (After Bowen 1999)

however, found at depths between 25 and 30 m along the north coast (Warren *et al.*,1984), probably represents Stage 6.

Deposits of the last, Ipswichian interglacial (Stage 5e, *c.* 130–115 ka) are poorly represented but a warm temperate mammalian fauna, including hippopotamus and elephant, is recorded in the Cefn and Gallfaenan caves and is assigned to this stage (Green and Walker, 1991). In the Conwy estuary reddish-brown silty clays, the Deganwy Formation, with included shells and pollen of probable marine and estuarine origin, are also probably of Ipswichian age and may equate with sands and muds sandwiched between the Llandudno Formation and Late Devensian Irish Sea diamicts at Morfa Conwy (Warren *et al.*, 1984).

The Early Devensian in Britain (Stage 5d-a; Stage 4, *c.* 115–50 ka) was characterized by a cold, but generally non-glacial climate. Deposition of breccia in the Pontnewydd Cave and the occurrence of a cold mammalian fauna in the Cefn and Gallfaenan Caves, including woolly rhinoceros and mammoth, probably represent this stage (Green and Walker, 1991). Thin but widespread, locally derived heads and screes, the Hiraethog Formation, overlying bedrock surfaces beneath Late Devensian diamicts, are a product of periglacial slope processes during this stage (Warren *et al.*, 1984). Glacial deposits of Late Devensian age (Stage 2, 10–25 ka) are divided into two groups reflecting the dominant ice-streams that generated them. The Welsh group forms an extensive sheet of coarse-grained, rubbly, blue-grey diamict crowded with ice-worn and striated clasts of Ordovician and Silurian rocks together with more far-travelled Welsh erratics derived from Snowdonia and the Arenigs, and is classified as the Meirion Formation. It crops out extensively across the Denbigh Moors, in the Conway valley and the central and southern Clwydian Hills. Along the eastern border a distinctive sub-division, the Ruabon Member (Thomas, 1985) consists of grey or brown, sandy and stony diamict containing a significant proportion of locally-derived Carboniferous rocks.

The Irish Sea group is represented along the northern coast by the St Asaph Formation. It consists of red or brown, clay-rich diamict and associated fluvioglacial sediment containing dominantly local clasts, especially Carboniferous limestones, sandstones and shales, together with Triassic sandstones and long-distance erratics including Criffel Granodiorite from the Southern Uplands and Eskdale Granite from the Lake District, plus numerous, usually broken or abraded marine shells. It overlies the Meirion Formation (Whittow and Ball, 1970; Embleton, 1961; Warren et al., 1984), indicating that Devensian Welsh ice reached the coast before Irish Sea ice. It also overlies the Tremeirchion caves which contain Middle Aurignacian implements (Garrod, 1926) and mammoth bones dated to 21 ka (Rowlands, 1971). Along the eastern border the Irish Sea group is assigned to the Stockport Formation and is divided into a number of members. The Dee Member is a stiff, red clay-rich diamict with far-travelled northern erratics and local Carboniferous rocks. It underlies the floor of the lower Dee valley and is correlated with the 'Lower Boulder Clay' of the tripartite sequence of Wedd et al. (1928). It overlies the Ruabon Member of the Meirion Formation, as in the north, confirming that Welsh ice also reached this area before Irish Sea ice. The Singret Member includes deposits of ice-front alluvial fan, sandur and ice-contact lacustrine origin deposited during uncoupling of the Welsh and Irish Sea ice-sheets around Wrexham (Thomas, 1985). The Wheeler Member includes sand and gravel formed in sub-glacial eskers and ice-front alluvial fan deltas in the Wheeler and Upper Alyn valleys (Brown and Cooke, 1977; Thomas, 1984 and 1985) and the Llai Member comprises sheets of diamict that intercalate with the Singret and Wheeler members. These are equivalent to the 'Upper Boulder Clay' of Wedd et al. (1928) and have been interpreted as resedimented debris flows by Thomas (1985).

Although numerous kettle basins occur in the Wheeler valley, around Padswood and on the surface of the Wrexham 'delta-terrace', none have been investigated stratigraphically and the regional fluctuations of Late Glacial climate are poorly recorded. A Late Glacial mammalian fauna indicative of cold, tundra conditions is, however, recorded in Pontnewydd cave. Holocene deposits (Stage 1, <10 ka) include the Tregaron Formation, an upland peat occurring across much of the high ground; the Gwynllwg Formation, consisting of marine and estuarine alluvium and peat, including a submerged forest (Tooley, 1974; Warren et al., 1984) exposed between Rhyl and Abergele; the Kenfig Formation, a series of sand dunes on the coast between Llandudno and the Point of Ayre and the Tywi Formation, comprising extensive river alluvium in all the major valleys. The Caerwys Formation records the most extensive calcareous tufa deposits in Britain and contains an exceptional fossil record including molluscs, leaf-beds and vertebrates (Preece et al., 1982; Pedley et al., 1996) ranging from the Late Glacial through to the Late Holocene.

Landforms of glacial erosion and deposition

North-east Wales lacks spectacular scenery associated with glacial erosion despite it having been submerged by ice at least during the last glacial stage. Embleton (1970) has summarized the evidence for glacial erosion and identified three zones of intensity. Zones of moderate erosional modification are restricted to the area west of the Conway valley where partial modification of the preglacial surface by cirques, over-deepened rock basins, breached watersheds and discordant tributary valleys are common. A zone of slight erosional modification includes the Clwydian Hills, which display a number of immature cirques and U-shaped valleys, and the Middle Dee valley which shows immature cirques at tributary valley heads, local over-deepening of the valley and partial truncation of spurs. The remainder of the region shows little or no modification by glacial erosion.

Glacial meltwater channels are common throughout the region (Embleton, 1970). They range from rock-cut gorges up to 40 m deep, such as the Alyn and the Hendre, north-west of Mold, to numerous short, shallow channels cut in glacial deposits. The great majority carry no current drainage and many, especially of the larger type, display 'up and down' long profiles suggesting that they were formed subglacially. Most of the channels occur below 200 m and the Denbigh Moors and the Clywdian Hills are largely free of them. In the Conwy valley a sequence of irregularly spaced parallel channels run along the eastern flank of the valley. Embleton (1970) interpreted them as ice-marginal and formed during successive stages in the retreat of the Conwy glacier but Fishwick (1977) considered the majority to be sub-glacial. In areas affected by Irish Sea ice the direction of channels is mostly to the south or south-east and a major system, up to 30 m deep and 100 m wide, runs for nine kilometres along the eastern flank of Halkyn Mountain (Peake, 1961; Derbyshire, 1963; Embleton, 1964 a and b and Thomas, 1984). In some instances the depth of erosion by meltwater incision has led to the adoption of new courses for some rivers since the end of the last glaciation. The most radical was the diversion of the River Alyn, whose preglacial headwaters flowed north to join the River Wheeler. It now turns eastwards across the Carboniferous Limestone via the deeply entrenched Alyn Gorge (Embleton, 1957). The diversion was accomplished by meltwater draining the margin of the Irish Sea ice-sheet at a time when it occupied the Wheeler valley and hence drained eastwards to escape. The adjacent Hendre Gorge cuts across the watershed between the Wheeler and Alyn river systems and was also probably cut by subglacial meltwater flowing eastwards at a time when the Welsh and Irish Sea ice-sheets were confluent. The lower course of the River Elwy was also diverted. Formerly flowing eastwards through the Carboniferous Limestone ridge flanking the western side of the Vale of Clwyd, it was diverted by the margin of Irish Sea ice occupying the Vale for some 3 km south to its present confluence with the River Clwyd (Embleton, 1961).

Landforms of glacial deposition are concentrated in the Wheeler valley, the middle Alyn and the Wrexham area. Elsewhere, depositional landforms are rare but extensive gravel terraces, probably former sandur surfaces fronting Welsh ice, occur in the Elwy valley (Livingstone, 1986) and in the Vale of Conway (Embleton, 1961). A series of gravel terraces lie to the rear of Colwyn Bay and rough terraces of gravel spread across the eastern slope of Halkyn Mountain from Holywell south; both are associated with deposition from the margin of Irish Sea ice. Major drumlin swarms, orientated in fields running north, north-east or east occur between the Conway and the Vale of Clwyd, south of Ruthin and along both flanks of the River Vyrnwy, respectively (Fig. 4.1). The largest attain heights of 30 to 40 m and lengths of up to a kilometre and the great majority are rock cored. They were all deposited by the passage of Welsh ice. Moraines are uncommon. Irish Sea ice, in particular, left few large moraine systems behind though the arcuate series of low hills running across the Vale of Clwyd from Tefnant to Bodfari (Fig. 4.1) has been interpreted by Embleton (1970), Rowlands (1971) and Livingstone (1990) as a retreat stage moraine. Retreat of Welsh ice has produced lateral moraine systems in the Vale of Conway (Embleton, 1961) and a small retreat stage moraine at Pwyll-glas at the southern end of the Vale of Clwyd. Large arcuate cross-valley moraines occur west of Welshpool, in the upper Dee north-east of Bala and west of Shrewsbury.

The Alyn-Wheeler ice-marginal drainage system
The most distinctive feature of Late Devensian glacigenic sedimentation in the region is the complex depositional system that runs in a narrow belt from Bodfari, eastwards along the Wheeler Valley, through the Hendre Gorge, south-east past Mold to terminate in a large, widening fan around Wrexham (Fig. 4.1). The system parallels the zone of confluence of Welsh and Irish Sea ice and is bordered by glacial deposits of predominantly Welsh origin to the west and Irish Sea origin to the east. Its surface is characterized by a suite of glacial landforms including esker ridges, irregularly distributed mounds and basins, pitted outwash surfaces, delta terraces, alluvial fans and a complex system of drift and rock cut channels. The system was formed as a consequence of the uncoupling of the Irish and Welsh ice-sheets during retreat of the last glaciation when it acted as the principal distributary for meltwater drainage (Thomas, 1985 and 1989). It can be divided into four interrelated, depositional provinces.

The Wheeler valley ice-marginal zone

The River Wheeler rises on the north-east slope of the Clwydian Range but turns sharply westwards to flow towards the Vale of Clwyd. Immediately to the east the Hendre Gorge crosses the watershed and connects the Wheeler catchment to the Alyn (Fig. 4.2). The upper Wheeler valley, between Nannerch and Bodfari, is deeply entrenched along the boundary between the Carboniferous Limestone plateau of Halkyn Mountain to the north and Silurian shales of the Clwydian Hills to the south. The flanks of the Clwydian Hills have a thin cover of diamict, almost exclusively of Welsh composition, whilst the plateau area is characterized by diamicts of mixed provenance including a dominant Carboniferous Limestone component together with more far-travelled northern erratics, indicating that the Wheeler valley lay along a zone of interaction between the two ice-sheets. West of Nannerch the valley floor is deep and narrow and landforms are characterized by a number of prominent linear ridges running parallel to the valley axis, numerous small, often irregularly distributed mounds, occasional kettle basins and small, fragmentary terraces, mostly sloping westwards (Brown and Cooke, 1977). To the east of Nannerch the valley is wider and shallower, mounds are larger and at least three terrace systems occur, the lowest of which, at Nannerch, is very extensive and is pitted with kettle holes. In this area the higher terraces slope eastwards, some of the intermediate terraces slope westwards, and the lowest is horizontal and at the same height as the adjacent inlet to the Hendre Gorge to the east.

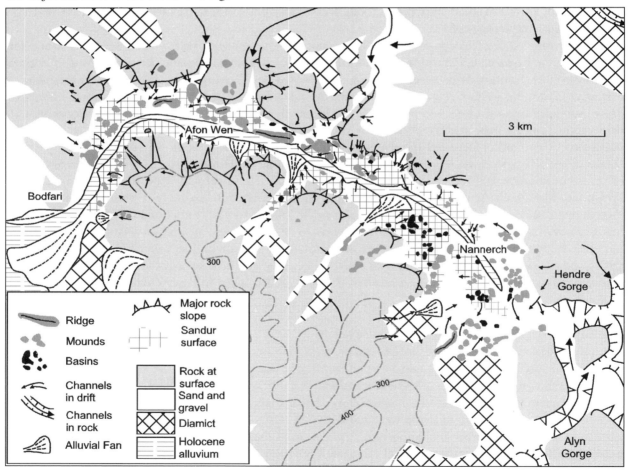

Fig. 4.2 Geomorphological map of the Alyn-Wheeler ice-marginal zone (After Brown and Cooke (1977), Thomas (1984 and 1985) and unpublished mapping)

According to Brown and Cooke (1977), glacial deposits in the Wheeler valley are in excess of 35 m thick at Nannerch and over 60 m at Maesmynan where the succession comprises 30 m of red sand, 12 m of red, sandy diamict and 18 m of coarse gravel intercalated with numerous thin diamicts. The red sands are derived from Triassic sandstones in the Vale of Clwyd, are thickest around Bodfari and progressively diminish in quantity eastwards, although they are recorded as far east as Mold (Strahan, 1890). Brown and Cooke (1977) considered the landforms and deposits of the Wheeler valley to be a product of the decay of a stagnating Irish Sea ice-margin retreating to the west or north-west. In the early stages of retreat the Wheeler valley was blocked by ice and streams within sub glacial tunnels drained the margin and formed short esker systems that fed ice-front alluvial fans and narrow valley floor sandur draining east. Many of these sandur surfaces, such as those around Nannerch, became pitted by the melt of underlying dead-ice. As the ice-sheet margin retreated drainage reversed and flowed westwards towards the Vale of Clwyd.

The Rhydymwyn Lake
East of the Hendre Gorge the depositional system widens to form an embayment around Mold (Fig. 4.2). At the northern end a series of prominent terraces descend the eastern slope of the Alyn valley and are bordered on their eastern side by Sarn Galed, a deeply entrenched, meandering channel that crosses the watershed between the Dee and Alyn rivers (Peake, 1961; Derbyshire, 1963; Embleton, 1964 a and b; Thomas, 1984) and is flanked by irregular mounds, kames and kettle basins associated with the margin of Irish Sea penetration from the adjacent Cheshire lowlands (Fig. 4.3 A).

At Rhosesmor (SJ 216670) a large quarry is cut in the surface of the highest terrace and has been described by Thomas (1984). It displayed two sedimentary packages, stacked one above the other, each consisting of a down-current transition, from north-east to south-west, from glacio-tectonised ice-marginal diamicts and fan gravels, through channeled sandur gravels, delta topset gravels, steeply-dipping foresets of upward-fining gravel to sand into laminated bottomsets of fine sands, silts and clays (Fig. 4.3B). Each package was interpreted as a response to a delta system fed by meltwater exiting the margin of the Irish Sea ice-sheet via the Sarn Galed subglacial channel into a temporary lake ponded between the Irish Sea ice-margin and the foothills of the Clwydian Hills (Fig. 4.3D). A lower delta, associated with a lake level at 180 m OD, built outwards from the ice-front for some 600 m into the eastern margin of the lake. At the delta margin water depth was probably no more than 10 m, increasing to over 90 m around Mold. At this stage part of the ice-margin was in contact with the lake water for the occurrence of dropstones in the bottomset sediments point to the existence of ice-bergs calved from the snout. An upper delta indicates that the lake rose some 12 m to form a water surface at 192 m, drowning out the lower delta and building another across it. The rise in lake level was associated with a minor readvance of the ice-margin for the proximal alluvial fan gravels of the lower delta are deformed, truncated and unconformably overlain by an assemblage of ice-marginal sediments including basal lodgement and flow till (Fig. 4.3B).

Peake (1961) identified seven major ice-marginal lake stages in the Alyn but provided no sedimentary evidence to support them and for this reason their existence was doubted by both Derbyshire (1963) and Embleton (1964 a and b). The delta sequences at Rhosesmor provide obvious support for at least two stages in this lake, termed the Rhydymwyn Lake by Peake (1961), and further evidence comes from boreholes across the lake basin floor (Fig. 4.3D). These show a passage from gravels overlying Irish Sea diamict to the north of Rhosesmor, through the delta sequences into extensive laminated mud and fine sand across the floor towards Mold. Fig. 4.3C attempts a reconstruction of the extent of the lake at its maximum stage of 192 m, based on an extrapolation of its calculated water level. At this height the lake drowned out the Hendre and Alyn Gorges and probably extended a little way up the Wheeler valley before abutting against the Irish Sea ice-front to the west. The lake also extended through part of the

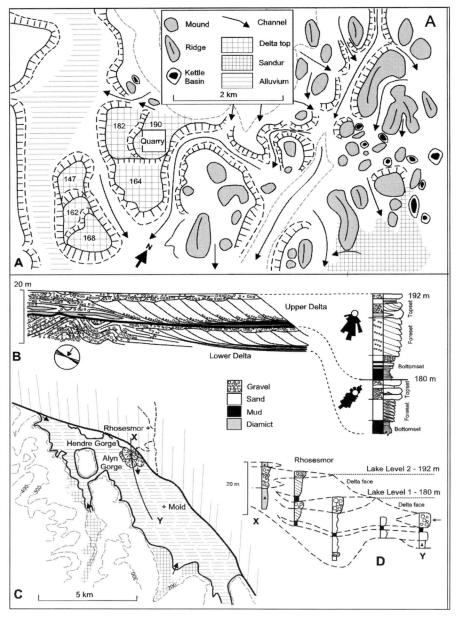

Fig. 4.3 Rhosesmor Quarry
A: The geomorphology of the area around the quarry, showing distribution of delta terraces and their relationship to the Sarn Galed sub-glacial drainage channel. For location of quarry see Fig 4.4.
B: Summary of the stratigraphic succession in the quarry showing the relationship between the two delta sequences together with a vertical log profile, inferred water height, palaeocurrents and strike of folds and thrusts.
C. Palaeogeographic reconstruction of the extent of Lake Rhydymwyn and its principal input points at its highest level of 192 m.
D. Cross-section north/south across the lake floor showing relationship between the deposits of the two lake levels

upper headwaters of the Alyn River towards Loggerheads, where a large delta plain fringes the margin of the lake at a height of 195 m. This is fed by an extensive valley sandur that extends some three kilometres upstream, suggesting that at this stage Welsh ice was actively feeding the lake from a retreating ice-margin to the south-west. Further evidence for the occurrence of Welsh ice in the area at this time is provided by a large fan-delta on the southern border of the lake, draining from the Afon Terrig. To sustain its water surface the lake can only have been dammed by an ice-barrier and the most likely scenario is that the extension of the Irish Sea ice-margin southward from Rhosesmor abutted against the steeply rising slope of Hope Mountain where the embayment around Mold narrows (Fig. 4.3C). Ice-dammed lakes are essentially temporary, fluctuate in level and frequently drain in response to movement of the ice-margin and changes in the internal drainage system. Rhydymwyn Lake seems no exception. On the assumption that the delta foresets at Rhosesmor are a response to the annual meltwater flood cycle, Thomas (1984) calculated a minimum life of the lake as some 90 years.

The Hope ice-marginal sandur trough
South of Mold the embayment narrows as the Alyn cuts across the strike of the Carboniferous Limestone. On the western margin slopes rise rapidly to Hope Mountain but on the east the valley floor is bordered by a complex area of ice-disintegration terrain running in a narrow zone parallel to the valley margin. This is especially well developed around Padswood where a series of parallel ridges, separated by wide, flat-floored channels and large, often irregular water-filled basins, irregular mounds and small terraces border the valley (Fig. 4.4).

Hope Quarry (SJ 360587), on the eastern border of the valley (Fig. 4.4), displays an upward coarsening succession from sands, through rapidly alternating sands and gravels into stacked sequences of massive gravel intercalated with thin diamicts. Large channel structures are common and current indicators show a predominant flow direction to the south-east, parallel to the present course of the Alyn. The sequence has been interpreted by Thomas (1984) as a braided sandur confined to a narrow trough running south-east down the Alyn valley between an Irish Sea ice-margin to the east and the rock margin to the west (Fig. 4.5A). The coarsening upwards succession indicates an increase in proximity to subglacial

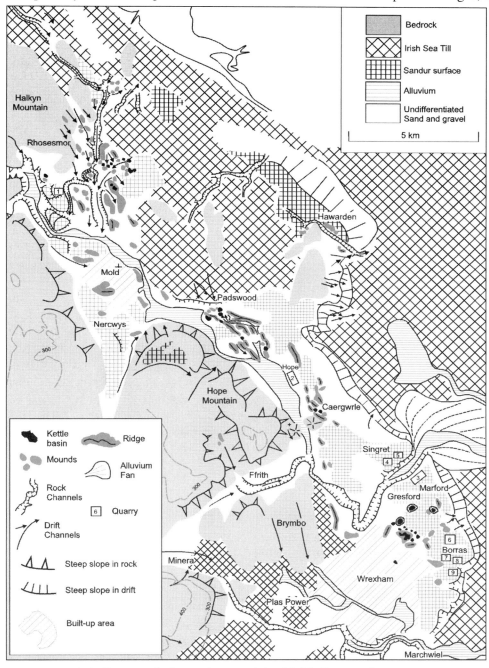

Fig. 4.4 Geomorphological map of the Mold-Caergwyle ice-margin sandur system and the Wrexham fan-delta (After Thomas (1985 and 1989) and unpublished mapping)

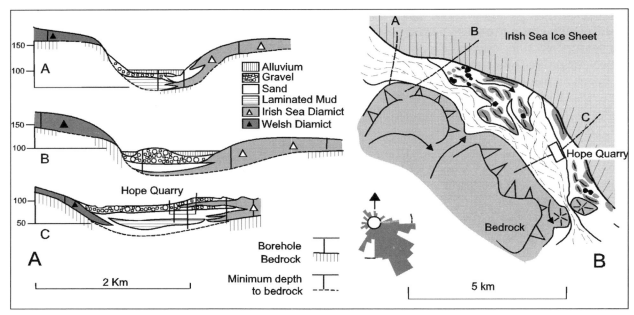

Fig. 4.5 Hope Quarry
A. Cross-sections showing relationship of major sediment assemblages across the floor of the Mold-Caergwyle ice-margin sandur system. Constructed from borehole records in Dunkley (1981).
B: Palaeogeographic reconstruction of ice-margin conditions around Hope

stream exit points as a response to minor ice-marginal readvance into the eastern margin of the valley. This is confirmed by the occurrence of intercalated diamict units amongst the higher parts of the succession interpreted as debris flow, released from dirt bands in the ice-margin, across the fronting sandur fan. The existence of much dead-ice terrain running parallel to this margin also supports this view. Boreholes across the valley floor (Fig. 4.5B) show that the eastern margin is underlain by thick Irish Sea diamict that breaks up rapidly westwards beneath the ice-disintegration terrain into a coarsening upwards succession from laminated mud through sand into coarse gravel. On the west a sheet of Welsh diamict overlies the lower bedrock slope of Hope Mountain and passes out eastwards into the coarsening succession. The sequence was deposited at a time when the middle Alyn valley acted as the main ice-marginal distributary into the Wrexham fan-delta.

The Wrexham 'fan-delta'
At Caergwrle, the Alyn river ignores a more direct route to the River Dee and, instead, flows in a narrow gorge between Hope Mountain and a Carboniferous outlier to its immediate east (Fig. 4.4). The gorge forms the apex of a large sloping fan, covering an area of more than 50 km^2, that runs south-east through Wrexham. Its east and north-east margin is marked by a prominent escarpment, up to 30 m high, overlooking the floor of the Dee valley and deeply entrenched by the modern Alyn River. The surface of the fan is very diverse and includes a prominent esker at Gwersyllt Park (SJ 323536), a large area of complex ice-disintegration terrain, including numerous water-filled basins, running parallel to the outer margin of the fan along its north-east boundary between Singret and Borras, and extensive areas of sloping sandur surface. Although the main feeder into the fan is via the Caergwrle gorge, other feeders enter from the Ffrith Gorge, from Brymbo and from the gap in the Carboniferous escarpment at Minera, all draining from Welsh source ice.

The first account of the Wrexham area was provided by Wedd *et al.* (1928) and the term 'Wrexham Delta-terrace' was coined by Lamplugh (in Wedd *et al.*, 1928) following his interpretation of the eastern escarpment as a delta front. Peake (1961 and 1981) regarded the feature as a 'composite prograding delta' built into a series of narrow ice-dammed lakes associated with easterly retreating Irish Sea ice. Poole and Whiteman (1961), in contrast, interpreted the feature as 'moraine outwash', built by proglacial outwash flowing westwards. In the first detailed, sedimentological investigation, Francis (1978) regarded the feature as an unconstrained alluvial fan fed by a dominant meltwater flow from Welsh ice; a conclusion supported by Dunkley (1981). Wilson *et al.* (1982) agreed but termed the feature a 'prograding outwash sandur' and argued that the outer escarpment was an erosional feature caused by removal of the distal portion by a major readvance of the Irish Sea ice-sheet. In a review Worsley (1985) noted the complexity of depositional environments, confirmed that the feature was essentially a sediment sink influenced by meltwater from two separate glacial sources and argued that it was one component in a large ice-marginal supraglacial sediment-landform system that ran from Wrexham across the adjacent Cheshire-Shropshire lowlands towards Whitchurch. To remove the confusing generic implications Worsley renamed the feature the 'Wrexham Plateau'. Thomas (1985) largely supported this interpretation but emphasised that the feature was a complex, diachronous ice-marginal feature that reflected rapidly changing depositional environments ranging from marginal ice-disintegration, ice-front alluvial fan, marginal sandur and pro-glacial and ice contact lake systems, fed predominantly by meltwater from Irish Sea ice sources running down the Hope ice-marginal sandur.

Reflecting the diversity of its surface, the internal composition of the delta-terrace is equally complex. Fig. 4.6 shows a series of cross-sections, derived from borehole data (Thomas, 1985), running

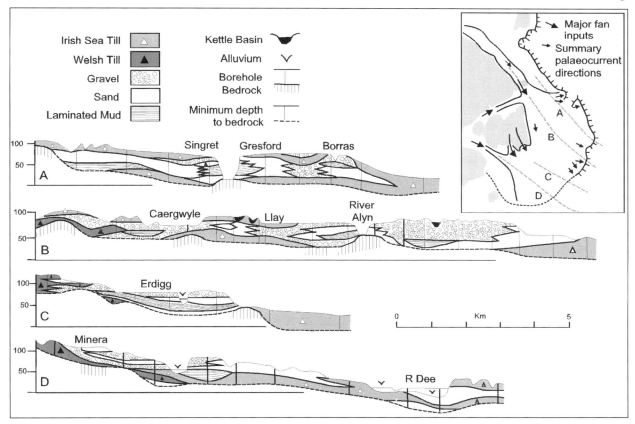

Fig. 4.6 Cross-sections through the Wrexham fan-delta. Inset shows location of section lines, location of major fan inputs and summary palaeocurrent directions. Constructed from bore records in Ball (1982) and Dunkley (1981)

radially down the fan slope. To the north and east of Wrexham the delta terrace is floored by a sheet of Irish Sea Till, the 'Lower Boulder Clay' of Wedd *et al.* (1928), that extends eastwards and thickens beneath the floor of the River Dee. The diamict thins westwards and is succeeded by a complex suite of sand and gravels, the 'Middle Sands' of Wedd *et al.* (1928), that coarsens upwards and outwards (Dunkley, 1981), Intercalated within the sands and gravels are localized units of laminated mud. Around Gresford the sands and gravels are overlain by further Irish Sea diamict, equivalent to the 'Upper Boulder Clay' of Wedd et al. (1928) and the 'Llai Till' of Peake (1961). In the west, especially around Minera, the 'Lower Boulder Clay' is locally underlain by Welsh diamict but the relationship between them is complex. Worsley (1985) has described 'interactive sedimentation' between the two diamicts in exposures at the former Plas Power opencast coal mine but Dunkley (1981) reports that sheets of Welsh diamict underlie, intercalate with and overlie both Irish Sea diamict and the succeeding sands and gravels. In the upslope, more proximal parts of the delta-terrace, particularly around Minera, both the Irish Sea diamict and the sands and gravels wedge out over thick Welsh diamict which, to the west occupies thick successions at the surface. Thomas (1985) and Thomas in Bowen (1999) have revised the stratigraphic succession in the area and referred it all to the Stockport Formation (Table 4.1). The 'Lower Boulder Clay' is now identified as the Dee Member of this formation; the 'Middle Sands' as the Singret Member; and the 'Upper Boulder Clay' as the Llai Member. The Welsh diamicts are identified locally as the Ruabon Member and form part of the more extensive Meirion Formation seen in the interior.

The Wrexham area is a primary source of aggregate and numerous quarries provide detail of the environments of deposition. Marford Quarry (SJ 358561, Quarry 3, Fig. 4.4) lies immediately to the rear of the outer escarpment on the southern side of the Alyn river and is the type-site for the Llai Member. The succession consists of 15 m of alternating gravel, sand and red-brown diamict showing rapid lateral and vertical variation, dipping north-east (Fig. 4.7B). Clast composition is predominantly local and includes Carboniferous sandstones, shales and coals, some limestone and a sprinkling of northern erratics, together with rounded and often armoured diamict balls. The diamict units are extremely variable in texture, thickness and lateral persistence but most consist of discrete, erosionally based sheets of massive or faintly bedded stony mud. Others display great internal variability and frequently grade upwards from stony diamict to matrix rich gravel. Palaeocurrents show transport from the north-west and are commensurate with deposition in the upper part of an alluvial fan draining from the entry point into the fan at Caergwrle towards the east and north-east. Sedimentation was dominated by coarse, longitudinal gravel bars deposited under conditions of high flood in wide but shallow braided channels. The regular intercalation of diamict may be interpreted as a response to frequent release of supraglacial debris flows from an adjacent ice-margin to the east into the margin of the fan and, in part, their subsequent reworking into braid bars (Francis, 1978; Thomas, 1984).

Singret Quarry (SJ 343559, Quarry 4, Fig. 4.4) displayed a complex stratigraphic succession that can be divided into a number of major facies assemblages. Across the floor is a sheet of hard, red diamict (assemblage 1, Fig. 4.7A), indicative of early passage of Irish Sea ice westwards across the area. This is succeeded by stacked sets of massive gravel fining upwards into parallel-laminated sand with intercalated muds and thin diamicts, disconformably overlain by gravel filled migrating channels (assemblage 2, Fig. 4.7A). Palaeocurrent indicators show flow to the north-east and the sequence is interpreted as a response to the progradation of an ice-marginal sandur system across the underlying diamict as a consequence of the retreat of the Irish Sea ice-margin. The initial upward fining and subsequent upward coarsening in the assemblage indicates significant facies shift caused either by migration of major distributary channels, cyclic variation in sediment supply or by fluctuations in the ice-margin position. The sandur deposits are disconformably overlain by a sheet of massive, brown or red Irish Sea diamict (assemblage 3). Francis (1978) interpreted this diamict as a debris flow, but its thickness, lateral persistence and uniform lithological composition compared with the thinner, discontinuous and varied diamicts at Marford Quarry, suggest an origin by subglacial lodgement caused by a readvance of the ice-margin across the proglacial

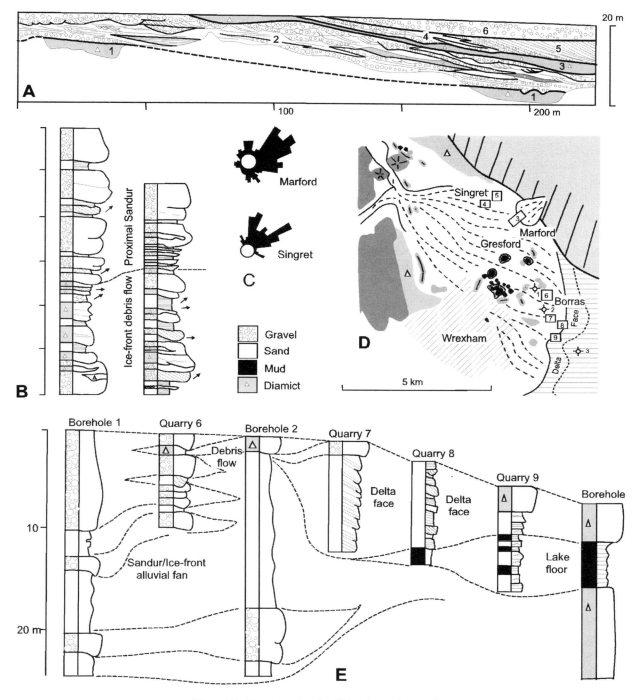

Fig. 4.7 Sections in the Wrexham fan-delta
A: Sections through the quarry at Singret (Quarry 4) showing distribution of major facies assemblages.
B: Vertical log profiles through Marford Quarry (Quarry 3).
C. Palaeocurrent distribution at Marford and Singret quarries.
D: Geomorphological map and location of quarries
E. Section north-west to south-east through the fan-delta. Based on quarry exposure and borehole data

area. The diamict is overlain by an assemblage of upward-fining massive gravels and parallel-laminated and trough and planar cross-bedded sands (assemblage 4), interpreted as a further response to ice-marginal sandur progradation. This assemblage passes eastwards into a thickening wedge of large-scale delta foresets (assemblage 5) dipping east. These are unconformably overlain by stacked sets of coarse gravel and sands (assemblage 6) which together are indicative of the eastward progradation of a large delta. This is confirmed by a further quarry at Singret (SJ349566, Quarry 5, Fig. 4.4), some 600 m down-current to the east, which displays large-scale sandy delta foresets in a prograding sequence 25 m thick. The water surface of the associated lake stood at c. 65 m. The overall succession at Singret clearly identifies the ice-marginal sedimentary response to complex local oscillation of the Irish Sea ice-margin and the consequent rapid and repeated shift in depositional environments from subglacial, through ice-marginal sandur into deltaic sedimentation.

A number of quarries around Borras confirm the complexity of the depositional sequence and have been described by Thomas (1985). To the north shallow pits (SJ 364531, Quarry 6, Fig. 4.4) display massive gravel intercalated with discontinuous lenses of stony diamict similar to those at Marford Quarry and were interpreted as a response to proximal alluvial fan or sandur sedimentation in which debris flow from an adjacent ice-margin was a significant component. A kilometre south another quarry (SJ 367521, Quarry 7, Fig. 4.4) shows repeated, tabular bedsets fining upwards from pebble gravel to coarse sand, dipping east-southeast between 18 and 28°. These are interpreted as simple coarse-grained proximal delta foresets prograding east-south-east into a shallow lake basin at least 8–10 m deep and with a surface elevation at c. 70 m. 200 m in the down-current direction an adjacent quarry (SJ 365522, Quarry 8, Fig. 4.4) displays a rapidly alternating series of sands dipping east at up to 12°. A metre-scale cyclicity occurs throughout the sequence and is initiated by parallel-laminated sand passing upwards into cross-laminated fine sand into silty drape ripples. Cycles are frequently separated by thick sets of trough or planar cross-bedded sands. Large, steep-walled channel structures occur at intervals together with lenses of granule gravel and extensive sheets of laminated and massive mud thinning up-dip. Dropstone structures are common and some channels display large, irregular lumps of open-textured stony diamict on their floors. Similar upward-fining cycles have been described by Clemmensen and Houmark-Nielsen (1981) and Thomas (1984) from low-angled, fine-grained delta foresets and have been interpreted as a response to seasonal variations in outwash streams entering a lake. The large channels that truncate the sequences are similar to those described by Cheel and Rust (1982) and were formed by high sediment density concentrations descending delta foresets from the mouths of topset distributary channels. The large diamict lumps were probably transported either as bed-load clasts of debris-rich ice or by floating bergs. A further quarry at Llan-y-pwll (SJ 365517, Quarry 9, Fig. 4.4) displays 8 m of very low-angled parallel or cross-laminated fine sands with thin impersistent layers of silt and clay, capped by 2 m of red diamict. The sands form upward-fining cycles from parallel-laminated sand into ripple-laminated fine sand, showing transport to the south-south-east. Francis (1978) interpreted the sequence as forming in the distal portion of a subaerial alluvial fan but Thomas (1985) regarded them as similar to those in Quarry 8 and formed at the toe of delta foreset slopes by lake directed density underflow.

Taken together the sedimentary evidence around Borras represents a transition from proximal, subaerial alluvial fan and sandur sedimentation through delta-foreset accumulation into proximal lake-floor deposition, prograding to the east and south-east and is well supported by the close correlation between section and borehole evidence (Fig 4.7E).The pattern of palaeocurrents, however, suggest that sections in individual quarries do not necessarily relate to the same depositional system but probably to independent, laterally coalescing, diachronous systems draining from the north-west into the western margin of Lake Bangor, a large ice-marginal lake system occupying much of the adjacent Cheshire lowlands (Thomas, 1989). Clearly, the development of the Wrexham delta-terrace as a whole was complex and neither of the two previously proposed models are adequate to explain the variety of depositional environments identified. Thus the structure is neither a simple delta (Peake, 1961 and 1981) nor a simple alluvial fan (Francis, 1978), but a much more composite and diachronous feature that reflects greatly varying sedimentary environments, including ice-front debris-flow and alluvial fans, sandur and proglacial and ice-contact lakes, formed at the margin of an Irish Sea ice-sheet that was itself repeatedly oscillating. As proposed by Thomas (1985), the Wrexham 'delta-terrace' may perhaps be more precisely described as a 'fan-delta' to reflect the principal processes of sandur fan and deltaic sedimentation.

The Severn Glacier
Although the existence of a Severn Glacier as a major distributor of Welsh ice draining eastwards has long been postulated (Wills, 1950), little is known about its maximum eastward penetration or its relationship to Irish Sea ice occupying the Shropshire lowlands. Pocock *et al.* (1938) showed that west of a line from Dorrington to Shawardine (Fig. 4.8) drift is exclusively of Welsh origin. East of this line, as far as Shrewsbury, Welsh diamict and outwash were recorded as either overlying glacial deposits of Irish Sea origin or intercalated with them. To the east of Shrewsbury, glacial deposits are derived exclusively from Irish Sea sources. Welsh diamict is also seen to overlie Irish Sea diamict at Ellesmere (Shaw, 1972 a and b), though glaciotectonics complicates the picture and the succession may be reversed (Worsley, 1970). A series of stratigraphic sections running parallel to the assumed flow direction of the Welsh ice (Fig. 4.8), confirms the pattern but Welsh diamict is nowhere seen to overlie Irish Sea diamict and only in one location (section C, Fig. 4.8) are Irish Sea diamicts seen to overlie Welsh diamicts. The ice-sheets therefore seem to have coalesced rather than overridden one another and the geometry of individual diamict sheets indicates that on subsequent retreat both margins fluctuated in a series of minor snout oscillations. Between the retreating ice-margins, outwash sedimentation was collected in narrow, ice-marginal sandur troughs, running to the south-east. These show a general upward coarsening succession from laminated muds, representing temporary lake basins ponded between the ice-margins, into sands and gravels representing subsequent sandur progradation.

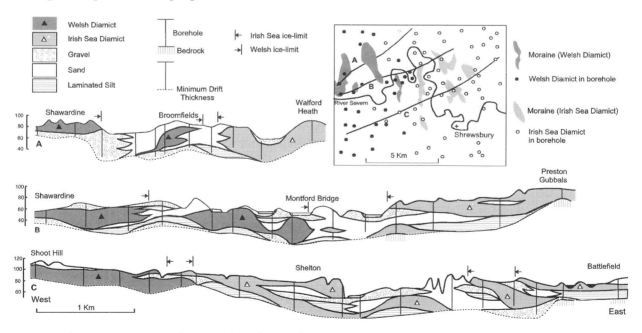

Fig. 4.8 Cross-sections showing the relationship between Welsh and Irish Sea diamicts in the area around Shrewsbury. Based on borehole records in Cannell (1982). Inset top right shows location of boreholes, presence of Welsh or Irish Sea diamict and distribution of major moraine ridges

A geomorphological map of the area between Welshpool and Shrewsbury is presented in Fig. 4.9 and identifies a number of major stages in the retreat of both the Irish and Welsh ice-sheets following uncoupling. Ice-flow indicators show that Welsh ice moved broadly north-east along the Severn and eastwards along the Vyrnwy as a series of lobes constrained by prominent bedrock topography. Thus, ice moving along the line of the Severn valley divided around the hill masses of Breidden, the Longmyndd and the Stipperstones to pass north-east along the through valleys that separate them. The north-east end of the valley separating Breidden Hill from the Longmyndd is blocked by a series of cross-valley moraines that pass, eastwards, into a large sloping sandur surface running east towards Shrewsbury. The wider valley separating the Longmyndd from the Stiperstones, is flanked on both sides by morainic ridges arcing across valley and complex ice-disintegration landforms and pitted outwash surfaces occur at the south-west end. The north-east end is partially blocked by a complex, multiple moraine system, more than 5 km wide, arcing across valley towards Pontesbury. To the rear of the moraine most of the valley floor is flat and boreholes indicate significant laminated clay, up to 30 m thick in the area around Marton Pool (SJ 295027).

The current exit of the River Severn into the Shropshire lowlands around Four Crosses is devoid of moraines and the valley floor is underlain by extensive Holocene alluvium bordered on the north and west by sandur terraces. This extends as far east as the moraines at Shawardine and Broomfield, both underlain by Welsh diamict. Any cross-valley moraines that did occur have probably been removed by melt-

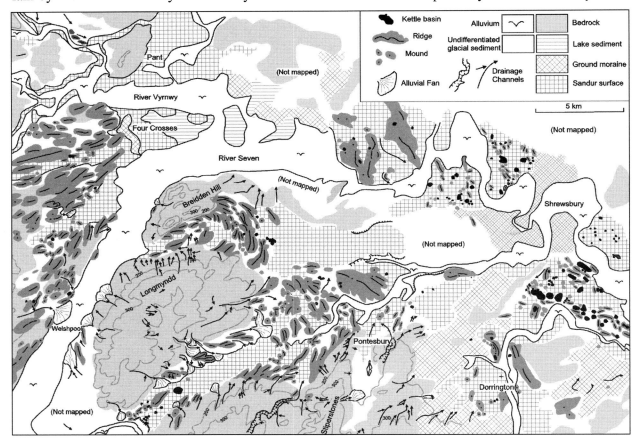

Fig. 4.9 Geomorphological map of the area between Welshpool and Shrewsbury. Based on author's unpublished field maps at a variety of scales. Note that some small areas are unmapped; others show only drift geology boundaries

water erosion and Holocene flood-plain incision. Beneath the valley floor east of Four Crosses borehole records show thick laminated clay (Pocock *et al.*, 1938) and suggest temporary ponding of meltwater in a proglacial lake basin during retreat. To the west, the area between Welshpool and the River Vyrnwy displays a complex zone of ice-moulded bedrock ridges, moraines and drift and rock-cored drumlins, running north-east.

The Irish Sea ice-margin is well marked by complex ice-disintegration terrain in a linear belt running north-west to south-east through Shrewsbury. In detail, the belt consists of a number of parallel sets of ridges, mounds and deep kettle basins separated by narrow, often pitted, ice-marginal sandur troughs. The diamicts underlying the ridges east of the Shawardine and Broomfields moraines are predominantly northern in origin but north-west of Shrewsbury the sands and gravels filling the intervening sandur troughs include a significant proportion of Welsh rock types. South of Shrewsbury, an outlying set of linear moraines and associated sandur systems around Dorrington are predominantly composed of northern rock types but with an admixture of local Welsh rocks. To the west glacial deposits are exclusively Welsh, suggesting that the Dorrington ridges represent the maximum penetration of Irish Sea ice into this area.

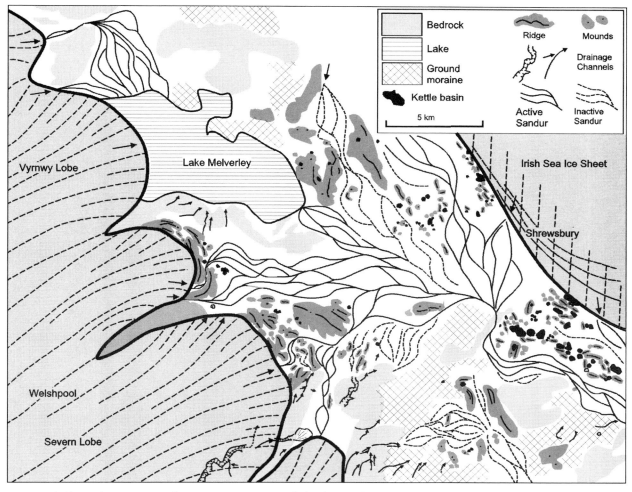

Fig. 4.10 Palaeogeographic reconstruction of the limit of the Severn glacier during an early retreat stage from its maximum and the extent of Glacial Lake Melverly. Limits of Irish Sea ice sheet and the Severn glacier are not necessarily contemporaneous

From the evidence presented here it is evident that towards the maximum of the Late Devensian glaciation, Irish Sea and Welsh ice coalesced along much of the border of north-east Wales and separated from one another only south of Shrewsbury where the Irish Sea ice-margin swings eastwards towards its terminal maximum north of Wolverhampton. At this stage the Severn glacial lobe seems to have been powerful enough to extend as far east as Shrewsbury and coalesce with the Irish Sea ice. During retreat the two ice-sheets progressively uncoupled and one stage in this separation is shown as a palaeogeographic reconstruction in Fig. 4.10. Of particular note is the occurrence of a temporary ice-marginal lake basin, the 'Melverly Lake', across much of the floor of the Severn valley west of Shrewsbury. This is based on the occurrence of extensive laminated clay, both at depth and at the surface. A number of earlier workers, especially Wills (1924) and Pocock and Wray (1925) postulated the existence of very large lake systems in the area, including Lake Buildwas that were instrumental in the permanent diversion of the River Seven south through the Ironbridge Gorge. Cannell (1982) and Thomas (1989), however, argue that the stratigraphy identifies only small, local and unconnected lake basins.

Sequence of events

Little is known of events preceding the last glaciation in north-east Wales but the region was probably glaciated by ice from both Welsh and Irish Sea sources during Marine Oxygen Isotope Stages 8 and 6. Except for cave sediments, buried Irish Sea diamicts and last interglacial estuarine sediment, little evidence of earlier glacial or interglacial stages has been preserved and the overwhelming imprint derives from the last, Devensian glaciation. At its maximum Welsh ice moved north, north-east and east to extend up to and a little beyond the borders of the country to meet Irish Sea ice moving south against the north coast and the borders. At the Devensian maximum the zone of coalescence between the two ice-sheets ran east from Conway, south-east through the Wheeler and Alyn valleys, past Wrexham, between Oswestry and Ellesmere towards Shrewsbury. South-east of Shrewsbury the two ice-masses separated.

The timing of the uncoupling of the Irish and Welsh ice-sheets is unknown. On the basis of weathering characteristics Boulton and Worsley (1965) distinguished two separate phases of Irish Sea ice advance; one terminating at the Wolverhampton limit, the other at the prominent Whitchurch moraine to the north. Recent work on the chronology of the Devensian glaciation by Bowen et al. (2002) has suggested that the former occurred at c. 40 ka and was associated with the Heinrich 4 climatic event, and the latter at c. 22 ka, soon after the Heinrich 2 climatic event. During uncoupling the ice-sheets effectively 'unzipped' on a line running north-west from Shrewsbury towards Wrexham. Although the details of Welsh ice involvement are not clear, Thomas (1989) has identified a number of stages in the retreat of the Irish Sea ice-sheet and noted the importance of a major ice-marginal sediment trough formed along the line of uncoupling by concentration of meltwater drainage from both the Irish and Welsh ice-sheet margins along the line of the Alyn valley into a major sediment sink in the Wrexham fan-delta.

5 The Cheshire–Shropshire Plain

by Peter Worsley

Introduction

This chapter concerns the relatively low ground in the counties of Cheshire, Shropshire and Staffordshire, bordered on the west by the uplands of Wales, and in the east by the steep west-facing flank of the Pennines. Beyond and northwards, the lowlands extend into southern Lancashire and the Wirral of northwestern Cheshire, while to the south, the limits encroach upon the area of varied relief which characterises the English Midlands. The modern drainage reaches the sea via four routes, the Rivers Dee and Mersey to the Irish Sea, the Severn to the Bristol Channel and the River Trent to the North Sea (see Fig. 5.1). The geology is dominated by a partially exhumed basin inherited from Permo-Triassic times, which is structurally a half graben flanked by mainly Carboniferous rocks (Worsley, 2002). In south Shropshire, older Palaeozoic and Pre-Cambrian rocks are present.

During the last Ice Age, these lowlands were invaded by extraneous ice from both Welsh and Irish Sea sources and the maximum extent of the latter provides a convenient envelope to the eastern and southern boundaries. There is an extensive array of glacial landforms, with widespread erosional features such as meltwater channels, and deposi-

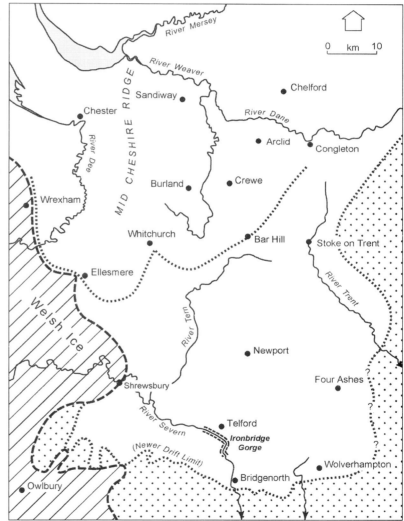

Fig 5.1 A map of the Cheshire–Shropshire lowlands showing the Last Glacial Maximum (Newer Drift) ice limit, drainage, and principal localities mentioned in the text. The stippled area beyond the glacial maximum ice limit was unglaciated during the Devensian

tional relief including extensive hummocky moraines, kames, eskers and kettle holes each displaying ice-contact slopes. W.B. Wright, the doyen British glacial geologist, promoted the notion of 'Newer Drift' for this kind of 'fresh' looking unmodified landform assemblage and contrasted it with 'Older Drift' which is usually degraded and dissected to such an extent that it crops out on interfluves rather than within valleys and seldom, if ever, displays recognisable glacial depositional landforms (Wright, 1914).

The field identification of the 'Newer Drift' maximum ice limit is not straightforward, as it is often poorly-defined, lacking any pronounced landform expression. Rather it is frequently a feather-edge feature to a thin veneer of glacial sediments or simply a concentration of erratic boulders. Difficulties are compounded where it abuts 'Older Drift' deposits related to ice moving from the same source, as the sediments can be very similar in composition. Accordingly, the precise position of the 'Newer Drift' border is often unresolved, and differences of opinion can amount to tens of kilometres *e.g.* the sector north-east from Wolverhampton through Rugeley towards Cheadle, Staffordshire, where the limit shown on the Geological Survey's 'Quaternary Map of the United Kingdom' of 1977, is much further east, contrary to the evidence given in the memoir (Stevenson and Mitchell, 1955). Therefore, it is more realistic to refer to an *approximate* ice advance limit.

The first national synthesis of the British Quaternary
An attempt to classify the British Quaternary following widespread consultation was published in 1973 (Mitchell *et al.*, 1973). This had been heralded by an earlier report (Shotton and West, 1969), which expounded the basic philosophy of sub-dividing the Quaternary into a sequence of cold and temperate stages, based on climatic cycles of glacial and interglacial status, with the glacials being divisible into alternating stadials and interstadials.

The Cheshire-Shropshire Plain was selected as the type area within the British Isles for the last cold stage, which was given the new name of 'Devensian'. The late Professor Shotton derived it from the 'Devenses', a British tribe which he alleged inhabited much of Cheshire. Such a tribe was entirely mythical but did have the sole merit of following earlier British tradition of using Celtic tribal names for divisions of geological time! More seriously, however, it departed from standard international practice of using geographical names for glacial events from rivers, examples of which are the classical alpine glacial stages of Gunz, Mindel, Riss and Würm from southern German rivers and Elster, Saale and Weichsel for the Fenno-Scandinavian ice-sheets of the north German Plain. In the literature immediately before 1970, authors had either used the terms Würm or Weichselian or simply 'Last' when referring to the latest major glaciation of the British Isles (Lewis, 1970).

Subdivisions of the Devensian
Rather than using classical bio and lithostratigraphy, Mitchell *et al.*, (1973) proposed the use of radiocarbon years as the basis for subdividing the Devensian. This decision to abandon standard practice was undoubtedly influenced by the lack of an ideal succession but it also meant that the subdivisions were defined from the top downwards because the radiocarbon timescale does not extend to the preceding Ipswichian interglacial. This caused a fundamental tenet of the classification scheme to be violated, namely that each stage be defined in terms of its base with the top being inferred by reference to the base of the following stage. Using what we now know to be uncalibrated radiocarbon years, the three subdivisions became: Late Devensian 10–26 ka BP, the Middle Devensian 26–50 ka BP, the Early Devensian being that part of the stage older than 50 ka BP extending back to an undefined terminal Ipswichian datum. This is not the place to engage in a discourse on radiocarbon dating other than to observe that any radiocarbon 'dates' of greater than 20 ka BP should be handled with particular reserve and are only indicative age estimates. When first published many 'old dates' had a greater precision ascribed to them than should have been the case, in part because the problems arising from contamination were not fully appreciated.

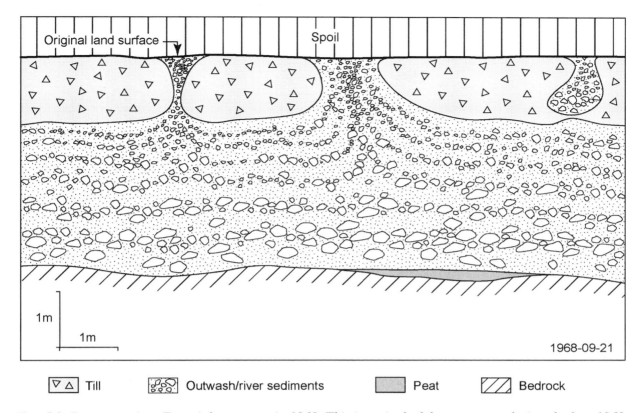

Fig. 5.2 Drawn section, Four Ashes quarry in 1968. This is typical of the exposures during the late 1960s and shows the characteristic elements of the stratigraphy at that time. Middle Devensian biogenics and similar-aged permafrost structures are not present and were only rarely exposed elsewhere in the quarry. Note the periglacial disturbances effecting the upper 2 m and the small ice-wedge cast to the right

Initially, Chelford in east Cheshire was selected as the Devensian stratotype locality, but the definitive 1973 report discarded it in favour of a newly discovered locality at Four Ashes in south Staffordshire, which was certainly beyond the imagined territory of the Devenses tribe (see Fig. 5.1). Nevertheless, as will be seen, the change of location had logic on its side.

Devensian stratotype—Four Ashes, Staffordshire

Four Ashes (SJ 916082) occupies a flat landscape on the north side of the shallow Saredon Brook valley, and is in many ways unique in the context of the British Quaternary. Its geological riches were revealed through sand and gravel quarrying, but unfortunately nothing has been conserved as a 'Site of Special Scientific Interest'. Indeed, the old pits were later buried beneath landfill. Nevertheless, a poorly exposed section of some 3 m of gravels forming the northern limit of the workings remains. A drawn section (Fig. 5.2) shows the typical stratigraphy exposed. In its late 1960s hayday (Fig. 5.3), the Four Ashes quarry revealed thin biogenic representatives of the Ipswichian interglacial and all three Devensian subdivisions—early, middle and late. Much of the immediate vicinity is underlain by 0.5–5 m of sandy gravels which overlie red sandstones of Permo-Triassic age. The gravels consist almost totally of quartzite clasts derived from the Triassic Kidderminster Formation conglomerate (Bunter Pebble Beds) which crops out in Cannock Chase to the east. Although they are strictly non-glacigenic they are nevertheless the products of river deposition in a cold stage environment. Discordant above the gravels is a patchy till up to 4 m thick (assigned to the Stockport Formation—see later), but an associated outwash facies is anomalously absent. Within the gravels over 50 discrete thin lenses of detritus peaty muds and silts are identified and astonishingly all came from a restricted geographical area within a radius of about 200 m. Their lens form and limited extent suggests that they are infills of shallow channels within a fluvial depositional system (see Fig. 5.3).

About half of the lenses were examined in detail and found to contain fossil beetles (A. Morgan, 1973) and flora (Andrew and West, 1977). Each contains an internally coherent assemblage, indicating minimal reworking of earlier deposited material as aggradation progressed, a surprising discovery in the light of the apparently low sedimentation rate. Four principal ecological groups were identified on the basis of the fossil assemblages in the lenses:

Fig. 5.3 Four Ashes Quarry September 1968 during a visit of the Quaternary Research Association. One of the basal organic lenses, of either interglacial or interstadial status, lies at the feet of the participants. The bedrock is exposed on the quarry floor and corresponds to the level of the ponded water. Note the prominent horizontal base to the Stockport Formation forming a paraconformity with the underlying non glacigenic (fluvial) gravels. (Photo: Peter Worsley)

   (a). Tundra, with Arctic insect faunas; high non arboreal pollen, grasses and sedges
   (b). Near tundra, with an unusual mixture of temperate and cold insect faunas; high non arboreal pollen, grasses and sedges
   (c). Boreal woodland, with boreal insect faunas; high arboreal pollen, of pine, spruce and birch
   (d). Temperate forest, with high arboreal pollen, with oak, alder and hazel; very few insects preserved.

At the basal unconformity of the sandy gravels, three biogenic lenses infilled shallow channels eroded into the bedrock beneath. Their biota revealed that trees were present in the landscape (groups c and d above), whereas those at higher levels in the gravel sequence indicate treeless environments (groups a and b above). Group (d) was identified on the basis of a single lens (site 44) but it contained both macro and micro floral fossils. These signified a temperate fen—deciduous woodland comparable with the last interglacial Ipswichian biozone Ip IIb, (*i.e.* the latter part of the early temperate sub-stage of the classical interglacial vegetation cycle). The other two lenses had beetle faunas of boreal character and one of these (site 10) had a compatible pollen biozone. These were strikingly similar to the interstadial deposits at Chelford and were correlated with them. It is important to note that sites 10 and 44 lay lateral to each other at the same horizon and it is only an assumption that the group (d) flora of site 44 antedates the group (c) biota of site 10. The biota in group (a) are standard arctic assemblages and occur in lenses at various levels in the gravel sequence. The most problematic assemblage is that of group (b). The flora is decidedly cool, but in contrast, the beetle fauna includes species which are tolerant of only a narrow environmental temperature range, with warm demanding ones in the ascendancy over their cold tolerant counterparts.

This anomalous mixture characterises the Middle Devensian 'Upton Warren Interstadial', which is described in Chapter Six, and lies in the middle of the gravel succession.

Eight radiocarbon assays were undertaken by the University of Birmingham radiocarbon laboratory. A single assay on site 44 group (d) interglacial wood was, as expected, infinite (>45 ka) and undertaken as a precautionary measure. The group (b) material yielded 3 dates bracketing 38–43 ka, consistent with other dates from the Upton Warren Interstadial event and finally a group (a) sample from below the group (b) horizon yielded an infinite age (>44 ka) whereas 3 others stratigraphically above it bracketed 30–37 ka.

Thus the lower fluvial sequence provides a sedimentary record extending from the last interglacial through to about 30 ka BP. The biota provide evidence of substantial climatic fluctuations in the order of interglacial—interstadial—stadial—interstadial—stadial prior to the site being enveloped by a glacier. Unfortunately, this diverse array of environmental change is recorded in an attenuated fluvial succession which must contain many hiatuses and hence only gives snapshots of what must have been a dynamic saga of change. An important implication arising from the Four Ashes stratigraphical evidence is that the Staffordshire regional 'Newer Drift' LGM (Last Glacial Maximum) ice advance probably occurred after some 30 ka BP and hence was likely to be of Late Devensian age (Shotton, 1967).

Chelford Sands Formation

Following Shotton and West, (1969), it was originally mooted that Farm Wood Quarry, Chelford would become the Devensian type site. This quarry came into prominence in the 1950s when a 1 m thick peat bed within a sand succession seemed to support the tripartite concept of an initial 'Lower Boulder Clay' glaciation, followed by a significant retreat with the deposition of 'Middle Sands' and a subsequent later 'Upper Boulder Clay' glaciation (Fig. 5.4). The biogenic Farm Wood Member was originally interpreted as having accumulated during the 'Middle Sands' glacial retreat interval. This member yielded a rich fossil micro and macro flora with spectacular tree stumps in the position of life and a beetle fauna. The biota pointed to a boreal forest environment with birch, pine and spruce, analogous to present day south-central Finland. This sequence defined the Chelford Interstadial (Simpson and West, 1958; Coope, 1959).

The detailed saga of radiocarbon dating the biogenics at Chelford and a similar site at nearby Arclid has been recounted earlier

Fig. 5.4 Oakwood Quarry (West), Chelford in 1978 when groundwater pumping enabled the interstadial deposits to be clearly seen in the middle of the Chelford Sands succession. They formed the fill of two coeval palaeochannels cut into the lower sands. The original peat probably filled the channels but due to later burial compaction by both sediment and ice loading, this has been reduced by almost an order of magnitude. Occasionally in situ *sub fossil tree stumps were rooted within the peat. Above the sands is the glacigenic Stockport Formation. This had to be removed before the sands could be extracted. (Photo: Peter Worsley)*

by Worsley, (1980). Suffice it to say here that the absolute age of the material is *now* considered beyond the capability of the method, and the finite radiocarbon dates are probably an artifact of low levels of contamination by more recent material. Application of thermoluminesence dating techniques to the sands adjacent to the Farm Wood Member has suggested that the absolute age is *circa* 100 ka BP (Rendell *et al.*, 1991) making the Chelford Interstadial a likely correlative of Marine Oxygen Isotope stage 5c.

A closer examination of the Farm Wood sedimentology revealed that the sands containing the Farm Wood Member were not 'Middle Sands' as had previously been understood, but rather Chelford Sands, a separate formation of fluvio-aeolian rather than fluvio-glacial origin, (Boulton and Worsley, 1965; Worsley, 1966a; Evans *et al.*, 1968). The Chelford Sands were subsequently shown to be of wide extent in east Cheshire and were interpreted as low-angle alluvial fans with the interstadial deposits being part of the fill of occasional incised palaeochannels (see Fig. 5.4). Aggradation of these sediments was associated with a sediment flux from east to west (away from the Pennine front) which progressively buried a pre-existing erosional landscape of mainly Permo-Triassic sediments.

Sub-drift surface and allied deposits
A partially buried landscape forms the unconformity between the Quaternary-aged sediments and the Mesozoic and Palaeozoic bedrock. Some exceptionally deep sequences of Quaternary materials have long been known, for example the Sydney borehole just north of Crewe, which was some 100 m deep and unbottomed. In an attempt to reconstruct the buried geomorphology of the Dee-Mersey estuary area, Gresswell (1964) identified four sub-parallel depressions trending north-west/south-east in the Dee–Mersey estuary area and coined the name iceways for them, invoking selective ice erosional processes along linear belts to explain their overdeepened characteristics. A decade later, Howell (1974) reported a much broader study of the buried surface throughout south Lancashire and much of Cheshire, using both borehole and seismic evidence. He was critical of the iceway concept and argued that the surface form was a function of sub-aerial fluvial erosion processes during periods of eustatically lowered sea level when an extensive network of incised valleys was cut. Howell acknowledged a problem with his reconstructed longitudinal valley profiles, as they were seemingly upgraded, and he explained this by suggesting that an absence of boreholes at critical locations was responsible. Ultimately he had to acknowledge an element of subglacial meltwater erosion to account for the over deepening. One such example was beneath the River Dee estuary where a bedrock threshold lay in the path of the postulated buried valley at the relatively shallow depth of some 30 m below modern sea level. Since 'upstream', the sub-drift surface descended to at least -88 m O.D., the hypothesis of grading/incision of sub-aerial rivers in direct response to eustatic sea level lowering was clearly not sustainable. The age of these thick buried valley fills remains uncertain and diachronism is likely to be common.

Although the broad outlines of the rockhead surface are known in the Mersey catchment, local morphological details are susceptible to the interpreter's preconceived ideas. This point was well illustrated by Johnson and Musk (1974), who gave students the task of contouring Howell's Mersey basin borehole data without revealing the location. The resulting maps were significantly different from those produced by Howell. In the other catchments the precise form of the rockhead surface remains obscure due to a paucity of key data; for instance, it has yet to be conclusively established that there is a continuous buried valley system extending from the present River Severn near Shrewsbury to the Dee.

A number of factors probably contribute towards the buried surface morphology and these include (a) limited eustatically controlled incision, (b) localised intensive sub-glacial meltwater erosion associated with tunnel valleys and (c) selective erosion by glacier ice flow. An added complication is the presence of Permo-Triassic saltfields. As Earp and Taylor (1988) have emphasised, salt dissolution, seepages of brine and resultant subsidence are not recent products of human activity but are of geological antiquity with natural flows of brine taking advantage of the deep gravel-filled channels that underlie the river valleys. One may speculate that in glacial environments these flows might have been enhanced, especially in pressurised subglacial drainage systems.

It appears highly plausible that there were a number of earlier glaciations but their identity is enigmatic due to the rarity of interstadial or interglacial biostratigraphic markers. However, some pre-Devensian tills are definitely identifiable, such as that at Burland (Bonny *et al.*, 1986) where probable Chelford Interstadial deposits overlie a deeply buried till, and the Oakwood Till near Chelford (Worsley *et al.*, 1983). Similarly, till occurs at Arclid beneath interstadial and interglacial deposits (Worsley, 1991, 1992) and boreholes near Crewe have proved organic deposits between tills (Rees and Wilson, 1998). Any earlier glaciations would have contributed towards the ultimate sub drift surface form.

The Late Devensian Glaciation

The LGM glaciation of the plains area involved ice coming from two distinct sources: eastward flowing ice from Wales and southward flowing ice from the northern Irish Sea Basin/Lake District. Above the bedrock sub drift surface and where present, an erosional unconformity cut into the Chelford Sands, is a complex sequence of tills and outwash facies. This forms a sedimentary unit which constitutes much of the land surface in the Cheshire–Shropshire Plain region and extends to the eastern and southern glacial limits. Most of the lowland is dominated by Irish Sea type glacigenic sediments which are called the Stockport Formation, (Worsley, 1967a; Chapter Two), with Lake District and Southern Uplands erratics, along with a derived marine molluscan fauna (Thompson and Worsley, 1966). A radiocarbon assay on just 2 derived marine shells from ice-marginal deposits at Sandiway suggests a late Middle Devensian age, and a selection of the same fauna from a range of Cheshire localities yielded amino acid D:L ratios consistent with a Devensian age (D.Q. Bowen, pers. com.). Along the western border the coeval glacigenic materials from Wales are named the Shrewsbury Formation (Worsley, 1991).

Importantly, this event should not yet be assigned to the Dimlington Stadial, which is often incorrectly assumed to be the equivalent of the Late Devensian (10–26 radiocarbon years BP as noted earlier in this chapter). At the east Yorkshire type site of Dimlington, 2 radiocarbon assays of around 18 ka from biogenics from beneath a thick till sequence define the base of the stadial. Hence the sequence does not embrace the early Late Devensian, and there is no certainty that the Irish Sea ice LGM is coeval with it, as discussed in Chapter Two. Indeed, Bowen *et al.*, 2002, have speculated that that the segment of the 'Newer Drift' ice limit in east Shropshire and Staffordshire is not of LGM age and might even fall into the Middle Devensian. However, until new data from the area are available it is best to keep an open mind on this issue.

Across the central parts of the lowland, a series of arcuate ridges defines the Wrexham–Ellesmere–Whitchurch–Bar Hill end moraine (see Figs. 5.1 and 5.5). Along the feature morphologically diverse

Fig. 5.5 A view of the north face of Wood Lane Quarry, Ellesmere as exposed in 1971 with two major thrusted till units in the central part (see Worsley, 1991 for a drawn section of the thrusts). The skyline curve reflects the morphology of the hummock landform into which the quarry has been excavated.
(Photo: Peter Worsley)

terrain units can be identified including till-cored ridges, hummocky ice-contact kame and kettle zones and frontal fans and deltas (Boulton and Worsley, 1965; Worsley, 1967b; Paul, 1983; Thomas, 1989; Paul and Little, 1991). This array emphasises that at the regional-scale end moraines are not at all comparable with the simpler cross valley moraines found in highland areas and appear to represent elements of a supraglacial landsystem associated with a sub polar glacier. Amongst other factors, the stratigraphic evidence from Four Ashes resolved an earlier debate as to the status of this end moraine by showing that it is either a readvance or retreat feature.

The richest archive pertaining to the Stockport Formation lies in the publications, maps and memoirs, of the British Geological Survey. However, the reader needs to be aware of the changes in glacial geological theory which have occurred through time. A 1: 10,000/560 scale re-mapping of much of the lowland area began in the 1930s and continued until the 1970s. The first two new memoirs to be published—Stockport (Taylor et al., 1963); Nantwich (Poole and Whiteman, 1966)—followed the lead given by the pre-war Wigan memoir (Jones et al., 1938) and later independently by Simpson and West, (1958), in maintaining the classical tripartite concept of glaciation, retreat and readvance as the conceptual basis for interpretation of glacial deposits. Since the first edition of Lewis (1970) went to press, three new memoirs have appeared and complete the planned revision of the region. The first was Macclesfield (Evans et al., 1968); although this publication was not released until 1969 despite being dated 1968. Although the authors identified the Chelford Sand as being non-glacigenic in origin and were clearly aware of the weaknesses of the tripartite model, the latter model persists in the text, resulting in an unsatisfactory hybrid terminology. The second was Chester; (Earp and Taylor, 1986) in which the tripartite notion is almost exorcised and finally the third, Stoke-on-Trent (Rees and Wilson, 1998) is totally free of its influence. Indeed, the latter work discusses the glacial sediments in a contemporary interpretative framework, emphasising the complex relationships in each case—till sheet (active ice), morainic (meltout), glacifluvial (outwash), glaciolacustrine and meltwater channel fills.

Apart from the standard Geological Survey mapping work, in the early 1980s a series of 1: 25,000 scale maps and reports devoted to sand and gravel potential were published under the auspices of the survey's Industrial Minerals Assessment programme. These covered much of the south-western part of the region, and apart from rapid revision of existing maps, this work was primarily based on hundreds of newly commissioned shell and auger boreholes, most of which did not reach rockhead, because of the great thickness of the glacial sediments encountered (in excess of 50 m).

Two independent attempts have been made to interpret this new data set. The first was by some of the officers of the Geological Survey, who had worked on the mineral assessment programme. Surprisingly, in view of the progressive abandonment of the tripartite glaciation model by memoir authors, Wilson et al., (1982) actually resuscitated it! They argued for two separate glacial advances separated by complete deglaciation and claimed that the Middle Sand was a prograding sandur. The second by Thomas (1989), in conjunction with his own reconnaissance level landform—sediment mapping, made a regional analysis with the objective of improving understanding of the depositional environments involved and elucidating the relationships of the Irish Sea and Welsh-sourced ice-sheets. He concluded that any simple order in the glacial sediments, such as the tripartite model would identify, was elusive. Finding no evidence to the contrary, he interpreted the whole succession as indicating one, albeit complex, glacial event. He advocated that a discontinuous basal lodgement till was deposited during the glacial advance, to be later overlain by the products of widespread supraglacial sedimentation during the retreat phases. In addition, he identified prolonged stillstands and localised frontal oscillations as part of the overall deglacierization process.

It has long been recognised that the Welsh and Irish Sea ice-sheets interacted along a 5–10 km wide zone running from Wrexham south to Shrewsbury. Borehole evidence suggests that the Welsh ice arrived first in the borderlands but later the contact shifted as the two ice-sheets jostled (Thomas, 1985). During

The Cheshire–Shropshire Plain

Fig. 5.6 The type Shrewsbury Formation succession as exposed in 1969 in the former Mousecroft Lane Quarry (south-west suburbs of Shrewsbury). This sequence, some 3 m thick, displays a classical coarsening upwards succession of outwash gravels (proglacial sandur), which is the fluvial signature of an advancing ice front, overlain by a till of Welsh derivation. (Photo: Peter Worsley)

Fig. 5.7 The northern face of Mousecroft Lane Quarry in 1973, showing the Stockport Formation (a complex of till and outwash facies) overlain by the Shrewsbury Formation. The contact between the two formations shows large scale low amplitude wave-like deformation attributed to the post depositional meltout of glacier ice blocks deeply buried within the Stockport Formation sediments. The meltout must have occurred after the Welsh derived ice had advanced across the site. In the local area there are many kettle holes e.g. Weeping Cross. (Photo: Peter Worsley)

the 1960s and '70s the extensive Mousecroft Lane gravel pit exposures in the south-west suburbs of Shrewsbury showed very clearly that the final ice-sheet present in that area was of Welsh derivation (Fig. 5.6). A complex of northern-derived till and outwash sediments (Stockport Formation) was unconformably overlain by a generally coarsening upwards proglacial sandur sequence which culminated in Welsh till (Shrewsbury Formation). This can be traced eastwards to the longitude of Uffington, (see later). The entire sequence in the vicinity of Shrewsbury is deformed by kettle holes and the Mousecroft quarry sections showed that the source of the subsidence was relict Irish Sea glacier ice within the Stockport Formation (see Fig. 5.7). Alas, landfill and housing have largely obscured this formerly well-exposed key area.

Meltwater Drainage

Glaciers normally have en and subglacial meltwater systems which emerge at the lowest parts of their margins. This generalisation applies during both the advance and retreat phases, with the greatest discharges occurring during the summer. We may speculate that when the Irish Sea ice advanced up the regional lowland slope into the Cheshire Plain it impeded the drainage which otherwise would have drained into the Irish Sea. As the ice approached the watershed between the Liverpool Bay basin and the Severn–Trent systems, it became less likely that internal and marginal drainage would have been able to cope with all the water being produced, so any proglacial ponded water would have first sought pre-existing cols in the confining boundary watershed, and beyond downstream flooding would have ensued, for example in north Staffordshire down the Churnet, Fowlea, Lyme and Meece Brooks of the uppermost Trent (Knowles, 1985). In this manner, the discharge of the Trent and Severn catchments would have been greatly augmented. However, these watersheds were eventually overwhelmed because the ice advance limit lay some distance beyond them. At this stage, the former sub-aerial flows probably became trunk subglacial systems draining through the cols, and augmented by subsidiary networks. These cols, and their downstream extensions, are characterised by deep infills below the modern floodplain, for example the River Sow near Stafford (Morgan, 1973), although here most of the aggradation may be attributed to the retreat phase. Impeded drainage also occurred during the ice retreat as watersheds became re-exposed. One such example lies in the Newport, Shropshire district where landform evidence shows lake overflow at Gnosall and esker–fan formation (see Fig. 5.8).

Undoubtedly the most spectacular glacial landform in the region, if not the whole of England, is the approximately 10 km long Ironbridge Gorge, though which the contemporary River Severn discharges much of the drainage of central Wales. It appears plausible that prior to the Devensian glaciation, the Welsh component of the River Severn drained northward to join the predecessor of the modern River Dee and hence flowed into Liverpool Bay. A 'preglacial watershed' between the modern Lower Severn and Irish Sea catchments extended across from Wenlock Edge to the ridge extending through the Wrekin to Lilleshall. The precursors of Dean Brook, Mad Brook and the River Worfe defined the headwaters of the Lower Severn because the Ironbridge Gorge did not yet exist. Rather optimistically, Shotton (1977 p.13) asserted 'There can be only one explanation for the Ironbridge Gorge. Drainage from the Upper Severn was blocked by ice to the north, impounding water which

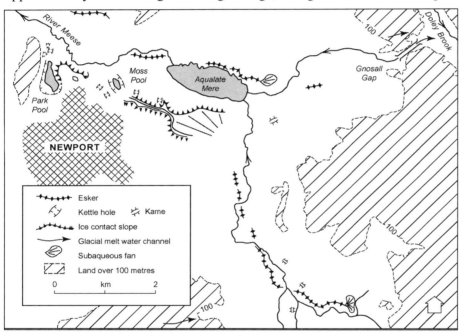

Fig. 5.8 A glacial geomorphological map showing the distribution of ice retreat landforms in the district around Newport, Shropshire and adjacent Staffordshire. The three lakes—Aqualate Mere, Moss Pool and Park Pool—occupy the sites of large kettle holes. Note that the eskers and their associated fans indicate that the sub-glacial drainage was towards the east and south-east. The Gnosall Gap breaches the Severn/Trent watershed and is thought to have been formed by ponded meltwater draining eastwards into the Trent basin before a connection with the Ironbridge Gorge drainage was established. Note that a small meltwater channel, which probably antedated the main phase of overspill, lies just north of the main overflow channel

poured across the lowest point it could find in the land ridge to the south, until the overflow gorge so produced was lowered enough to give permanent diversion of the river'.

In fact the genesis of the Ironbridge Gorge remains an unsolved mystery and apart from lake overspill, a number of other mechanisms have been proposed, including pure subglacial drainage, ice-marginal drainage, and exhumation of an infilled older gorge. Within the bedrock-floored gorge itself, there are virtually no superficial deposits so sedimentary clues are effectively absent. Therefore, it is necessary to examine the areas both up and downstream of the gorge, and the terrain on either side, for clues.

The notion of a glacially determined diversion of drainage was first proposed by Charles Lapworth in the late nineteenth century. Wills (1924), undertook the detailed glacial geological fieldwork to test this hypothesis. He suggested that the initiation of the drainage diversion was due to a sequence of small overspilling meltwater lakes impounded against the western side of the preglacial watershed, with a controlling role played by ice-marginal oscillations. His final published views are beautifully illustrated by a series of cartoon maps in Wills (1950) but some years later he expressed the view that parts of his interpretation were 'badly out of date' (pers com.1964). Until around 1960 very few British workers were aware of the possibilities of subglacial meltwater drainage mechanisms, so glacial lake overflows were often erroneously postulated: indeed, knowledge of contemporary glacial environmental processes was very rudimentary.

Subsequent to Wills, the only thorough published study of the Ironbridge Gorge area glacial history is that undertaken by Richard Hamblin for the Geological Survey, (Hamblin, 1986; Hamblin and Coppack, 1995). This work was part of a special investigation of the Telford New Town district. A wealth of new detail pertaining to the glacial materials on either side of the gorge was documented, aided by new exposures in open cast coal mining operations and borehole data. A generally complex stratigraphy was revealed, although in many places two tills of similar character were identifiable separated by waterlain deposits. However, Hamblin considered that all the glacigenic sediments were of Late Devensian age and the two tills were attributed to lodgement and meltout processes. Further, he concluded that the gorge was probably initiated as an ice-marginal meltwater drainage channel adjacent to a decaying ice-mass as evidenced by the meltout till sheet, but he did not rule out a possible subglacial component. Clearly, once established, drainage through the gorge would be maintained and meltwater from both Wales and that sector south of the displaced watershed between the Severn and Dee/Weaver basins would flow through it.

To the north of the Ironbridge Gorge an infilled glacial meltwater eroded valley, the Lightmoor Channel, was defined running roughly parallel to and of similar dimensions to the Ironbridge feature. Hamblin believed that this channel had functioned during the last glacial advance phase. This valley was the upstream continuation of a similar feature previously identified by Hollis and Read (1981) but the latter workers preferred an interpretation with an advance- retreat- advance cycle for a two stage infill. Struck by the similarity of the Lightmoor Channel buried valley and the demonstrably pre-Devensian Seisdon-Stourbridge Channel (Morgan, 1973), it was suggested by Worsley, (1991) that the Lightmoor Channel was originally formed in an earlier, possibly Anglian-aged glaciation.

Much effort has been expended on linking the proglacial outwash in the lower Severn valley with the ice limit in the vicinity of Bridgnorth (Wills, 1950). The principal outwash feature is a valley sandur known as the Main Terrace and from below the Ironbridge Gorge and in the tributary valley of the River Worfe to the east, this can be traced down to the Bristol Channel. In several places, tills have been found to interdigitate with the outwash facies but this is the norm in the complex ice terminal environment, where marginal oscillations are common. Similarly the outwash terrace can be expected to show some variations in height in the proximal zone. To date, no local biogenics have been discovered which can directly confirm a Late Devensian age for the outwash although in the past deposits in non-glaciated downstream tributaries such as that of the River Salwarpe have unduly influenced chronological thinking. Worsley (1976) made a critique of linking the Upton Warren Interstadial sequence to the Main Terrace

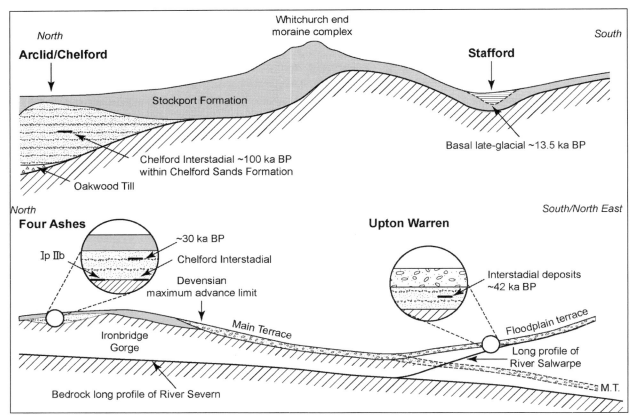

Fig. 5.9 A schematic north/south cross section to show the regional relationship of the non glacigenic Chelford Sands and the glacigenic Stockport Formation, to the River Severn Main Terrace and the key sequences at Four Ashes and Upton Warren (modified after Worsley, 1977)

on the basis of geomorphological continuity, as was done by Coope *et al.*, (1961) and exposed the assumptions involved. Despite this, Hamblin appears to accept the morphological linkage and in doing so overlooks the major unconformity with the River Salwarpe terrace at Upton Warren, otherwise he would not persist in the notion of a pre LGM age for the ice limit at Bridgnorth (see Fig. 5.9 for a diagrammatic explanation).

The highest terrace which is claimed to be present both upstream and downstream of the gorge section of the river is named either the Uffington (upstream) or Worcester Terrace (downstream). As might be expected, downstream of Ironbridge, the Worcester Terrace is lower than the Main Terrace and inset within it. It is doubtful if the terrace fragments classified by the Geological Survey as Uffington Terrace, were once a continuous feature extending to Ironbridge. It may well be an outwash landform of limited extent in the area immediately east of Shrewsbury due to the presence of ponded water between that point and the Ironbridge Gorge. This would be coeval with a glacial advance event, when Welsh ice reached the site of Shrewsbury (Shrewsbury Formation). Considerable uncertainty prevails over the details of the deglacial environment between Shrewsbury and Ironbridge at this time and the extent and role of residual ice in the landscape (Shaw, 1972 a, b; Worsley, 1975).

A structurally determined north/south orientated discontinuous ridge of Helsby Sandstones protrudes above the Cheshire Plain to a maximum height of 150 m. Numerous dry channels and networks cut into bedrock and the flanking glacial deposits are common. An example of their form and distribution in the southern part of the ridge is shown in Fig 5.10. Although many of these are likely to be the product of glacial meltwater erosion, conclusively proving this is difficult. Some of the channels might be fluvial or nivational in origin formed within a permafrost environment which was established after deglacierization (see next section). In this context, it is important to note that none of the channels are known to extend over watersheds or possess 'up and down' longitudinal profiles. Some of the channels have been previously reported in the Geological Survey memoirs but all were viewed as being cut by sub-aerial meltwaters. Following 1:10,630 scale field mapping, Worsley (1967b, 1970) suggested a sub-glacial chute origin, and this view was later supported by Sambrook Smith and Glasser, (1998); Glasser and Sambrook Smith, (1999) after preliminary air photograph reconnaissance and field checking and survey. The latter workers attempted to map the entire ridge area and a total of 134 channels divisible into four sub-groups were identified based upon the assumption that all were of glacial meltwater origin. Small scale 'p' forms (rounded subglacial drainage channels eroded into bedrock) have also been identified at Lyme, 20 km to the north-west (Leviston, 2001) and by the writer near Frodsham.

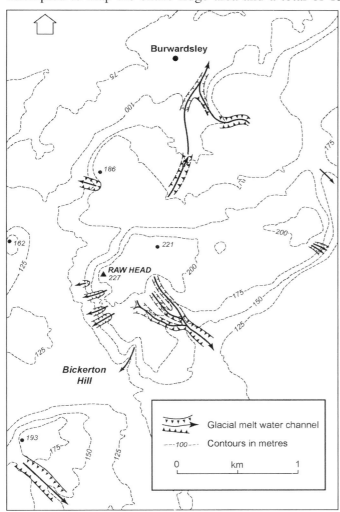

Fig. 5.10 Map showing the distribution of some typical bedrock floored glacial drainage channels associated with part of the southern Mid-Cheshire Ridge around the Bickerton Hills. The overall pattern of the channels suggests that the ice was generally stagnating and that sub glacial conduits eroded into the Helsby Sandstone were cut by meltwater flowing down both sides of the ridge

Periglaciation

Although traditional accounts of cold stage environments from within an area of glaciation tend to have a strong glacial emphasis, an equally important element ought to concern the effects of permafrost development. Indeed, with the increasing attention given to glacier thermal regime, the realms of glaciology and permafrost science merge.

The initial recognition of permafrost related structures in the region came with the reports by Worsley, (1966a; 1966b). First, examination of the east Cheshire stratigraphy showed that intraformational ice wedge casts occurred in both the Chelford Sands and Stockport Formations. In addition interformational ice wedge casts occurred at the unconformity between the two formations. It was clear from this evidence that permafrost was a recurrent phenomenon both in the 'preglacial' Devensian and after the ice had started to retreat from the LGM. Second, at Congleton, a surface polygonal network was serially excavated and this showed wedge structures penetrating a thin till sheet lying unconformably on Chelford Sands. Importantly, the wedge infills were of clast-free exceptionally clean sand. The absence of any evidence indicating the derivation of the fills from the host till during permafrost decay, suggested that the sands had infiltrated open thermal contraction cracks, hence these were true 'sand wedges' (as such the first to be identified in Europe). It appeared that sand, probably derived from exposed Chelford sands, had been transported by the wind to the site. Ventifacts, testimony to aeolian transport, are commonly associated with the Devensian succession, especially at or just below the modern land surface (Thompson and Worsley, 1967). [At the proof stage, after approval by the referees, the Congleton manuscript was shown by Anders Rapp (the journal editor) to Troy Péwé, who had previously described sand wedges in Antarctica. Having the preconceived notion that British sand wedges were implausible, Péwé persuaded Rapp that the title be changed from sand to the much less specific frost wedges. Péwé's published comments clearly show that he had failed to grasp the essential field characteristics described in the paper!]

These findings demonstrated that beyond the LGM glacial limit in the extra glacial area, ice wedge casts did not necessarily relate to the LGM as had previously been assumed but could equally antedate or postdate that event. Indeed, it is now clear that most, if not all of the Devensian ice recession, occurred before there was any significant climatic amelioration from a persistent polar climate. Devensian glacial retreat was probably primarily a function of a prolonged negative mass balance caused by aridity rather than climatic amelioration.

The east Cheshire findings were later to be corroborated by evidence from Four Ashes, where three horizons of sedimentary structures, interpreted as ice wedge casts signifying the former presence of permafrost, were identified (A.V. Morgan, 1973). The lowest horizon is enigmatic, consisting of one wedge structure penetrating bedrock at an unusually oblique angle from the basal unconformity of the gravels. Unfortunately, its age is indeterminate and could be either pre Ipswichian or Devensian. Within the gravels, a few intraformational ice wedge casts defined a single horizon which, on biostratigraphic grounds, postdates the Upton Warren horizon, but antedates the emplacement of the till. Post ice retreat ice wedge casts and allied involutions were common across the site.

Finally, through the work of the late Edward Watson, numerous relict cryogenic mound sites have been identified within Wales. Following air photograph reconnaissance, Watson was aware of a potential site just inside Shropshire at Owlbury, near Bishops Castle but it was only recorded by him as a dot on a small map of Wales (Watson, 1977). This was later located and investigated by Gurney and Worsley, (1996) who showed that a number of enclosed basins, some with shallow ramparts, occupied a valley bottom underlain by lake sediments. These were interpreted as permafrost related mineral palsas and since they clearly post dated the deglaciation, they may signify localised permafrost growth during the Dimlington or Loch Lomond stadials.

Conclusion

Since the 1970s global understanding of Quaternary environmental change has been revolutionised by the results from analyses of oceanic and ice-cap cores. In particular, the record of glacier and marine stable oxygen isotope variability has heavily influenced the chronological frameworks through which the impacts of the last cold stage have been interpreted. Unfortunately, terrestrial records are normally found to be fragmentary and this is undoubtedly the case for the Cheshire–Shropshire Plain. Yet nationally, the Four Ashes and Chelford successions are probably the most informative for understanding the pattern of environmental change within the Devensian. Landscapes are usually dominated by erosional processes, rather than those of deposition, so that much of the time lapsed is represented by gaps in the sedimentary record, in other words unconformities. The intensive geomorphological and sedimentary processes crammed into the few thousands of years spanning the LGM when the Irish Sea and Welsh ice-sheets invaded the lowlands largely determined the character of today's landscapes.

6 The lower Severn valley

by Darrel Maddy *and* Simon G. Lewis

Introduction

The Pleistocene development of the lower Severn valley is recorded in the fluvial sediments of the Mathon and Severn Valley Formations. Glaciation of the basin is known to have occurred on at least three separate occasions with deposition of the Wolston/Nurseries (Marine Isotope Stage 12), Ridgacre (MIS 6) and Stockport (MIS 2) Formations. The most complete stratigraphical record is that of the Severn Valley Formation, which post-dates MIS 12 and comprises a flight of river terraces, the highest of which is *c*. 50 m above the present river. The terrace sediments are predominantly composed of fluvially deposited sands and gravels, largely the result of deposition in high-energy rivers under cold-climate conditions. Occasionally towards the base of these terrace deposits low-energy fluvial sediments are preserved. Often these contain temperate faunal remains and yield geochronological age estimates that support their correlation with interglacial conditions. The chronology of terrace aggradation in the lower Severn seems to correspond with the Milankovitch 100 ka climate cycles. Similar terraced sequences occur in the tributary valleys of the Warwickshire Avon (Avon Valley Formation) and Wye (Wye Valley Formation).

The evolution of the Severn valley is closely related to crustal instabilities. The initiation of the Severn drainage line is believed to result from tectonic reactivation of the Triassic basin during the Tertiary. More recent crustal uplift is implied by the formation of the terrace staircase that indicates progressive incision of its valley during the post- MIS 12 period. Superimposed on this long-term uplift are periods of complex localized crustal movements associated with glacio-isostatic adjustments. Complex terrace sequences develop in response to rapid incision during periods of glacio-isostatic rebound, with large incision events reflecting the rebound adjustment to late glacial stage isostatic depression.

Recently studies of river sedimentary archives, with their wealth of stratigraphic information, have enjoyed a revival of interest (Bridgland, 2000). This chapter discusses one such archive, the sediment/landform assemblages deposited and shaped by the River Severn (Fig. 6.1). The River Severn, together with its tributary, the River Wye, currently drains a large proportion of Central Wales and the English West Midlands. However this drainage pattern represents only the latest configuration of drainage of this region. Throughout the Pleistocene this area has undergone repeated drainage reconfiguration as a consequence of successive river basin invasion by glacial ice from the Welsh uplands.

This chapter presents a synthesis of our existing knowledge concerning the Pleistocene evolution of the Lower Severn Valley (defined as Bridgnorth to the sea). In any synthesis we have to achieve at least three goals, each of which represents a stage in interpretation:

A) To briefly describe the available landform and sedimentary evidence.
B) To place the sediment/landform evidence within a chronological framework so that we can establish an evolutionary story. Traditionally frameworks would be established using biostratigraphical correlation. Increasingly the availability and use of independent dating methods has allowed more meaningful chronological frameworks to be developed. Here independent age estimates are used together with the biostratigraphy to attribute

deposits to the now globally accepted Pleistocene framework of the Marine oxygen Isotope Stages (MIS: Shackleton and Opdyke, 1973; Shackleton *et al.*, 1990).

C) To discuss the causal mechanisms for the changes observed. These are perhaps the most interesting and yet most intractable questions.

In essence these stages can be thought of as the 'what', 'when' and 'why' questions. The 'what' and 'when' constitute a classical geological approach, this approach can be synthesized under the heading Stratigraphy. The why, represents more of a process-based geomorphological approach that allows a more holistic view of landscape development. Some of the 'Why' questions will be addressed later in the section entitled River Severn Drainage Development: Causal Mechanisms.

Previous Models of Quaternary Stratigraphy

The earliest descriptions of the terraced gravel deposits in the Severn valley are by Murchison (1836, 1839), Symonds (1861) and Maw (1864). These authors believed that these gravels were of marine origin based upon the presence of marine shells (Symonds, 1861). However, later investigations by Lucy (1872), mainly in the tributary Avon valley, ascribed many of these gravel deposits to a terrestrial origin. The acceptance of the 'glacial theory' in the latter half of the nineteenth century led to the recognition that large areas of the Severn basin had been glaciated (*e.g.* Harrison, 1898) introducing a new, complicating factor into the understanding of this sedimentary record. It was soon realized that the marine shells in the terrace gravels were introduced by glacial incursion into the basin and therefore did not reflect a marine depositional environment.

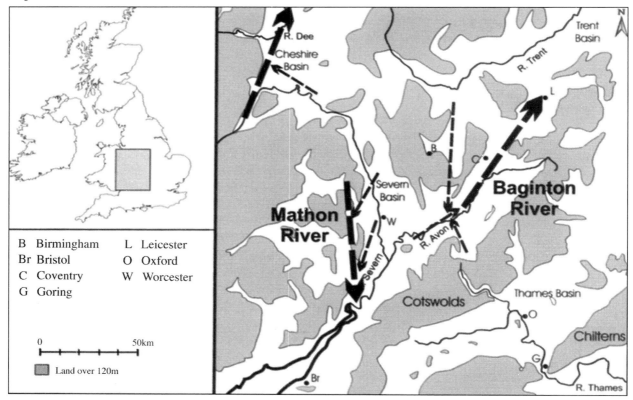

Fig. 6.1 Pre- MIS 12 Drainage in the area of the current lower Severn river basin. Bold arrows indicate the probable main Pre-MIS drainage line. Mathon river (after Barclay et al., *1992), Baginton river (after Maddy, 1999b)*

Later in the nineteenth century, advances in the understanding of river system behaviour demonstrated that large-scale landscape evolution could result solely from fluviatile action. This conclusion was reflected in the early twentieth century work of Gray (1911, 1912, 1914, 1919). He was not convinced by the wholesale application of a glacial origin to the deposits of the Severn basin, and maintained that many of the deposits could be explained by the fluvial redistribution of glacial sediments.

The most significant advances in understanding of this sequence followed the detailed mapping of the region by Wills (1924, 1937, 1938). The work on the Severn was also enhanced by detailed work on the record of its major tributary, the Warwickshire Avon (Tomlinson, 1925). Wills recognised a series of river terraces at progressively higher levels above the present river and was able to demonstrate that the terrace deposits archived a record of complex interactions of fluvial processes and repeated glacial encroachment into the basin. Perhaps most significantly he noticed that the present catchment area is a relatively recent phenomenon, being the product of Late Devensian (MIS 2) glacial diversion. Prior to the Late Devensian the Severn watershed did not extend much upstream of Bridgnorth (Fig. 6.1). Wills (1948) extended his work to include a comprehensive assessment of the development of the Severn valley over the whole Quaternary, including long-term tectonic factors.

Subsequent work on the Severn valley deposits (*e.g.* Hey, 1958; Maddy *et al.*, 1995) has tended to build upon the work of Wills, providing progressively more detail and applying technological advances which in turn are allowing new questions to be asked. Similar recent work has looked also at the Avon valley (Maddy *et al.*, 1991) and Wye valley sequences (Hey, 1991). Taken together this body of work can now be summarized.

Lower Severn Valley Quaternary Stratigraphy: Current Understanding

Little is known of the early Pleistocene drainage in this area as no deposits of this age have yet been observed. The oldest known Quaternary sediments in this region comprise the fluvial deposits of the *Mathon Formation* (Barclay *et al.*, 1992: Fig 6.1: Table 6.1). These sediments are buried beneath younger glacigenic sediments correlated with the *Wolston Formation*. Exposed only in gravel pits, these sediments have no surface morphology.

The Mathon Formation comprises predominantly coarse gravels, high-energy fluvial sediments representing drainage along a north/south drainage line west of the Malvern Hill axis (Barclay *et al.*, 1992: Fig 6.1). Barclay *et al.* (1992) argue that the presence of large quantities of Bunter pebbles from the Kidderminster Conglomerate found in the Birmingham area suggests that much of the northern reaches of the lower Severn basin drained via this route during the time represented by their deposition. As no equivalent age fluvial sediments have been identified in the present lower Severn, this drainage line is

MIS	Formation fluvial	Members	Formation glacial	Formation fluvial	Members	Formation glacial	Formation fluvial	Members	Formation glacial
2-1		6. Power House			?Bretford				
2		5. Worcester							*Herefordshire*
			Stockport		Wasperton				
5e-2		4. Holt Heath			Eckington			Bullingham	
	Severn Valley			Avon Valley			Wye Valley		
6		3. Kidderminster Station	*Ridgacre*		Cropthorne			Hampton	
7					Strensham				
9-8		2. Bushley Green			Pershore			Holme Lacy	
11-10		1. Spring Hill							
12		Woolridge	*Wolston*			*Wolston*			*Risbury*
pre 12	Mathon								

Table 6.1 Lithostratigraphic nomenclature for the lower Severn valley (after Bowen, 1999)

shown as the main trunk river in Fig. 6.1. It remains possible however that any equivalent age Severn sediments have been removed by subsequent valley incision and erosion, thus an early Severn drainage, similar to that of the present day, could have co-existed with the Mathon drainage.

No geochronology is available from the Mathon Formation although biostratigraphical correlation between organic beds within this sequence and organic channel fill sediments at Waverley Wood Farm (Shotton *et al.*, 1993) have been suggested (Allan Brandon pers comm.). These latter sediments have been correlated with MIS 15 on the basis of their aminostratigraphy (Bowen *et al.*, 1989) and to the late Cromerian complex using biostratigraphy (Shotton *et al.*, 1993). If correct, this would suggest these sediments are at least 500,000 years old.

Without doubt the most significant event recorded in the sedimentary record of the lower Severn valley is the extensive Middle Pleistocene glaciation (recognized by Wills) that deposited the widespread Wolston/Nurseries Formations (Table 6.1). This glaciation is believed to be time-equivalent to MIS 12 (Anglian) (Bowen, 1999) and therefore *c.* 450,000 years old. Within the lower Severn valley the Anglian glaciation is represented by the *Woolridge* and *Thurmaston Members* of the *Wolston Formation* (Table 6.1). The Woolridge member forms the Woolridge Terrace of Wills (1938), high-level flat surfaces which can be traced on isolated hills above 80 m OD downstream of Tewkesbury (Hey, 1958). Typically the deposits underlying the terrace are less than 2 m in thickness and consist of sandy clayey coarse gravels. At Woolridge fluvial gravels lie directly upon a thin Triassic till correlated with the *Thrussington Member* of the Wolston Formation (Maddy, 1989) and thus these sediments are interpreted as outwash from the Anglian ice-sheet responsible for its deposition. Thick sequences of equivalent glacigenic sediments can be found upon the Ridgeway (Fig .6.2), the ridge of high ground that separates the Severn from the Warwickshire Avon catchment to the east. A probable ice-margin for the MIS 12 ice-sheet is shown in Fig. 6.2. This glaciation was responsible for major drainage rearrangement, especially in the Lower Severn valley. Prior to MIS 12 drainage to the east of the main Severn axis flowed north-east and deposited the Baginton Formation (Maddy, 1999a, b : Fig. 6.1). Glaciation obliterated this route and reversed the drainage to form the present southwest draining catchment of the Warwickshire Avon (Shotton, 1953: Fig 6.2), thus substantially enhancing the drainage area of the Severn system.

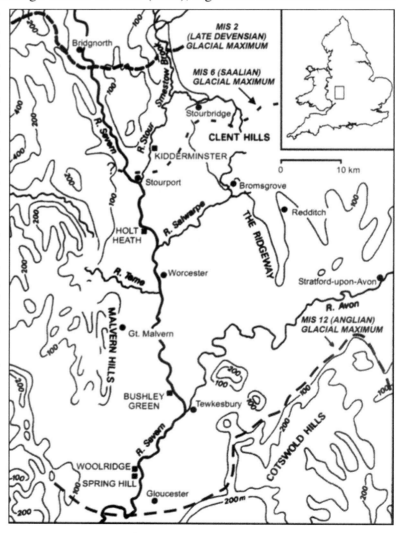

Fig. 6.2 The Lower Severn Valley

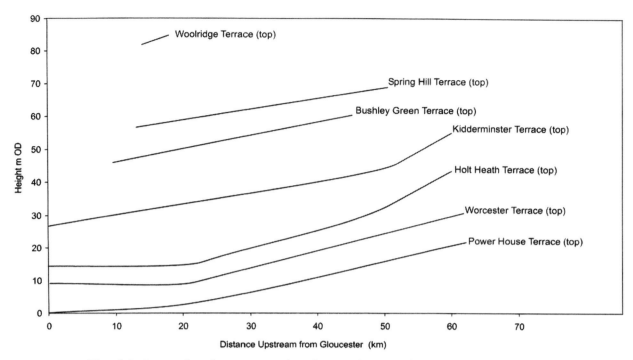

Fig. 6.3 Interpolated terrace surface longitudinal profiles after Maddy (2002)

Following glaciation in MIS 12 the remaining Middle-Late Pleistocene of the Lower Severn valley is recorded in a well-developed staircase of river terraces (Severn Valley Formation *and their equivalent Avon Valley Formation* and *Wye Valley Formation* in the respective tributaries) and extensive spreads of later glacigenic sediment (*Ridgacre Formation* and *Stockport Formation*) (Maddy, 1999a, b; Maddy, 2002).

The post- MIS 12 river terrace sequence of the lower Severn valley is discussed in detail by Maddy *et al.* (1991, 1995). In summary six terraces are recognized in the Severn valley (Fig. 6.3) underlain by discrete sediment bodies which comprise the members of the Severn Valley Formation. Terraces confluent with some of these can be recognized within the main tributary valleys, especially along the Warwickshire Avon and Wye rivers.

1. The highest terrace so far identified in the Lower Severn which appears to represent a modern Severn drainage line is the *Spring Hill Member* which can be mapped from just south of the present Stour / Severn confluence downstream to Gloucester (Fig. 6.3). The sediments beneath the terrace comprise predominantly gravel and range in thickness up to 7.2 m but more typically they are no more than 2–3 m in thickness. No geochronological evidence is available directly from this terrace but Maddy *et al.* (1995) suggest correlation with MIS 10 (*c*. 350 ka) on the basis that it pre-dates the Bushley Green Member (see below) but succeeds the MIS 12 glaciation represented by the Woolridge Member of the Wolston Formation. As yet no equivalent unit has been identified in either the Avon or Wye valleys.

2. The *Bushley Green Member* represents the sediments that comprise the Bushley Green Terrace of Wills (1938). This terrace can be mapped from Lower Broadheath to Apperley with an upper surface *c*. 45 m above the present river (Fig. 6.3). A thickness of 6.9 m has been recorded at the type locality but generally the sediments of this unit are less than 4m in thickness. The basal sediments (Fig. 6.4) include a molluscan fauna indicative of a temperate environment. Amino acid ratios from this lower level suggest

Fig. 6.4 Sediment exposures at Bushley Green 1986

time equivalence to MIS 9 (Bowen *et al.*, 1989). The vast bulk of the sediments at Bushley Green comprise the overlying high-energy sandy braided river sediments. The presence of an intraformational sand wedge at the bounding surface is interpreted as a cryogenic structure, this suggests deposition for the braided river sediments under cold climate conditions. These cold climate sediments are thus attributed to MIS 8 (Bridgland *et al.*, 1986). At the top of the Bushley Green exposures Wills (1938) described a diamict deposit, which he suggested represented a till. This interpretation was rejected by Bridgland *et al.* (1986) as they preferred to interpret this capping unit as a solifluction deposit derived from the higher Spring Hill level deposits nearby.

The Bushley Green terrace is confluent with the *Pershore Member* (formerly fifth terrace of the Warwickshire Avon identified by Tomlinson, 1925) of the Avon Valley Formation (Maddy *et al.*, 1991). A sequence similar to that of Bushley Green *i.e.* basal temperate sediments overlain by cold climate deposits, is recorded from Allesborough Hill Pershore (Whitehead, 1989). The *Holme Lacy Member* (formerly third terrace of Brandon and Hains, 1981) is thought to represent the Wye valley equivalent (Hey, 1991).

3. The *Kidderminster Station Member* represents the sediments that comprise the Kidderminster Terrace of Wills (1938) that can be traced along the whole length of the Lower Severn and up the Stour Valley beyond Stourbridge (Fig. 6.3). This Member is particularly well developed in the Stour Valley (Fig. 6.2) where it has been little affected by subsequent erosion. Sequences of gravels up to 7.9 m in thickness have been recorded; although downstream the deposits are generally much thinner. The sediments contain a suite of lithologies, including local Permian Clent Breccias, not seen in the higher terraces. More exotic clasts include large Welsh volcanic boulders (Wills, 1938), similar to Welsh boulders associated with surface glacigenic sediments east of the Stour valley around Churchill. Rock exposure age estimates using ^{36}Cl methods from several of these boulders suggest time equivalence with MIS 6 (Maddy *et al.*, 1995). It is probable that the new lithologies may be the result of glacial erosion in the Clent Hills area (Fig. 6.2). Although no till sequences have been directly recorded in association with this fluvial unit it is probable that this inferred glaciation is equivalent to the glacial event represented by the upper tills in the Birmingham area *i.e.* the Ridgacre Formation. A possible ice-margin for the MIS 6 glacial event is shown in Fig. 6.2.

The Kidderminster Station Member is equivalent to the *Cropthorne Member* (formerly fourth terrace of Tomlinson, 1925) of the Avon Valley Formation. Organic sediments at the base of the Cropthorne

Member at Ailstone (Maddy *et al.*, 1991) are believed, on the basis of their biostratigraphy and amino acid geochronology, to be time-equivalent to MIS 7, with the bulk of the overlying cold-climate fluvial gravels assigned to MIS 6. No MIS 7 deposits have been identified within the main Severn Valley. In the Wye valley, the *Hampton Member* (formerly Second Terrace of Brandon and Hains, 1981) has been correlated with the Kidderminster Station Member by Hey (1991).

4. The *Holt Heath Member* is the most easily mapped of all the lower Severn terraces and comprises the sediments identified by Wills (1938) as the Main Terrace. This member can be mapped from Bridgnorth where it lies approximately 30 m above the present river to Gloucester where it descends beneath the modern floodplain (Fig. 6.3). Typically the deposits range in thickness up to 10 m. Last interglacial sediments (OIS 5e) may be present at the base of this Member at Stourbridge (Boulton, 1917) in the Stour valley, although this faunal record of *Hippopotamus* is of uncertain value given the likelihood of reworking. Younger sediments within this member include organic channel fill sediments attributable to the Upton Warren Interstadial Complex at Upton Warren (SO 935673) in the tributary Salwarpe valley (Fig. 6.2). Radiocarbon age estimates of 41, 500 ± 1200 and 41,900 ± 800 suggest time equivalence with MIS 3 (Coope *et al.*, 1961). However, amino acid ratios of 0.066 ± 0.007 suggest ages of >50 ka and thus may suggest these dates are underestimates (Bowen *et al.*, 1989).

The majority of the sediments within the main valley were deposited by outwash from the Late Devensian (MIS 2) ice-sheet as they contain characteristic Lake District and Scottish erratics. A probable ice-margin for the MIS 2 event is shown in Fig. 6.2. Advance of the Late Devensian ice-sheet into this area, represented by thick sequences of glacigenic sediments to the north of Bridgnorth (the Stockport Formation), is believed to postdate 28 ka based upon the exposures at Four Ashes (Morgan, 1973). The arrival of Late Devensian ice had a profound effect on the drainage system of the Severn. After deglaciation drainage from Central Wales, which had previously drained northwards into the Irish Sea, was diverted southwards into the Lower Severn Valley (Wills, 1924). Thus the modern drainage line was established.

One confusing observation is the outcrop of a probable till deposit beneath the Holt Heath Member at Stourport (Dawson, 1988; Goodwin, 1999). This deposit was attributed a pre MIS 2 age by Dawson (1989) and later tentatively correlated with the MIS 6 by Maddy *et al.* (1995). Evidence is now available elsewhere in the UK of a possible MIS 4 glacial advance. At present this remains the best candidate for such an advance into the Lower Severn, although an early MIS 2 advance also remains a possibility.

The Holt Heath Member is equivalent to the *Wasperton Member* (formerly second terrace of Tomlinson, 1925) of the Avon Valley Formation. Organic sediments at the base of the Wasperton Member at Fladbury (Coope, 1962) are believed, on the basis of a radiocarbon age estimate of 37, 000 ± 700, to be time-equivalent to MIS 3, with the bulk of the overlying cold-climate fluvial gravels assigned to MIS 2. In the Wye valley, the *Bullingham Member* (formerly First Terrace of Brandon and Hains, 1981) has been correlated with the Holt Heath Member by Hey (1991).

Uniquely the Avon valley also preserves sediments from the last interglacial (MIS 5e) within the sediments of the Eckington Member (formerly third terrace of Tomlinson, 1925). A rich molluscan fauna together with the presence of *Hippopotamus* confirm their biostratigraphical attribution to the last interglacial (Keen and Bridgland, 1986). Amino acid ratios of 0.116 ± 0 (2) also support this age attribution (Bowen *et al.*, 1989).

5. The *Worcester Member* comprises the sediments identified as the Worcester Terrace by Wills (1938). The upper terraced surface lies approximately 8 m below the Holt Heath Member and is traceable from Bewdley to Tewkesbury (Fig. 6.3) where it continues below the level of the current alluvium (Beckinsale

and Richardson, 1964). The deposits consist of sands and gravels that are often coarse. Although no direct geochronology is available, this member is thought to have been deposited after the Late Devensian reached its maximum extent at *c.* 18 ka and given the age of the Power House Member (see below) it must pre-date 13 ka. The sediments are therefore most likely to result from high-energy flows during deglaciation (Dawson, 1989; Dawson and Bryant, 1987).

No direct equivalents to the Worcester terrace have been identified either in the Avon or the Lower Wye valley.

6. The *Power House Member* comprises the sediments identified as the Power House Terrace of Wills (1938), although several discontinuous terraces are most likely present. The sediments are up to 12m in thickness and can be mapped from Bridgnorth to Worcester (Williams, 1968). Below Worcester the deposits are correlated with sediments beneath the modern alluvium (Beckinsale and Richardson, 1964). Organic sediments at Stourport have been dated through radiocarbon age estimation to 12,570 ± 220 BP (Shotton and Coope, 1983). Brown (1982) considered these sediments to be the post-glacial valley fill.

The Power House Member has been correlated with the Bretford Member (formerly first terrace of Tomlinson, 1925) of the Avon Valley Formation (Sumbler, in Maddy, 1999b). However, the Bretford Member is only present in the upper Avon valley and therefore cannot be mapped at the confluence. Such a correlation is therefore speculative. No equivalent unit has yet been identified in the Lower Wye valley.

River Severn Drainage Development: Causal Mechanisms

Valley creation requires a process that both excavates and removes sediment. The only plausible candidates for this process in the Lower Severn valley are fluvial and glacial action. The position of the Lower Severn valley at the margins of the big Pleistocene ice-sheets in the lowland areas promotes glacial deposition rather than erosion. Indeed the thick sequences of glacigenic deposits relating to MIS 2 (Stockport Formation), MIS 6 (Ridgacre Formation) and MIS 12 (Wolston/Nursery Formations) tend to support the predominant constructional topography associated with ice-sheet margins in this area. This does not however, rule out the localized importance of glacial erosion, it does however, suggest that considerable valley excavation by glaciation is unlikely.

A more plausible process for valley excavation is the progressive incision of the River Severn. Progressive incision is manifest by the terrace sequence. Adjustments in river longitudinal profile, leading to floodplain abandonment and terrace formation occur in response to a variety of mechanisms both internal and external to the fluvial system, furthermore these changes occur over widely differing temporal and spatial scales. The terraces described above are the result of basin-wide fluvial incision events which most probably reflect response to changes in external controlling variables such as base-level changes, climate change or crustal instabilities. This picture is further complicated by glacially induced drainage diversions and the superimposition upon these basin-wide events of significant reach-scale incision resulting from localized catchment changes either through river capture or channel avulsion/meander cut-off etc. Unfortunately, during the Quaternary these variables would have driven fluvial activity collectively, making the identification of the relative importance of a single cause difficult. Despite these problems some observations do allow plausible causes of incision to be evaluated.

Base-level change: A model of progressive sea-level lowering during the Quaternary, proposed by Fairbridge (1961), was an often-cited mechanism (*e.g.* Clayton, 1977) for the progressive incision of river valleys *i.e.* where a lowering of base level of a river is not balanced by lengthening of its course, incision will result in changes in profile immediately upstream. Such incision could, given enough time and a readily erodible substrate, be transmitted throughout the whole basin via knickpoint recession. Thus low glacial stage sea levels could have resulted in downstream incision with subsequent transmission of this incision upstream via knickpoint recession.

Little is known about the downstream extension of the Severn through the Bristol Channel and beyond during low sea level stands. However, it is likely that the river simply extended its course offshore, incising only in the very downstream reaches close to the continental shelf-break hundreds of kilometres offshore. Schumm (1993) and Leopold and Bull (1979) have suggested that the effects of base-level change will tend not to be propagated far upstream, particularly in large rivers, which have more scope to accommodate adjustment via other channel variables. Thus sea-level controlled incision is an unlikely cause of the incision in the lower Severn valley.

Climatic change: Climate-driven changes have undoubtedly been important in governing aggradation-incision cycles (Bull, 1991; Bridgland, 1994; Bridgland and Maddy, 1995; Maddy *et al.*, 2000; Maddy *et al.*, 2001). Climatic control exerts influence on the sediment/discharge ratio with incision promoted when low sediment availability is concurrent with high discharge and aggradation being promoted in times of high sediment supply. Leeder and Stewart (1996) have suggested that incision may be initiated upstream, when increasing discharge outpaces increasing sediment availability. This can migrate downstream, a process they referred to as discharge-controlled incision or 'kinetic incision', thus providing a mechanism for the transmission of upstream initiated incision throughout the whole basin.

The conditions necessary for the initiation of incision might, for example, be anticipated at the beginning of warm events when hill-slope stabilization by vegetation reduces sediment supply. The occurrence of interglacial sediments at the base of river terrace sediment packages in the Severn *e.g.* Bushley Green (MIS 9) and Stourbridge (MIS 5e) tends to suggest that incision has indeed occurred immediately prior to the interglacial *i.e.* on the glacial-interglacial transition. This timing of incision also occurs when global sea-levels are rising, providing further evidence for the relative unimportance of sea-level change as a driving mechanism in this part of the valley.

Although climate control is important in governing the likely timing of some incision events it cannot provide a mechanism that leads to long-term progressive valley incision. This problem is discussed in relation to terrace formation in the Thames valley (UK) by Maddy (1997) and Maddy *et al.* (2000, 2001). In these papers it is suggested that climate change governs the timing of aggradation-incision cycles, in synchrony in the Thames, as here in the Severn, with the Milankovitch glacial-interglacial cycles, but progressive valley incision and the consequent development of terrace staircases must require long-term crustal uplift. This is a conclusion supported by recent modelling exercises (*e.g.* Veldkamp and van Dijke, 2000).

It is also of note that not all terrace formation in the Severn can be attributed to a climate-driven mechanism, for example the incision between the deposition of the Holt Heath and Worcester Members. This occurs during deglaciation prior to the onset of any major vegetation changes associated with ameliorating climates. Furthermore, this terracing event does not occur in the downstream tributaries suggesting a localized control to terrace formation. As base level control is also unlikely (as base levels were probably rising at this time) an alternate mechanism must be sought for this event.

Crustal Instability

Three elements of crustal instability can be considered important in the development of the Lower Severn Valley:
 1. Tectonic deformation of the Mesozoic cover and the initiation of Severn drainage
 2. Regional crustal uplift throughout the Quaternary
 3. Glacio-isostatic crustal rebound associated with basin glaciation/deglaciation.

1. Tectonic deformation of the Mesozoic cover and the initiation of Severn drainage
The precise antiquity of the Severn drainage line is unknown. It has been assumed that rivers developed on an uplifting landscape during the Eocene (Wills, 1948). These rivers drained across the

uplifting Cretaceous Chalk. As the Chalk surface was tilted west-east, the result of greater uplift of the Welsh Massif, it is assumed that drainage developed down-dip. Tectonic movements, the result of the Alpine Orogeny (lasting into the Miocene), resulted in the downwarping of critical areas such as the Cheshire Basin, the Needwood and Knowle Basin and the Severn Basin (Fig. 6.5). It has been argued that this warping may have resulted in a disruption of the down-dip drainage pattern and the formation of a Severn drainage guided by the underlying reactivation of the Severn Basin (Wills, 1948; Maddy, 1997), southwards into the Bristol Channel. Similar movements along the Knowle Basin may have controlled the early drainage of an extended Thames river system (Maddy, 1999a).

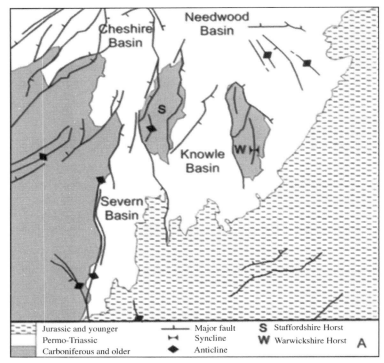

Fig. 6.5 Geological structure of the lower Severn river basin

2. Regional crustal uplift throughout the Quaternary
The concept of uplift control on valley incision has had a long history in the geomophological literature, embodied in the ideas of rejuvenation. Mechanisms for crustal uplift in Southern England are discussed by Maddy *et al.* (2000) who list a number of mechanisms including: direct plate-tectonic stress at, or near, plate margins; intraplate stress (*e.g.* Cloetingh *et al.*, 1990) and large-scale glacioisostatic adjustment (*e.g.* Lambeck, 1993; 1995).

Using the altitude of the base of each terrace Maddy (2002) calculated time-averaged incision rates for the post- MIS12 of *c.* 0.15 m ka^{-1}. This compares with rates of 0.07 m ka^{-1} calculated from the terrace record of the Upper Thames (Maddy, 1997) and rates of *c.* 0.10 m ka^{-1} in the Lower Thames valley (Maddy *et al.*, 2000). This higher incision rate is consistent with the west-east tilting of the Jurassic/ Cretaceous strata to the east of the Severn valley, suggesting this pattern may be the continued response to a long-term crustal adjustment, perhaps even to doming (Cope, 1994). However, Maddy (1997) and Westaway *et al.* (2002) suggest that increased uplift rates begin immediately prior to the onset of lowland glaciation *c.* 2.6 Ma and argue that this uplift is largely therefore a Quaternary phenomenon. This conclusion is based not only on the identification of the onset of valley incision in the UK but also much further afield in the Netherlands (Van den Berg, 1996; Westaway, 2001) and France (Antoine, 1994).

Westaway (2001) and Westaway *et al.* (2002) present a model for uplift driven by flow in the lower continental crust in response to the repeated pressure/thermal gradient changes induced by the loading/unloading cycles of sea-level change on the continental shelf and ice-sheet waxing and waning on the continents. An alternative mechanism involving lithospheric flexure in response to erosion of the Late Triassic and Early Jurassic clays and marls from areas surrounding the Severn valley has also been suggested by Watts *et al.* (2000).

The evidence for long-term uplift is not restricted to the Severn valley. Perhaps the most obvious expression of uplift control in this region are the incised meanders of the river Wye (Miller, 1935). These meanders are incised *c.*185 m below a pronounced planation surface which caps the Forest of Dean at an altitude of approximately 200 m OD. Using the uplift model similar to that of the Lower Severn this would require *c.* 1.4 Ma, assuming once again that the sole control on incision is the uplift. This would place the change from predominantly landscape planation to valley formation in the early Quaternary.

3. Glacio-isostatic crustal rebound associated with basin glaciation/deglaciation

The isostatic depression of crust under ice-sheet loading is a widely recorded phenomenon. On deglaciation the weight of the ice is removed and the crust isostatically readjusts by uplifting (rebounding). During rebound river sediments deposited in previously depressed areas will be uplifted and incised. As these conditions can be very localized the effect would be to produce anomalously high sediments relative to similar age sediments downstream where no isostatic movement has occurred. Thus terrace profiles become distorted and more complex terrace sequences can form during rebound in the affected areas. Examination of the terrace long-profiles (Fig. 6.3) can thus be a powerful tool in determining the possible effects of glacio-isostatic adjustments in river systems.

As already noted the incision from the Holt Heath/Worcester Members is associated with deglaciation at the end of MIS 2 (Devensian) prior to 13 ka. As no major climate change is apparent at this time this terracing event may be associated with glacio-isostatic compensation, with greater amounts of localized incision, up to *c.* 30 m, closer to the ice-front (Fig. 6.3). The pause in incision during Devensian deglaciation, indicated by the deposition of the Worcester Member, may reflect a short-lived re-advance of the Devensian ice-sheet, or more likely a catastrophic phase of ice-sheet down-draw resulting in substantial outwash deposition.

Similarly the MIS 12 Woolridge Member is incised by *c.* 30 m prior to deposition of the MIS 11–10 Spring Hill Member. As this area was completely covered by MIS 12 ice (Fig. 6.2) it is reasonable to anticipate that this region may also have been glacio-isostatically depressed until deglaciation and hence this unusually high amount of incision may be a response to glacio-isostatic rebound during deglaciation.

In both these cases the incision levels associated with deglaciation are of the same order of magnitude to the levels of incision in front of the MIS 12 ice-sheet in the Thames valley where Maddy and Bridgland (2000) have argued for a glacio-isostatic component.

There is also a downstream decrease in terrace gradient of the Holt Heath Member, perhaps reflecting increased warping of the terrace towards the ice-sheet margin, perhaps the result of differential rebound. A similar picture emerges with the Kidderminster Station Member perhaps suggesting similar warping of the terrace profile during late MIS 6, although evidence from this event is much more ambiguous.

Conclusions

The Pleistocene development of the lower Severn valley is typical of other rivers in Southern England in that the long-term incision appears to be driven by crustal uplift. Similarly, the stratigraphy of the terraced sediment bodies appears to support the timing of terrace aggradation and incision being principally driven by sediment/discharge changes consequent upon climate change, a phenomenon already noted from the Thames valley (Bridgland, 1994; Maddy *et al.*, 2001). The terrace staircase therefore appears to directly reflect the Milankovitch-driven 100 ka climate cycles, with incision concentrated during the cold-warm transitions (although not exclusively limited to these periods) and aggradation during warm-cold climate transitions and to a more limited extent within the cold climate episodes themselves. Base level controls on this system are believed to be insignificant.

Although the overall level of valley incision is governed by regional uplift, complex terrace sequences can develop in response to localized changes. In the case of the Severn, a complex Late Devensian deglacial history has resulted in multiple terraces forming during a phase of glacio-isostatic rebound. The greatest incision rates experienced during this deglaciation reflect adjustments in river response to this rebound together with accommodation of the uplift experienced during the cold stage.

7 West Wales

by James L. Etienne, Michael J. Hambrey, Neil F. Glasser *and* Krister N. Jansson

Introduction

West Wales, for the purposes of this volume, is defined as the area from Tal-y-Llyn at the foot of Cadair Idris to Mynydd Preseli to the southwest of Cardigan, and extending eastwards as far as the watershed (Fig. 7.1). Embracing a fine rocky coastline, rolling green hills, verdant river valleys and highland moors, the region has much to offer geologists and geomorphologists. Yet West Wales is not immediately associated with classic glacial phenomena, as are Snowdonia and the Brecon Beacons, and many of its excellent coastal Quaternary sections have been neglected until recently, at least by comparison with those on the Lleyn Peninsula and in South Wales. The region is of key importance in understanding the glacial history of western Britain, and especially the relative roles played by the Irish Sea Glacier as it flowed south from north-east Ireland, south-west Scotland and the Lake District, and local Welsh ice which formed a cap over the Cambrian Mountains. The relative neglect of the area, glaciologically speaking, means that many issues remain unresolved, with abundant opportunities for further research.

Many aspects of the glacial and periglacial history of this region have been described by Watson (1970) and it is not the intention here to simply revise his work. Rather, this chapter focuses on recent research that has led to unravelling the glacial history of the coastal tract, and highlights some of the more interesting phenomena to be found in the inland region, and how they might be interpreted.

The region has been affected by local Welsh ice-masses, from which ice-streams flowed coastwards, and Irish Sea glaciers which flowed southwards. The interaction between these ice bodies remains uncertain and they may not have been coeval. The evidence suggests that west Wales was glaciated during the Late Devensian, although there may have been earlier Quaternary glaciations.

Geographical and Geological Background

West Wales is characterised by hilly country, dissected by broad open fluvial valleys, deeply incised glacial troughs and narrow valleys cut by glacial meltwater (Fig. 7.1). Relatively high ground of peaty or grassy moorland, over 600 m a.s.l (above sea level), forms the northern boundary of the region, with the rolling plateau of the Dyfi (Dovey) hills, dissected by cirque basins and glacially eroded troughs. Prominent summits include Tarrenhendre (633 m) and its un-named neighbour (667 m) to the northwest of Machynlleth, and the Waen Oer group (670 m) enclosed by the A470 and A487 roads, north-north-east of Machynlleth. From Machynlleth southwards, a glacially dissected plateau at around 500 m marks the watershed between west- and east-flowing rivers, and continues to the highest summit in mid-Wales, Pumlumon Fawr (752 m; SN 789869). Progressing southwards, the Cambrian Mountains remain as a dissected rolling plateau, but with few summits exceeding 600 m. The range trends southwest and west, terminating in the Mynydd Preseli (468 m) to the southwest of Cardigan (Aberteifi).

These upland areas enclose a hilly coastal belt, some 25–30 km wide, of agricultural and forestry land up to 300 m elevation. The boundary between the two regions is essentially defined by the course of the Afon Teifi, the region's longest river (Fig. 7.1). The Teifi rises in the Cambrian Mountains at Teifi

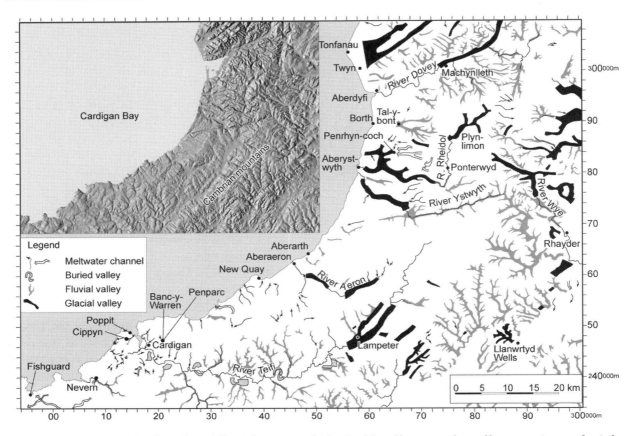

Fig. 7.1 Large scale glacial and fluvial geomorphological landforms and smaller, prominent glacial meltwater channels in west Wales. The map is based on 1: 50, 000 scale digital tiles of UK Ordnance Survey maps sourced from EDINA Digimap. Inset: shaded relief map illustrating incised fluvial and glacially eroded valleys and major bedrock structures trending north-east/south-west, based on the Land-Form PANORAMA™ Digital Terrain Model, sourced from EDINA Digimap

Pools (450 m), falling rapidly to 200 m before continuing gently through a wide valley trending south-west and west towards the coast near Cardigan. Other prominent rivers occupy ice-fashioned valleys trending westwards towards the coast, and include the Dyfi which flows through Machynlleth, the Rheidol (Fig. 7.2a) and Ystwyth which join the coast at Aberystwyth, and the Aeron which meets Cardigan Bay at Aberaeron (Fig. 7.1). In addition to the major valleys, deeply incised glacial meltwater channels cross the area (Fig. 7.1).

The coast, trending south-south-west to west, is mainly of the slope-over-wall type, with irregular, heavily indented cliffs capped by smooth vegetated slopes with angles of about 30–40°. Small coves and extensive sandy beaches (*e.g.* Aberdyfi, Borth, New Quay and Poppit) add to the attraction of the coastline. The southern part is easily accessed on the Pembrokeshire Coast Path, a popular long-distance walk, but much of the rest of the coastline is equally impressive, although less frequently visited.

The landscape is strongly influenced by the bedrock geology of the region. The rocks were deposited in a Lower Palaeozoic marine basin, and represent a classic area of British geology. Broadly-speaking, the rocks have a north-east/south-west trend, reflecting the structure developed during the Caledonian Orogeny, ~400 million years ago. The Afon Teifi follows this trend, while other rivers follow major faults, of which the Bala Fault, which runs through Tal-y-Llyn, is the most impressive

(Fig. 7.2b). Most of the rocks are of Silurian age, but Ordovician rocks occur in the south around Cardigan, and in the area north-west of Machynlleth, with smaller inliers around Pumlumon Fawr. These rocks are typically interbedded mudstones and sandstones, deposited largely as a result of turbidity currents feeding the floor of the marine basin. During the Caledonian orogeny these rocks were extensively mineralised, and the legacy of a now defunct lead, zinc, copper and silver mining industry is seen in the spoil heaps, adits, abandoned railway routes and polluted rivers of the area. The orogeny was also responsible for the development of a strong cleavage in Ordovician mudstones, providing good slates, notably at Corris, near Machynlleth. No longer exploited, the neighbourhood in this area is characterised by deep hillside excavations and waste heaps. In the south of the area, basic igneous rocks of Ordovician age crop out on Mynydd Preseli, including the famous 'bluestone,' used to construct Stonehenge on Salisbury Plain, and responsible for the prominent tors which cap the Preseli hills.

Sitting unconformably upon the Lower Palaeozoic rocks are Quaternary glacial, periglacial and interglacial deposits. Most of the former are envisaged to be products of the Late Devensian glaciation, which peaked about 18,000 years ago. These deposits include diamicton (often interpreted as till), sand and gravel of glaciofluvial origin, and glaciolacustrine muds. Together with periglacial deposits, these drape the hillsides and valley floors. The thickest sequences amount to several tens of metres and are dominated by glacial lake deposits, proven by boreholes in buried valleys in the Cardigan area. Still thicker sequences may be preserved in the Dyfi estuary, but these are yet to be drilled. Inland, Quaternary deposits are typically poorly exposed, but there are some excellent coastal sections. The glacial deposits provide a valuable economic resource, notably sand and gravel currently being extracted in quarries around Cardigan.

There is also evidence for one or more earlier glaciations, recorded in meltwater channel morphology and sedimentology, initially suggested by Hambrey et al., (2001), and further developed by Glasser et al., (2004). More recently, localised glaciers are thought to have developed in cirque basins on Cadair Idris, and in the Brecon Beacons during the Younger Dryas (11–12,000 years ago); however, across much of the West Wales region the evidence available suggests periglacial environments dominated at this time.

Parts of the area have recently been re-surveyed by the British Geological Survey (BGS), including the production of map sheets at 1:50,000 scale, and memoirs or summaries for Cadair Idris (Sheet 149, (BGS, 1995; Pratt et al., 1995)), Aberystwyth (Sheet 163, (BGS, 1989; Cave and Hains, 1986)), Aberaeron (Sheet 177 (BGS, 1994)), Llanilar (Sheet 178, (BGS, 1994; Davies et al., 1997)) and Cardigan (Sheet 193 including part of sheet 210, (BGS, 2003; Davies et al., 2003)), the latter of which is one of the most rigorous Quaternary sheets ever published by the Survey. This chapter aims to provide an overview of some of the principal sedimentological, geomorphological and structural attributes of Quaternary deposits and landforms in the region, and for convenience is divided into three main areas: (i) the Cambrian Mountains, including the Dyfi Hills and Mynydd Preseli, (ii) the lower Afon Teifi basin, and (iii) the Ceredigion coast from Cardigan to Aberystwyth, and through Powys and Gwynedd to Tonfanau near Twyn.

The Cambrian Mountains and Adjacent Uplands
The upland region extends from the Dyfi Hills in the north, through Pumlumon Fawr, and east and south of the Afon Teifi to Mynydd Preseli. This section also embraces the hilly country towards the coast, which is an integral part of the same Welsh ice-cap story. In terms of glacial reconstructions the region is poorly understood and would benefit from detailed investigation. In contrast, considerable work has been undertaken on the periglacial history, a comprehensive review of which is provided by Watson (1970).

Glacial erosional features
The largest scale erosional landforms in West Wales are glacial troughs (Fig. 7.2a, b), of which the largest are the Dyfi and Teifi valleys (Fig. 7.1). The Dyfi rises in a cirque on Aran Fawddwy and flows through a narrow, then open, trough to Machynlleth and finally into an estuary characterised by tidal flats and aeolian sand dunes. Its cross-sectional profile is parabolic, rather than U-shaped, as is true for most glacial valleys. To the south, the Teifi drains the Teifi Pools plateau, occupying a broad open trough in its upper part (*e.g.* around Tregaron Bog), but shows little sign of significant glacial erosion downstream of Lampeter. Many of the details of the preglacial fluvial landscape survive, although extensive glacial deposits suggest a complete ice cover at some stage in the evolution of the river basin.

Numerous glacial troughs meet the coast from the east, similar in form to the classic troughs of Snowdonia, but much smaller in scale. Excellent examples with parabolic cross-sections include those around Aberystwyth, (*e.g.* the valleys above Talybont and Penrhyncoch), and around the Rheidol and Ystwyth valleys. Upland areas also display shallow open troughs, accessible examples of which occur around Nant-y-Moch reservoir (SN 755867) and the area to the east of Cwm Ystwyth (SN 789741). Many low-level troughs terminate up-valley against steep slopes called 'trough-heads,' and include the Afon Leri above Talybont (SN 730880), Craig Maesglase (SH 830140) in the Dyfi Hills, the area around Nantyrarian, on the A44 road east of Aberystwyth (SN 718814), and around Pumlumon Fawr (Fig. 7.2c).

Well-formed cirque basins are few in number, and only one has a tarn, the dammed Llyn Llygad, on the north flank of Pumlumon Fawr (SN 792876). Less convincing cirques occur on other flanks of Pumlumon Fawr, on the Dyfi Hills, and on the Tarrenhendre group.

Glacial meltwater erosion
The most obvious products of glacial meltwater erosion are narrow, incised meltwater channels that occur across much of the region. These glacial meltwater channels fall generally into three groups. The first is a group of small (typically <50 m wide) channels incised into the western flanks of the Cambrian Mountains. These channels trend in a roughly north/south direction. Many, but not all, of these channels cut across cols and other local topographic watersheds. Good examples occur around Tal-y-bont (SN 655893) (Fig. 7.1). The orientation and morphology of these channels suggests that they were formed close to the margins of an Irish Sea glacier abutting the coastal lowlands. Some of these channels may also represent drainage of small ice-dammed lakes in the valleys, impounded where the Irish Sea glacier impeded westward drainage. In some cases, these channels exploit the structural trend within the bedrock geology.

Second is a group of much larger (generally >50 m wide) channels which drain westwards from the Cambrian Mountains (Fig. 7.1). These channels form entire valleys or occur as part of a composite meltwater channel/glacial trough continuum. Many of these channels are partially infilled with glacigenic sediments and Holocene alluvium. The orientation suggests they were formed under a westward-flowing Welsh ice-sheet, but it is not clear if they formed during periods of ice-cap growth, at the ice maximum, or during recession, or whether they represent subglacial or proglacial meltwater erosion. The third group is a series of glacially and meltwater-eroded valleys around Cardigan (Fig. 7.1). These channels represent repeated modification of the drainage system of the lower Afon Teifi by successive glaciations during the Quaternary (Glasser *et al.*, 2004).

A final example of the influence of glacial meltwater erosion is in the upper Afon Rheidol. Here the river is incised into a southward-deepening gorge, which includes the spectacular "incised meanders" between Ponterwyd (SN 749808) and Parson's Bridge (SN 749791) (Challinor, 1933; Jones and Pugh, 1935). The origins of the incised meanders are unclear but possibilities include glacial meltwater erosion, a glacial diversion of the Afon Rheidol, or superimposed drainage from a formerly drift-covered valley floor.

Fig. 7.2 Glacial erosional features in the west Wales region. a) The Vale of Rheidol. b) The glacial trough at Tal-y-Llyn, exploiting the structural trend of the Bala Fault. c) The Cwm Gwerin trough-head. d) Ice-moulded bedrock in the Nant-y-Moch area, again with a strong structural control. Ice flow was from left to right. e) Striated surfaces at the foot of Craig y March, Cwm Gwerin (SN 906883)

Medium-scale features of glacial erosion

Glacial erosional landforms, on a scale of several to tens of metres, such as roches moutonnées, are generally uncommon in the valleys of West Wales, but may be observed on the high ground in the vicinity of Teifi Pools (SN 795685). These are not the classic stoss-and-lee forms, but rather they closely reflect deformation structures such as anticlinal folds in the Silurian bedrock. On the plateau itself, extensive areas of abraded bedrock are indicative of scouring by ice. Similar features occur in the area around Nant-y-Moch (Fig. 7.2d).

Small-scale glacial erosional features

Striated bedrock surfaces are well exposed in a few places, especially where peat and grass has recently peeled away from the bedrock. The best striated surfaces are probably those in Cwm Gwerin on Pumlumon Fawr, especially at the foot of Craig y March (SN 906883; Fig. 7.2e) and a nearby riegel (transverse bedrock ridge or rib) (SN 915888). Chattermarks are also well preserved on these surfaces. In

Textural description	British Geological Survey description	Location	Interpretation
Clast-rich muddy diamiction	Boulder clay	Hillsides and valley floors	Deposition from glacier ice followed by reworking through slope and frost processes
Boulder/cobble gravel with variable matrix	Moraine drift	Valley-fill	Glacial deposits reworked to a greater or lesser extent by meltwater
Angular gravel with mud matrix	Head	Hollows and patches on hillsides	Deep physical weathering of bedrock and slight movement downslope
Angular gravel with no matrix	Scree	Below crags	Frost-shattered debris; some downslope movement; in places grades into 'head'
Sand (well-sorted)	Glacial sand	Rare terraces along valley sides	Glaciofluvial deposition
Gravel (well-sorted)	Glacial gravel	Rare terraces along valley sides	Glaciofluvial deposition
Laminated silt/clay	Glacial lake deposits	Valley floor; low terraces	Seasonal sedimentation in glacial lakes (varves)
Isolated sandstone/mudstone subrounded/subangular striated boulders	—	Intermediate hill tops and valley slopes	Subglacially transported erratics of relatively local origin
Well-sorted pebble/cobble gravel	River terraces, alluvial fans, alluvium	Lower slopes; valley floors	Postglacial stream-transported sediment
Peat	Peat	Blanket bogs on mountain tops and in high valleys; raised bogs adjacent to estuaries (notably Borth Bog)	Postglacial wet/humid climate

Table 7.1 Principal Quaternary surficial deposits in West Wales and their interpretation

the valleys, these surfaces are commonly associated with smooth ice-moulded bedrock forms. Current exposures of striated surfaces occur up to altitudes of 690 m near the summit of Pumlumon Fawr, indicating the presence of thick trunk glaciers in the surrounding valleys.

Glacial depositional features

As noted in the British Geological Survey maps and memoirs referred to above, drift deposits are widespread, especially on the lower slopes and valley floors throughout the region (Table 7.1). Given the extent of glacially influenced sedimentation in the region, one would expect to find a variety of depositional landforms, such as moraines, drumlins and kame terraces. This is true of the upper Teifi Valley between Tregaron and Lampeter, where a series of cross-valley moraines retain infilled proglacial lake basins, and are associated with glaciofluvial plains which have been incised by several metres by the modern river. The moraines are composed of typical ice-contact deposits, and, where exposed in small gravel pits, consist of sand, pebble/cobble gravel, muddy cobble/boulder gravel and diamicton. Large (>100 m diameter) kettle hole depressions occur in some ice-contact deposits (*e.g.* near Tregaron). In addition, a fine set of drumlinoid features (probably rock-cored) occur for several kilometres below Tregaron, where they form elongate mounds on the valley floor and flanks. At higher elevations, moraines have been observed near the lip of cirques. There is a small moraine in a poorly defined cirque on Gribin Fawr in the Dyfi Hills (SH 802158), while the BGS Aberystwyth Memoir (Cave and Hains, 1986, Plate 260) illustrates moraine near Rhiw-gam, south-south-east of Machynlleth (SH 793944). The cirques on the north side of Pumlumon Fawr and surrounding high-level valleys lack well-defined moraines and other depositional landforms, as do the west-facing glacial troughs in the vicinity of Aberystwyth. However, moraines and kettle-hole basins occur in the Dyfi Valley to the north. Thomas *et al.*, (1982) report a number of moraines and associated kettle basins at Minllyn and below Mallwyd in the Dyfi Valley, and moraines around Cwm Cewydd in the Afon Dugoed tributary. The mounds of debris that impound Tal-y-Llyn, once thought of as a terminal moraine, are, in fact, a post-glacial landslide complex which detached from the hills to the southeast (Watson, 1970; Hutchinson and Millar 2001).

Style of glaciation in the mid-Wales uplands

The overall morphology, as depicted by glacial troughs on the uplands and adjacent hills, together with striated surfaces, basally transported erratics and extensive glacial drift, provide clear evidence of wet-based glaciation throughout the region. It has been assumed (*e.g.* by Cave and Hains, 1986) that the deposits are of Late Devensian age, but none has been dated. There is little to suggest that the deposits are coeval with the Late Devensian deposits of the Irish Sea Glacier along the coastal tract (see next section), so the timing of the last Welsh Ice Cap remains uncertain.

The nature of the landform record does not indicate well-defined valley glaciers, but rather near-total ice cover, with fast flowing 'ice-streams' occupying the valleys, separated by near stationary ice on the interfluves (see Jansson & Glasser, 2004). The drumlins in the upper Teifi valley may indicate formation by fast flowing ice. It is conceivable that, although the ice-cap was warm-based throughout at its peak, thinning led to freezing of ice on the interfluves, so overall the ice was polythermal. The absence of glacial depositional landforms in some valleys is puzzling but, if during recession the ice switched from wet-based to cold-based state, most deposition may have ceased. This is the situation with some polythermal glaciers in Svalbard (Glasser and Hambrey, 2001). Although terminal moraines would have been formed during the ice-maximal positions, in the northern areas they would only be preserved seawards of the present coastline.

The Lower Afon Teifi Basin

The lower Afon Teifi region and adjacent areas are crucial for reconstructing past glacial environments in West Wales, not least because it is here that the Late Devensian Irish Sea glacier overrode the Ceredigion and Pembrokeshire coasts and advanced for some distance inland (*e.g.* Charlesworth, 1929; Jones, 1965; Waters *et al.*, 1997; Hambrey *et al.*, 2001). The direction of ice flow is thought to have been approximately from north to south (Bowen and Henry, 1984; Campbell and Bowen, 1989), and this has recently been quantified by measurements of clast fabrics, striation orientations and fold structures in deformed periglacial (head) and glacial sediments (Hambrey *et al.*, 2001). Throughout the region, glacial deposits attributable to the Irish Sea glacier are characterised by a number of features, principally erratic clast content, reflecting source areas in northeast Ireland, southwest Scotland and material reworked from the floor of the Irish Sea and Cardigan Bay. The Irish Sea and Cardigan Bay have also acted as a supply for shells, fragments of which occur widely in Irish Sea glacial deposits.

Glacial Lake Teifi

Charlesworth (1929) hypothesized that during the recession of the Irish Sea glacier from this area, a series of large ice-dammed lakes formed, drowning coastal river valleys (Fig. 7.3a, b). The largest of these, glacial Llyn (Lake) Teifi is thought to have developed in the Teifi valley near Cardigan, fed by meltwater from highland-derived Welsh ice, the Irish Sea glacier, and by overspill from another lake body, Llyn Aeron to the northeast (Charlesworth, 1929). Similar lake systems developed to the west and southwest in the Moylgrove (SN 118446) and Nevern (SN 083400) areas. Charlesworths' ideas have fuelled a debate on depositional environments in this area that has lasted over 70 years.

Jones (1965) suggested that the level of Llyn Teifi was regulated by a series of cols that gave rise to successively lower lake levels during glacier recession. In this model, deep channels were eroded during periods of lake overspill. However, subsequent research in the Fishguard area led Bowen and Gregory (1965) to suggest a subglacial origin for some of the channels. Recent geomorphological and sedimentological investigations of meltwater channels in the area, for example the Cippyn channel (SN 135480), (Fig. 7.3b), demonstrate incision prior to the Late Devensian advance of ice across the region (Hambrey *et al.*, 2001; Glasser *et al.*, 2004). Erratic clast assemblages within pre-Late Devensian sediments indicate at least one earlier stage of Irish Sea glaciation (Glasser *et al.*, 2004). This is consistent with the form of the Teifi valley, which provides evidence for at least two periods of downcutting, infilling and course diversion (Jones, 1965). Part of this evidence lies in the recognition of two buried valley systems, investigated by geophysical means (Allen, 1960; Francis, 1964; Nunn and Boztas, 1977; Carruthers *et al.*, 1997) (Fig. 7.1). Such downcutting is a likely result of base-level grading in response to depressed sea levels during successive periods of Pleistocene glaciation (see Blundell *et al.*, 1969).

Arguments in favour of the Llyn Teifi hypothesis have, until recently, relied on silt, sand and gravel deposits exposed in a series of quarry pits in Penparc, near Banc-y-Warren (SN 204475). Several authors have interpreted these deposits as remnant deltaic sediments that built out into Llyn Teifi (*e.g.* Jones, 1965; Helm and Roberts, 1975; Fletcher and Siddle, 1998). Allen (1982) alternatively interpreted the complex as part of a subaerial outwash plain, presenting evidence of large angular blocks of soft sand and silt within coarse gravel beds, indicating reworking of frozen sediments. Eyles and McCabe (1989) subsequently re-interpreted the complex as part of a glaciomarine delta, supposedly developed when glacio-isostatic depression allowed an extensive marine transgression throughout the area. However, this interpretation is not supported by reconstructions of sea level change since the last glacial maximum (Lambeck, 1993, 1995, 1996; Lambeck and Purcell, 2001). Recent investigations of the Banc-y-Warren deposits support a subaerial provenance as envisaged by Allen (1982), but not on downwasting ice (Hambrey *et al.*, 2001).

An investigation into a landslip situated at Llandudoch (St. Dogmaels) saw the development of renewed research interest in the area, including a mapping project undertaken by the British Geological Survey (Fletcher, 1994; British Geological Survey, 1997, 2003; Waters *et al.*, 1997; Wilby, 1998; see also Fletcher and Siddle, 1998; Hambrey *et al.*, 2001; Walker and McCarroll, 2001, and references therein).

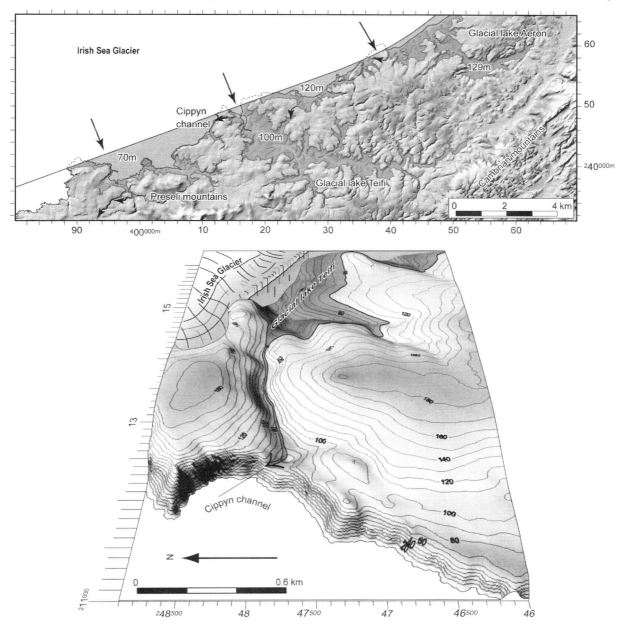

Fig. 7.3 a) Simplified two-dimensional reconstruction to show the location and possible extent of glacial Lake Teifi and peripheral lake bodies dammed in coastal river valleys by the Irish Sea glacier during advance. b) Three-dimensional reconstruction to show the profile of the Cippyn channel, and its potential as a control on the level of glacial Lake Teifi during early stages of advance. Terrain models based on the Land-Form PANORAMA™ Digital Terrain Model (DTM), sourced from EDINA Digimap

Fig. 7.5 a) Recumbent fold structure at Traeth-y-Mwnt (SN 194519). Figures for scale.
b) Glaciotectonically-deformed valley-fill sediments at Gilfach-yr-Halen (SN 435614). Hardhat for scale (circled)

As part of the project, detailed geophysical investigations were undertaken in order to characterise the nature and thickness of drift filling buried valleys in the area as a target for a drilling programme (Carruthers *et al.*, 1997). Rotary-drilled core was recovered from four localities, namely Llwynpiod Farm (SN 17684764), Pen-y-Bryn (SN 17614285), Penparc (SN 20124844) and Llandudoch (SN 15854549).

Sedimentological analysis of core from these sites reveals a complex glacial depositional record, involving two distinct phases of sedimentation. The first equates to the formation of a thick (~75 m) and laterally extensive, fine grained lacustrine sequence. This generally coarsens upwards into, or is abruptly overlain by a second silt, sand and diamicton-dominated succession reflecting ice-marginal and subglacial sedimentation (Fig.7.4 [rear cover]; Hambrey *et al.*, 2001). This stratigraphy, although supporting the formation of an ice-dammed lake, is more consistent with damming of the lower Teifi valley during ice advance, not during recession as originally proposed by Charlesworth (Waters *et al.*, 1997; Fletcher and Siddle, 1998; Hambrey *et al.*, 2001). Lake level reconstructions based on topographic constraints suggest that Llyn Teifi may have extended up to 25 km inland (Waters *et al.*, 1997).

The estuary mouth and adjacent area
Elsewhere in the region, sections at Poppit (SN 144490 to 148489) near the estuary mouth have traditionally been considered to provide evidence for marine erosion and beach sedimentation during periods of elevated sea level during the Ipswichian interglacial (John, 1970; Bowen, 1977; Bowen and Lear, 1982; Lear 1986, 2003a, b). Here the cliffs reveal a flat-lying bedrock platform exposed up to 3 m above the contemporary high water mark, overlain by variably cemented conglomerate, sand and gravel which are further overlain by reworked periglacial deposits and are transitional upwards into shelly Irish Sea diamicton. The interpretation of this section has recently been questioned by Etienne *et al.*, (2001, 2003) and Hambrey *et al.*, (2001) who argued that the nature of the platform, and the character of the overlying sediments may be products of subglacial erosion and glaciofluvial deposition. Unlike the *Patella* beach of George (1932), for example (Bowen, this volume), the sand and gravel in this section do not appear to contain any shell debris. However, Bowen and Lear (1982) reported a clay horizon within the succession, thought to bear *in situ* marine foraminifera supporting an Ipswichian age (~125, 000 yrs BP). In the light of recently published material (Bowen *et al.*, 2002), and the recognition of pre-Late Devensian glacial deposits in the nearby area (Glasser *et al.*, 2004), it is plausible that these fauna were reworked from material offshore during an earlier Devensian expansion of the Irish Sea glacier.

To the east of the estuary, at Traeth-y-Mwnt (SN 194519), a thick sequence of diamicton and muddy gravel sheets show impressive deformation, with a large recumbent (overturned) fold structure comprising part of the cliff section (Fig. 7.5a). The spectacular deformation within these sediments has been interpreted in a number of ways. Davies (1988) suggested that the fold structure formed as a result of downslope failure of gravel beds, sliding along a décollement based on till. Eyles and McCabe (1989) suggested a similar process, but occurring in a subaqueous (glaciomarine) environment. An alternative interpretation is that the fold structure relates to subglacial deformation (Campbell and Bowen, 1989). In a recent detailed study, Rijsdijk (2001) argued that the recumbent fold, and other features in the section including pod-like gravel bodies reflect deformation induced as a result of density contrasts between saturated depositional units during sedimentation, and re-adjustment of the sediment pile.

Further to the north, along the coast, a fold structure deforming sand and gravel deposits filling a buried valley at Gilfach-yr-Halen (SN 435614) is similar in scale to that at Mwnt, but slightly different in morphology. Here, sand and gravel beds thin from the lower limb into the fold hinge, becoming further attenuated and ultimately sheared out on the upper limb (Fig. 7.5b). The orientation of this fold is consistent with glaciotectonic deformation by an ice-mass advancing from the north.

Throughout the lower Afon Teifi region a consistent stratigraphy is recognisable, recently documented by Hambrey *et al.*, (2001). A pre-Late Devensian period of glaciation saw radical landscape modification by meltwater, and the accumulation of glacial sediments. Following a more temperate climatic regime, deep periglacial weathering of the Ordovician bedrock resulted in the local accumulation of periglacial head deposits. Excellent examples of blocky head can be observed in a track-side quarry at Allt-y-Goed (SN 136493) near Poppit. During the Late Devensian, the Irish Sea glacier advanced inland, damming coastal river valleys and leading to the formation of large ice-contact lakes, best observed in cored buried valley sediments. Continued advance of the glacier resulted in displacement of Llyn Teifi *via* pre-existing channels (*e.g.* at Cippyn), localised deformation of lacustrine sediment and deposition of widespread diamicton interpreted as basal till (*e.g.* at Gwbert Caravan Park (SN 162492)). Beyond the

Teifi valley, pre-existing head deposits were deformed by advancing ice (*e.g.* at Lleine (SN 093437)), and locally reworked into heterogeneous ice-marginal and subglacial deposits. During glacier recession, a thick sequence of ice-marginal glaciofluvial sediments accumulated (*e.g.* Banc-y-Warren, Monington (see Owen, 1997), (SN 138437), and Pantgwynmawr (SN 123424)).

The Coast

To the north of the Teifi estuary, the Ceredigion, Powys and Gwynydd coasts form a broad sweeping arc trending northeast to southwest, enclosing Cardigan Bay (Fig. 7.1). The coast is dominated by high sea cliffs (locally over 100 m) cut in rocks of lower Palaeozoic age (primarily Silurian and Ordovician marine mudstones and sandstones), and locally by cliffs composed of variably consolidated Quaternary deposits. Low-lying fan-shaped bodies form gently sloping platforms in the Llanrhystud, Llan-non and Aberaeron areas.

Along this stretch of coast, and further to the north, are three linear gravel ridges, Sarn Gynfelyn, Sarn y Bwlch and Sarn Badrig, each of which extends for several kilometres offshore. The ridges are composed primarily of boulder gravel, and trend northeast to southwest (Foster, 1970), sub-parallel to the orientation of the Lleyn Peninsula. These sarns have been interpreted as medial moraines, thought to delimit the line of confluence between Irish Sea and Welsh glacial masses (*e.g.* Foster, 1970). Offshore sampling programmes between the sarns have recovered glacial sediments indicative of Welsh provenance; beyond their westerly extent only drift of Irish Sea glacial provenance has been reported (Garrard and Dobson, 1974). The sarns are visible at low tide, and comprise mixed clast lithologies, including some material sourced from Anglesey. It is curious that none of the sarns appear to have any landward continuation.

Glacigenic sediments

Numerous exposures of glacigenic sediment occur along the Ceredigion coast. Watson (1970) noted the occurrence of Irish Sea glacial deposits as far north as Llanrhystud (SN 532701). However, sections at Tonfanau, near Towyn (Gwynydd) (SH 560040), comprise material that can only originate from offshore. Here, a range of glacial sediments occur, including diamicton and a thick sand, gravel and muddy laminite

Fig. 7.6a) Ice-marginal sand and gravel deposits at Tonfanau, Gwynydd (SH 560040), b) Corresponding stratigraphic log shows sediment characteristics

succession, interpreted as the products of subglacial and ice-marginal sedimentation respectively (Fig. 7.6a, b). Abundant fossiliferous Jurassic, and Permo-Triassic erratics within this succession indicate reworking of Mesozoic strata from the floor of Cardigan Bay to the west of the Tonfanau fault system, and herald an Irish Sea glacial provenance. This confirms the observations of Cave and Pratt (1995), and is contrary to the view that glacial deposits exposed along this stretch of coastline were deposited by highland-derived Welsh ice-masses (D. Morris, *unpublished*; reported in Bowen, 1977). Given the local proximity of Sarn y Bwlch and the stratigraphic evidence offshore, this site requires either an earlier advance of an Irish Sea ice-mass (*cf* Bowen *et al.*, 2002), or a radical re-appraisal of the accepted configurations for Late Devensian Welsh and Irish Sea ice-masses in this part of Cardigan Bay. Further data are required to resolve this issue.

To the south of the Dyfi estuary, a recurring stratigraphic sequence is evident and is comparable to that recorded in the Cardigan area. Bedrock is frequently transitionally overlain by blocky and/or crudely stratified periglacial deposits (*e.g.* Morfa Mawr (SN 496654), Llanrhystud (SN 532701), and New Quay (SN 392595); Fig.7.7a). These head deposits are locally reworked, and are often laterally or vertically transitional with a poorly sorted 'rubble drift.' Rubble drift deposits are polygenetic in character, typically comprising a mixture of clast lithologies and variable clast roundness. These deposits occasionally pass upwards into subglacial tills of Irish Sea glacial provenance. Examples of this relationship are exhibited in the sections to the east of the lifeboat station in New Quay. Here, Irish Sea tills contain a wide variety of rock types, including locally derived mudstone, sandstone and coarse grit, cone-in-cone nodules from the Silurian Aberystwyth Grit Formation, mineralized brecciated sandstone and vein quartz of probable local derivation. Erratic clasts include striated New Red Sandstone, fossiliferous Carboniferous limestone, yellow-weathering siltstone, chert and a variety of igneous and metamorphic rocks including olivine basalt, acid tuff and quartzose gneiss. Isolated examples of fragmented shells occur within this succession. The tills at New Quay are overlain by a series of sands and gravels.

Other examples of shelly deposits occur at Drefach, near Aberaeron (SN 468636) and at Llan-non (SN 507669). At Llan-non these deposits overlie subglacial tills of Welsh provenance, and are incised by later-

Fig.7.7 Some periglacial phenomena in west Wales. a) Stratified head deposits underlying glaciofluvial sediments at Morfa Mawr, near Llan-non (SN 496654). b) Large-scale involution structures deforming sediments in the Llan-non platform (SN 507669). c) Sorted periglacial stripes on Pumlumon Fawr. d) Cwm Du in the Ystwyth Valley, an example of a nivation cirque (according to Watson, 1970). e) 'Pingo U' in the Cledlyn Valley near Cwrt Newydd—evidence for the former presence of discontinuous permafrost

ally extensive sand and gravel sheets (Watson, 1965, 1976). Mitchell (1960, 1972) noted a zone of weathering at the transition between the tills and the overlying gravel deposits, which he interpreted as a product of Ipswichian interglacial weathering. Mitchell termed this deposit the Llansantffraid Interglacial soil. The sediments at Llan-non are locally deformed by excellent examples of involutions, interpreted as cryoturbation structures (Fig.7.7b; see below). Watson believed these structures to represent at least two different stages of periglacial weathering; however, more recent research suggests that the entire sequence can be explained in terms of a single transition from full glacial to periglacial conditions during the Devensian (see Harris, 2001a). A more comprehensive review of this site is presented in Campbell and Bowen (1989).

At Llanrhystud (SN 532701), as at Llan-non and Aberaeron, sand and gravel dominated fan-shaped bodies occur. In all cases, thick (>2 m) sheet-like gravel bodies within these fans are laterally extensive, commonly over hundreds of metres, and may represent terrigeneous fan build out during a number of high discharge flood events, possibly as a result of drainage of unstable high-level glacial lakes dammed in the valley interior of the hilly coastal belt. This mechanism could explain the apparent lack of preservation of moraines in the lower reaches of many of the coastal valley systems (see above).

Periglacial phenomena
Throughout West Wales, the effects of periglacial activity are widespread. Many hillsides, at all elevations, have a few metres thickness of broken angular rock fragments, described as 'head', resting transitionally on bedrock reflecting *in situ* weathering as a result of freeze-thaw processes. The thinly bedded and cleaved nature of the Upper Ordovician and Silurian rocks of the region is likely to have been an important factor in the physical weathering processes, leading to rock fracturing along both bedding and cleavage planes. Throughout the area, numerous road cuttings and minor farm quarries expose such sediments. Some of the largest excavations occur at the head of the Tal-y-Llyn valley, where 18 m of head and scree have been recorded (Watson, 1970). The head is clearly *in situ* in places, but elsewhere shows signs of downslope gravity sliding, slope wash and solifluction. These deposits indicate a cold periglacial climate and have been widely recognised throughout the British Isles (Ballantyne and Harris, 1994). Prominent solifluction terraces have been documented by Watson (1970) in the Nant-y-Moch valley, east of Aberystwyth. Sorted periglacial features including stripes occur on Pumlumon Fawr (Fig. 7.7c).

On ground approaching 500 m in elevation, nivation cirques may be observed (Watson, 1970). These are incipient cirques, and are shallow, outward-sloping hollows that collected drifting snow, especially when blowing on the extensive plateau areas made snow available to feed permanent firn patches. Watson (1970) suggested that Cwm Du, in the upper Ystwyth valley is a nivation cirque; however, the large volume of sediment suggests that other processes may have been responsible for its formation (Fig. 7.7d).

At lower elevations, in the hilly coastal belt, smaller-scale features such as ramparted ground-ice depressions occur. These depressions are up to tens of metres in diameter, and have ramparts two or three metres in height, typically enclosing bog sediments which preserve good pollen records. Perhaps the most famous examples are those in the Cledlyn valley near Cwrt Newydd (SN 491480; Fig. 7.7e), highlighted by Watson (1971) and Watson and Watson (1972). These authors suggested that the mounds represent open-system pingos; however, Gurney (1995) alternatively proposed that the features are lithalsa (resulting from segregated ground-ice). Both interpretations are thought to indicate the prior existence of discontinuous permafrost. Watson and Watson (1972) attributed the formation of these features to the Late Devensian deglaciation (Dimlington Stadial); however, Harris (2001b) and Walker and James (2001) have suggested that a Younger Dryas age is more likely. Other relatively poorly preserved ground-ice depressions occur in the Teifi valley, below Tregaron, which may also represent the remnants of pingos. Open system pingos, such as those interpreted from the Cledlyn Valley, typically occur in contemporary environments with mean annual air temperatures ranging from -1° to -5.6° C (Holmes *et al.*, 1968; Péwé, 1969). Temperature reconstructions based on coleopteran assemblages from

Llanilid in south Wales suggest that during the Younger Dryas, summer month temperatures were in the order of 10–12°, with winter month temperatures of -3° to -10° C (Walker *et al.*, 2003). These mean annual air temperatures are therefore generally consistent with those required for open system pingo development.

Along the coast, abundant evidence exists to support periglacial conditions. Some of the best evidence for periglacial processes of sediment accumulation occur in the sections at Morfa Bychan (SN 561763), to the south of Aberystwyth. Traditionally, muddy gravels within this succession have been interpreted as glacial in origin (Keeping, 1882; Reade, 1896; Williams, 1927). During the late 1950s, Wood (1959) alternatively proposed that the section comprises in part tills, deposited during an earlier (pre-Late Devensian) glacial event, and subsequently reworked by periglacial processes towards the end of the Devensian. Watson and Watson (1967) later re-interpreted the entire section as periglacial slope and aeolian deposits, demonstrating gravel clast orientations consistent with downslope deposition.

Bowen (1973a,b; 1977) argued that the occurrence of striated clasts within parts of these sections supports deposition from a Devensian ice-mass, with penecontemporaneous and de-glacial redistribution by solifluction processes ('paraglacial' sedimentation, Ballantyne, 2002). This interpretation was later supported by the work of Vincent (1976) and more recently by Ballantyne and Harris (1994). Micromorphological analyses of the deposits at Morfa Bychan have since been presented by Harris (1996, 1998), who concluded that the till-like sediments probably accumulated fairly rapidly as a mudflow-type deposit. Some consensus therefore exists that the sections here comprise both primary periglacial deposits and paraglacial sediments. Similar sediments dominate the cliffs at Aberarth between Dolauarth (SN 479641) and Graig Ddu (SN 491649), and may have accumulated under similar conditions.

In addition to the large-scale deformation structures described from sites such as Mwnt and Gilfach-yr-Halen (see above), many coastal sections exhibit deformed glacigenic sediments with a variety of complex involution structures. Regularly-spaced, and relatively small-scale involutions occur in head deposits between Clarach (SN 585842) and Wallog (SN 589856), in scarp-face exposures of rubble-drift deposits near Careg Yspar (SN 101450), at Drefach (SN 468636), near Aberaeron, and at Llan-non (SN 507669), and provide good evidence to support periglacial conditions following the Late Devensian deglaciation. Such cryogenic phenomena have long been recognised within Late Devensian maximum ice limits (*e.g.* Watson, 1970; John, 1973; Ballantyne and Harris, 1994; Worsley, 1987).

At Llan-non, the occurrence of vertically-oriented gravel clasts within flame-like vertical intrusions of sandy gravel may be indicative of frost heave, while larger-scale structures are more readily explained by loading. Watson (1965) similarly interpreted some of these features as resulting from cryogenically-induced density contrasts. More irregular structures may be related to a combination of both loading and sediment liquefaction resulting from cryohydrostatic pressures (Fig. 7.7b). At Aberaeron, Watson (1965) described flame-like intrusions of gravel separated by fine-grained bulbous structures. Such features can still be observed in plan view on the foreshore at low tide, where they form small patches of circular (~50 cm diameter) patterned ground. Vertically-oriented gravel clasts and cryoturbation involutions also occur widely throughout the Drefach sections near Aberaeron. Fine-grained angular muddy and sandy gravel deposits occurring stratigraphically above Late Devensian glacigenic sediments are also thought to indicate periglacial environmental conditions.

Large-scale soft-sediment deformation structures also occur at Tonfanau to the north; however, the large vertical scale of these features may be more consistent with a provenance from thermokarst (resulting from permafrost degradation and subsidence), rather than seasonal active layer processes. In some instances, deformed gravel beds contain a fill of horizontal planar laminated silt and sand, which may reflect thaw-lake development.

The wide range of soft sediment deformation structures observed supports a range of different cryogenic processes, with evidence for cryoturbation resulting from intense frost action, and possible thermokarst development resulting from permafrost degradation. The preservation of ice wedge casts in the

Aberystwyth area is well-documented (Watson, 1981), which, coupled with the preservation of ground-ice depressions in the local area, and the widespread geographical distribution of involution structures, supports periglacial conditions characterised by at least local permafrost development. Providing some time-constraints on the generation of these features is problematic. It has been demonstrated that ground-ice depressions in the Cledlyn Valley probably relate to cold climatic conditions during the Younger Dryas; however, Watson (1965) attributed both these and cryoturbation structures to the Late Devensian deglaciation. In Pembrokeshire, John (1973) similarly interpreted thermal contraction and cryoturbation structures as Late Devensian in age. Perhaps the most convincing evidence to support his interpretations is ice wedge casts described from Mathry Road (John, 1973). Since these casts occur within aggradational terrace deposits, permafrost conditions must have prevailed during the waning phases of deposition. John (1973) also mentioned the occurrence of ice wedge casts in the Banc-y-Warren deposits at Penparc.

Sedimentological evidence to support frozen ground conditions during the Late Devensian deglaciation of this area is present in the form of soft-sediment clasts of sand and silt which occur within ice-marginal successions at Tonfanau, Clyn-yr-Ynys (SN 168504) and Penparc. Arguably, the best interpretation for these is for erosion of frozen channel banks during the onset of the melt season, as originally proposed by Allen (1982) for the deposits at Penparc (Banc-y-Warren). This sedimentological evidence supports at the very least seasonally frozen ground conditions, and it is plausible that many of the cryoturbation structures discussed above are attributable to periglacial conditions following deglaciation in the latter part of the Dimlington Stadial. Further work is required to elucidate the exact timing of the formation of these structures.

Interpretation of Data on Irish Sea Glaciation
A combined analysis of the geomorphological and sedimentological features of Quaternary sites in the area of the lower Afon Teifi, and the Ceredigion coastal tract, reveals a consistent, although locally complex stratigraphy. Correlating specific deposits is inhibited by the lack of chronological data, and the likely diachronous nature of the depositional systems in which they accumulated. Localised reworking within some of the successions further complicates this. High-resolution studies applying dating techniques such as optically-stimulated luminescence of sand grains within outwash sediments, and surface exposure dating by cosmogenic nuclides may, in time, help to resolve some of these issues.

The sedimentological and stratigraphic evidence throughout this part of the Irish Sea basin suggests terrestrial conditions of sedimentation, with little evidence to support the glaciomarine depositional environments envisaged by Eyles and McCabe (1989). Evidence for at least two periods of glaciation is present, the first of which resulted in radical modification of the landscape. The large volumes of meltwater associated with this system indicate either a temperate or polythermal ice-mass. Finally, the altitudinal distribution of glacigenic deposits in the lower Afon Teifi region requires a Late Devensian Irish Sea glacier at least a few hundreds of metres in thickness. The large volume of glacigenic sediment attributable to this glaciation, particularly basal tills, indicate a temperate or polythermal character, but given the evidence for widespread periglacial conditions following deglaciation, the latter may be more appropriate.

Conclusions
West Wales has been affected by local (Welsh) and by invasive (Irish Sea) ice. The dates at which glaciation took place are uncertain, but the most obvious glaciation is generally assumed to have been of Late Devensian age. There is limited evidence of earlier glaciations. The interactions between Welsh and Ice Sea ice-masses have yet to be determined in detail. West Wales therefore continues to offer opportunities for further research, particularly in the upland regions of the Cambrian Mountains, and in determining the interaction between the Irish Sea glacier and outlet glaciers which drained the last ice-cap over mid-west Wales.

8 The upper Wye and Usk regions

by Colin A. Lewis *and* G. S. P. Thomas

Introduction

Although it is impossible to prove the precise morphology of the pre-glacial landscape (Brown, 1960; Clarke, 1936–7; Thomas, 1959) it is apparent that the basic outline of the Usk and Wye valleys, and of their tributaries, as well as of the uplands of the region, originated prior to the onset of Quaternary glaciation. The ice-sheets that developed over the region, the valley glaciers, the cirque glaciers and the periglacial features (such as protalus ramparts) thus formed on a landscape whose outlines already existed and which influenced glacial developments.

The mid-Wales ice-sheet

The maximum known extent of Quaternary ice in the Wye-Usk region is evidenced by the altitude to which erratics from areas further north have been deposited on the Black Mountains and the lack of comparable erratics on the Brecon Beacons. Presumably such erratics do not occur on the Brecon Beacons because those uplands nurtured their own ice cover, which prevented their invasion by mid-Wales ice.

Howard (1903–4) recorded Silurian slates, sandstones, quartz and dark basic igneous clasts at altitudes up to 400 m beneath the Black Mountains escarpment. Such erratics may be seen south of Hay-on-Wye on the interfluve between the Dulas and Cilonw Brooks in the vicinity of Twyn y Beddau (SO 245386), at altitudes around 380 m (Fig. 8.1).

The erratics discovered by Howard were derived from the Builth Wells-Aberedw area of the Wye valley and indicate that ice flowed generally southwards down the Wye valley to impinge upon the northern escarpment of the Black Mountains. The gathering grounds for this ice were presumably the uplands of Pumlumon and their southerly extensions.

Howard noted that Silurian erratics occur in the Rhiangoll valley, indicating that mid-Wales ice passed through the Pen-y-genffordd col at the head of that valley. The floor of the col is at little more than 320 m, some 240 m above the floor of the nearby Wye valley. Howard also considered that ice from the Wye valley entered the Usk valley by flowing up the Llynfi valley and passing over the low ground in the vicinity of Bwlch and near Tal-y-llyn station. On the basis of the evidence available to him Howard concluded that 'none of the Wye drift passed over the...escarpment between the Golden Valley and the Rhiangoll.' Dwerryhouse and Miller (1930) suggested that Wye ice entered the head of the Golden Valley.

North of the Wye valley, near Llanbella in the vicinity of Huntington (SO 240530), on the interfluve between the Arrow and Gladestry Brook valleys, there are hummocky glacial deposits. These contain Silurian clasts derived from the region to the north and west of Llanbella. The deposits exist at altitudes up to about 340 m and, like the erratics recorded by Howard beneath the escarpment of the Black Mountains, evidence the former existence of an ice-sheet over the region.

Erratics of Builth Wells olivine dolerite have been recorded in the Tywi valley (Bowen, 1970, this volume). Although it is not known when these erratics were deposited, they are believed to be of glacial origin (Griffiths, 1940), indicating that ice from the Builth region spilled south-westwards across the

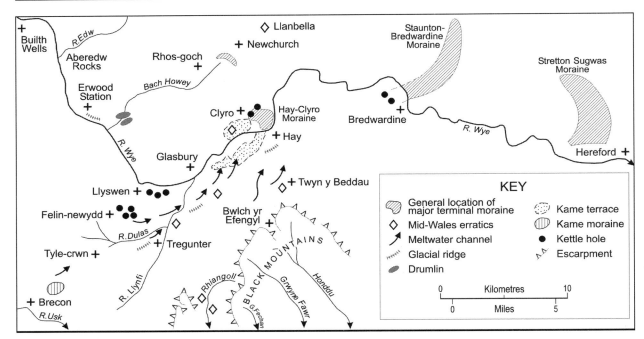

Fig. 8.1 The location of major terminal moraines in the Wye valley downstream of Glasbury and of erratics, meltwater channels and other indicators of glaciation north of the Black Mountains and south of Builth Wells

watershed between the Tywi and rivers that flow towards the Wye. The watershed is at an altitude of 290 m.

Mid-Wales, north of the Black Mountains-Brecon Beacons uplands, was therefore inundated by an ice-sheet that was at least 240 m thick in the Glasbury-Talgarth area when ice passed over the col at the head of the Rhiangoll valley, between the Wye and Usk drainage basins. This ice-sheet must have been continuous with ice over the Brecon Beacons-Fforest Fawr uplands, otherwise it would have impinged upon those uplands and deposited indicator erratics. The main escarpment of the Black Mountains apparently projected above the ice-sheet. Glaciers existed in the dip slope valleys of the Black Mountains, as is evidenced by glacial deposits and landforms in the Honddu, Grwyne Fechan and other valleys, although the correlation between them and the mid-Wales ice sheet has not been established.

M'Caw (1936) suggested that mid-Wales ice breached the Black Mountains escarpment at Bwlch yr Efengyl (the Gospel Pass; SO 235351; Fig. 8.1), at the head of the Honddu valley. This valley displays features of glaciation, being trough-shaped and having a remarkably well-developed terminal moraine at its southern end, at Llanfihangel Crucorney (SO 325208; Fig. 8.4). The floor of Bwlch yr Efengyl is at an altitude of 542 m so that, if the interpretation of M'Caw (1936) is correct, the mid-Wales ice-sheet on the northern side of the Black Mountains must have been at least 220 m thicker in the Glasbury-Talgarth area than the evidence presented by Howard (1903–4) suggests, attaining a depth of at least 460 m. Until mid-Wales erratics have been found in the upper Hondddu valley, however, it cannot be proved that mid-Wales ice flowed through the Gospel Pass.

The age and eastern limits of the mid-Wales ice-sheet as evidenced by erratics deposited beneath the Black Mountains escarpment and in the Rhiangoll valley, remains uncertain. In the Usk valley, downstream of the general vicinity of Crickhowell, ice of assumed Late Devensian age extended as a valley glacier almost to Usk. This may have been an outlet glacier of the mid-Wales ice-sheet that Howard (1903–4) had discovered.

In the Hereford basin mid-Wales ice of presumably Late Devensian age (Richards, this volume) spread out as a piedmont lobe, leaving well developed features at Stretton Sugwas, just west of Hereford, and in the Kington-Orleton region (Luckman, 1970). Older, probably Anglian Cold Stage (Middle Pleistocene, Oxygen Isotope Stage 12) glacial deposits exist north-east of Hereford (Richards, this volume) and contain Welsh erratics. Whether the mid-Wales ice-sheet, identified by the erratics recorded by Howard (1903–4) in the Black Mountains region, correlates with the older or younger glacial deposits in the Hereford basin is unknown. If, however, the hummocky glacial deposits near Llanbella (SO 240530) are of the same age as the deposition of mid-Wales erratics on the flanks of the Black Mountains, it is likely that the mid-Wales ice-sheet that they indicate was younger than the Anglian Stage, since glacial deposits of that antiquity are unlikely to have retained their morphology.

The maximum ice extent, as recorded by Howard (1903–4), was therefore probably not that responsible for the extension of glacier ice north-eastwards of Hereford. Consequently it is likely that there was an earlier and larger ice-sheet in mid-Wales, the evidence for which, within mid-Wales, has yet to be identified.

High-level meltwater channels
A series of now-dry channels occur at altitudes around 380 m on the commons near Twyn y Beddau (SO 240387) and cut across the summit of the interfluve between the Digedi and Cilonw valleys to the south of Hay-on-Wye, beneath the escarpment of the Black Mountains (Fig. 8.1). These channels lead north-eastward and probably originated as glacial meltwater channels utilised by meltwater flowing essentially parallel to the escarpment of the Black Mountains as it drained towards the lower ground of the Hereford basin.

The Wye glacier
Major terminal moraines
At least three major terminal moraines exist in the Wye valley downstream of Glasbury: at Stretton Sugwas (SO 465426) near Hereford (Luckman, 1970; Richards, this volume); Staunton-on-Wye/Bredwardine (SO 330445; Luckman, 1970; Richards, this volume); and between Hay-on-Wye and Clyro (SO 220435; Fig. 8.1). All three features contain mid-Wales erratics and were formed by ice from that area.

The Stretton Sugwas moraine
This moraine formed at the edge of a piedmont lobe as the mid-Wales ice-sheet extended onto the lower ground of the Hereford basin. Both Luckman (1970) and Richards (this volume) have described the moraine in detail.

The Staunton-on-Wye/Bredwardine moraine
The Staunton-on-Wye/Bredwardine moraine formed at the end of a valley glacier that extended eastwards down the Wye valley from accumulation areas in Powys. Richards (this volume) has described how the moraine extends from Brobury Scar (SO 353444), where a section is exposed, through Staunton and Norton Canon to Hyatt Sarnesfield (SO 380500). Up-valley of Bredwardine, especially in the vicinity of Weston Farms (SO 455320), well developed kame and kettle holes exist in the glacial deposits and some of the kettle holes contain minor lakes. There was thus a dead ice area upvalley of the terminal moraine, where the glacier stagnated and melted in situ.

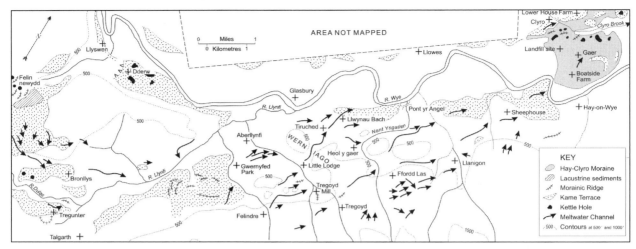

Fig. 8.2 Meltwater channels and other glacial features in the Talgarth- Hay region. (Redrawn, with minor additions and alterations, after Williams, 1968a)

The Hay-Clyro moraine
The Hay-Clyro moraine (Williams, 1968a; Fig. 8.2) forms a ridge, partly plastered on bedrock, that is followed by the B4351 road. In 1990 a landfill site adjacent to this road (SO 219434) exposed 4 m of diamicton containing lenses of sand and gravel. Some of the clasts within the deposit were of fossiliferous Silurian rock, presumably derived from outcrops in the Aberedw area, further up the Wye valley, near Builth Wells. Similar clasts, some of which bear shallow striations, exist in the lateral moraine/kame terrace between Llowes and Clyro that leads into the terminal feature. They have been exposed by road works beside the A438 opposite the point where a meander of the River Wye lies immediately below the road. The striae on the clasts indicate that the clasts were transported within the glacier, where clasts rubbed against and scratched each other.

The minor road that leads from the B4351 past Boatside Farm (SO 432225) and a house named "Gaer" to the A438 at Lower House Farm (SO 222443) traverses the moraine. Kame and kettle holes exist between Gaer and Lower House Farm as well as what appear to be isolated perforation kames. Morainic ridges, aligned approximately parallel to the axis of the Wye valley, occur between Clyro village and the minor road upstream of the point where that road crosses the Clyro Brook that flows past Lower House Farm. Kame and kettle topography also exists adjacent to the right bank of the Clyro Brook below the point where it is crossed by the minor road. Occasional meltwater channels run across the moraine, being most prominent in the vicinity of Clyro. An outwash terrace extends downstream of the moraine, towards Clifford. Symonds (1872) reported that the remains of boar, deer and ox had been discovered in this terrace, although these may have been part of the post-glacial, possibly Holocene, fauna of the valley.

Kame terraces and meltwater channels associated with the Hay-Clyro moraine
Lateral deposits, mapped by Williams (1968a) as kame terraces, exist on both sides of the Wye valley downstream of Glasbury and terminate at the Hay-Clyro moraine (Fig. 8.2). They are best exhibited on the right bank of the valley, erosion having obliterated much of the kame terrace/lateral moraine material on the Radnor side of the valley. Cuttings in these features, as already noted for the lateral moraine/kame terrace immediately downstream of Llowes, expose occasional Silurian clasts as well as the remains of the local country rock, indicating that the glacier responsible for these features flowed down the Wye

valley from the vicinity of Builth Wells. Shallow meltwater channels are cut into the terraces, especially in the vicinity of Sheephouse (SO 210414), about 1.4 km upstream of Hay-on-Wye. These channels are aligned essentially parallel to the axis of the valley.

A complicated series of meltwater channels exists higher up the right hand side of the valley, especially between Felindre and the vicinity of Llanigon (Fig. 8.2; Williams, 1968a; Lewis, 1970a). The Wye glacier apparently spread out as a piedmont lobe downstream of the funnel-like valley above Llyswen to form a mer-de-glace between Hay-on-Wye and an unknown area in the Llynfi valley upstream of Talgarth, where the Wye ice might have been in contact with ice from the Brecon Beacons and adjacent uplands. These channels allowed meltwater from the decaying ice to drain north-eastward into the Wye valley beyond the limits of the Hay-Clyro moraine.

Some of the meltwater channels, such as that which cuts across the Wern Iago (Tregoyd Wood) ridge between Little Lodge Farm (SO 187382) and Tiruched (SO 185389), appear to have been superimposed upon and thus cut through ridges. The Little Lodge-Tiruched channel starts beside Little Lodge Farm, cuts through the Wern Iago ridge, crosses the Tiruched valley that opens into the Wye at Glasbury, and proceeds across the ridge between the Tiruched valley and Llwynau Bach (SO 395187) to emerge on the kame terrace on the right bank of the Wye valley. The intake of the channel at Little Lodge lies down-valley of a morainic ridge, containing clasts from the Builth/Aberedw area, on the southern side of the minor valley occupied by the Tregoyd Mill Brook some 200 m down-valley of the hamlet of Tregoyd Mill (SO 188378). The ridge is best seen in the fields west of the minor road from Felindre to Tregoyd Mill. The channel was therefore not of ice marginal origin. Instead, it was probably initiated subglacially, as glacial meltwater drained north-eastwards to escape along the kame terrace at the edge of the main Wye glacier downstream of Glasbury.

A different origin apparently applies to the deeply incised valley of Nant Ysgallen, which forms a gorge between Heol-y-gaer (SO 195391) and the point where the stream debouches onto the kame terrace near Pont yr Angel (SO 199403). Upstream of Heol-y-gaer the stream drains a flat area between Tregoyd and Ffordd Las (SO 198380–207393) in which there are sediments that appear to be of lacustrine origin (Williams, 1968a). An ice margin ending immediately west of this flat area would have blocked drainage off the Black Mountains and could have ponded a pro-glacial lake. As the ice retreated towards the lower land of the Wye valley floor the lake apparently drained, via Nant Ysgallen, cutting the gorge near Heol-y-gaer as it did so.

Other, and much less impressive channels, exist on the southern flank of the Wye valley between Gwernyfed Park (SO 175373) and Glasbury. Some of these are semi-circular, as if one side of the channel is missing, as is the case with a number of small channels north of Gwernyfed Park school that apparently drained towards the Tregoyd Mill Brook. These may have developed as ice-marginal channels, allowing meltwater to drain down-valley along the ice-front, with one side of the channel being the ice-front itself.

The age of the Hay-Clyro moraine
This moraine is a spectacular and obvious feature in the landscape and on morphological grounds it is likely that it dates to the Late Devensian, although its age has not been established numerically.

The Llyswen-Dderw kame moraine
An area of kettled sediments extends downstream from a steep slope, that is up to 8 m high, located on the eastern side of the stream that flows past the former Nonconformist chapel (SO 133378) on the eastern side of Llyswen (Fig. 8.2). Sand and finer sediments exist in the flattish area on the Llyswen side of the steep slope. Kettle holes, some of which are occupied by ponds, exist on the Dderw Farm (SO 140376) and extend down-valley to the vicinity of Boxbush (SO 154377). The steep slope near the former chapel at

Llyswen may well be a former ice-contact slope, while the sands and gravels that form the kettled area were probably deposited by outwash streams that flowed out of the melting ice. The kettle holes presumably occupy areas where blocks of ice, carried onto the outwash plain by meltwater streams and covered or partly covered in fluvioglacial sediments, subsequently melted. The flattish area up-valley of the steep slope may have been occupied by a pro-glacial lake, in which fine sediments were deposited, as the ice-front melted back from the steep slope but still blocked drainage down the main Wye valley.

The Wye valley between Llyswen and Builth Wells
The River Wye flows within an U-shaped gorge between Llyswen and Builth Wells (Fig. 8.1), the valley floor being some 100 m or more below the level of the uplands on either side of the valley. Silurian bedrock is particularly well exposed down-valley of the confluence of the Edw and the Wye, at Aberedw Rocks (SO 080460), where some of the outcrops resemble roches moutonnee and may owe their shape to glacial plucking. Well developed terraces, some of which may be of glacial outwash origin, occur on the valley floor and lower valley sides between Llyswen and the vicinity of Builth Wells, as between Erwood and the site of the former Erwood railway station (SO 089439). Linear ridges downstream of the road bridge across the River Wye near Erwood station may be the remnants of a former ice-margin.

Drumlins exist at altitudes around 210 m in the Bach Howey valley, which is a left bank tributary of the Wye, between Erwood and Painscastle, in the vicinity of the Castle Mound (SO 125450). Terminal moraines cross the Bach Howey valley east of Painscastle, as between Rhosgoch Common (SO 195485) and Newchurch. Sections in glacial till at altitudes around 300 m in the vicinity of Llanbella (SO 240530) expose striated Silurian clasts and evidence the former movement of ice eastward over the uplands of the region from an ice centre further west.

An area of moundy terrain on the southern flank of the Duhonw valley west of Glanwye (SO 066498), and the till exposed on the wooded east bank of the Wye adjacent to a house and garage upstream of Llanfaredd (SO 064512), may be the remains of a terminal moraine of the Wye glacier.

The Vale of Irfon
The Wye valley changes its nature upstream of Builth Wells as it crosses a strike vale north of the north-facing escarpment of the Eppynt-Carneddau uplands. The River Irfon, which with its tributaries drains the mid-Wales uplands between the Elan-Claerwen and Tywi valleys, flows southwards off those uplands before turning east near Llanwrtyd Wells and flowing as a strike stream along the Vale of Irfon to join the Wye at Builth Wells.

North of Llanwrtyd Wells, particularly upstream of Abergwesyn, the Irfon flows in a trough-shaped valley. Deposits of glacial till occur at various sites within the valley, a section on the east side of the valley downstream of the site of the former church of Abergwesyn, exposes some 8 m of till derived from the local rocks of the upper Irfon valley (SN 864525). Ice from the mid-Wales uplands therefore drained down the Irfon valley, which it imprinted with the shape of a glacial trough in the region upstream of Llanwrtyd, before debouching onto the lowlands of the Vale of Irfon between Llanwrtyd Wells and Builth Wells.

The Vale of Irfon contains many features of glacial deposition. Landforms resembling drumlins are well developed in the region between Cilmery (SO 005515) and Builth Wells, with their long axes predominantly aligned parallel to the long axis of the Vale. Till is exposed on the flanks of one such feature between Cilmery railway station and Glan-Irfon (SO 003509). Upstream of Cilmery, between there and Llanfechan House (SN 979504), kames occur on the floor of the Vale, as in the field (SN 989506) beside the A483 between the road and a house adjacent to the track that leads north to Cilmery Farm. Interpretation of many landforms in the Vale, including those that resemble drumlins, is hindered by the paucity of exposures.

Further west in the Vale, as at Aber-Dulas-uchaf (SN 918468), there is blue-coloured till, containing Ordovician clasts that indicate that the ice responsible for its deposition flowed southwards from the central Welsh uplands and into the Vale. The till at Aber-Dulas-uchaf, which is exposed to a depth of 2 m, exists under outwash and younger (possibly Holocene) gravels.

Between Llanwrtyd Wells and Llawr dre-fawr (SN 858450) exist kames, with the boggy area of Waen Rydd occupying a basin with steep faces that resemble ice-contact slopes to north, south and west. Immediately further west, at Erwbeili (SN 853447), beside the A483 road, there is a delta deposit with well-marked foreset bedding dipping south-westward. This indicates that a pro-glacial lake was formerly trapped between an ice-front at Erwbeili and the present watershed between the Vale of Irfon and the Tywi catchment to the west. Presumably the lake drained over that watershed.

The complicated glacial features in the Vale of Irfon and elsewhere imply that the Vale was formerly inundated under an ice-sheet derived from mid-Wales but that ice subsequently only occupied the lower portions of the Vale. Whether the mid-Wales ice-sheet, that covered the Vale and abutted against the Black Mountains, melted entirely but was succeeded by readvance of ice from the mid-Wales mountains into the Vale, is unknown. The drumlin-like landforms between Cilmery and Builth indicate subglacial drainage towards the Wye valley during deglaciation. Tills in the western Vale, extending across the lower part of the Eppynt and into the south flowing Cilieni valley, indicate that mid-Wales ice also flowed into the upper Usk catchment. The pro-glacial delta at Erwbeili indicates that although ice formerly spilled into the upper Tywi catchment, across the Sugar Loaf col (SN 845440), its western margin was later restricted to the Irfon Vale. Whether the Erwbeili deposits relate to a withdrawal stage of the main ice-sheet, or a readvance of a glacier down the Irfon valley, is presently unknown.

The Wye valley upstream of the Vale of Irfon
Hummocky topography exists upstream of Builth Wells in the vicinity of Builth Road railway station, where there are kame-like landforms, suggestive of the former presence of stagnant ice. Thicknesses of up to 6 m of till are exposed in stream side sections and in cuttings beside the abandoned railway line some 200 m north-west of Builth Road. Kame-like mounds occur on the floor of the Wye valley between Builth Road and Penmincae (SO 004543), where there is a wooded esker-like ridge. Unfortunately the lack of sections renders interpretation of these features uncertain.

North of Newbridge-on-Wye, where the Wye emerges from its Doldowlod gorge section, a gravel pit near the junction of the A470 with the A4081 exposes bedded sands and gravels that were deposited by meltwater from a glacier issuing out of the gorge. Similar deposits near Llanyre (SO 037614) indicate that ice formerly covered the Ithon lowlands. At Caerfagu abattoir (SO 045657), in the Dulas valley, an esker, or possibly delta deposit, is exposed, indicating that meltwater from an ice-mass in the Nantmel-Rhayader area drained eastwards as the ice melted.

On the western side of the Wye valley, opposite Doldowlod House where Nant Cymrun joins the main valley, a well marked lateral moraine exists at altitudes around 300 m (SN 985623). The right bank tributary further north, that joins the Wye at Llanwrthwl, is trough-shaped and carried diffluent ice from the Elan valley as it spilled towards the Wye. The valley sides are markedly plucked while boulders, probably derived from the Elan valley uplands, are in evidence on the valley floor.

A lateral moraine exists on the south side of the Elan valley above the confluence of that valley with the Wye near Rhayader. Terminal moraines exist on the floors of valleys draining off the uplands between Rhayader and the Elan valley reservoirs, as in the valley of Glanllyn (SN 948690), where a small lake nestles below a trough's end over which a stream tumbles in waterfalls. Similar trough's ends exist in other valleys that carry drainage off the central Welsh uplands, indicating the importance of those uplands in nurturing the mid-Wales ice-masses that formerly developed into ice-sheets that extended into the Welsh border-lands and the Hereford basin.

Fig. 8.3 Pingo remnants in the valley of the Nant Cae-garw, near Llanidloes, photographed when they were being examined by Pissart in 1961. The circular ramparts formed as the ice of the active pingos melted, allowing sediments incorporated within that ice to sludge down the sides of the pingos. The ill-drained basins within the ramparts represent the central portions of the formerly ice-cored landforms. The formation of pingos is described by Ballantyne and Harris (1994). (Photo: C. A. Lewis)

Periglacial features, in the form of solifluction deposits, ice-wedge casts and pingo remnants, exist in the isolated areas north of Rhayader. Pingos near Llangurig, in the Nant Cae-garw valley (SN 950798; Fig. 8.3), which is a tributary of the Dulas that flows northward towards Llanidloes, have been described by Pissart (1963) and are thought to be Late Glacial in age. They evidence the existence of at least sporadic permafrost in the area following its deglaciation. Ice-wedge casts in the Claerwen and other valleys of the central Welsh uplands (Potts,1971) are also indicative of the existence of permafrost in those areas following ice-sheet deglaciation.

The Usk valley

The glacial features of various parts of the Usk valley have been mapped by Williams (1968a), Lewis (1970a, b), Ellis-Gruffydd (1972, 1977) and Thomas (1997). South of Abergavenny a large morainic structure runs in a broad arcuate band across the valley of the middle Usk. This moraine has long been recognised as marking the limit of the last glaciation in the area and has been repeatedly incorporated into maps of the Late Devensian glaciation (Charlesworth, 1929; Bowen, 1973, 1981; Campbell and Bowen,1989). Except for the work of Thomas (1997) and the excellent but unpublished geomorphological maps of Williams (1968a), however, very little work has been undertaken in the area.

The Usk valley may be divided into three regions: the middle Usk between Talybont and Usk; the lower Usk valley, downstream of Usk; and the upper Usk and Brecon Beacons-Fforest Fawr uplands and the Eppynt. Each of these regions, as well as the adjacent Black Mountains and Grwyne Fawr/Cwm Coed-y-Cerrig trough, will be described following presentation of a glacial sediment-landform assemblage system and a general statement on drift characteristics.

Sediment-landform assemblages

The middle and upper Usk valley is a type example of a glacial valley depositional system that includes at least four major sediment-landform assemblages. These assemblages are mappable units in which relatively homogenous morphologic, stratigraphic and lithologic characteristics occur, as shown on Fig. 8.4. Similar assemblages have already been described from the Wye valley, although they were not then related to a sediment-landform assemblage model.

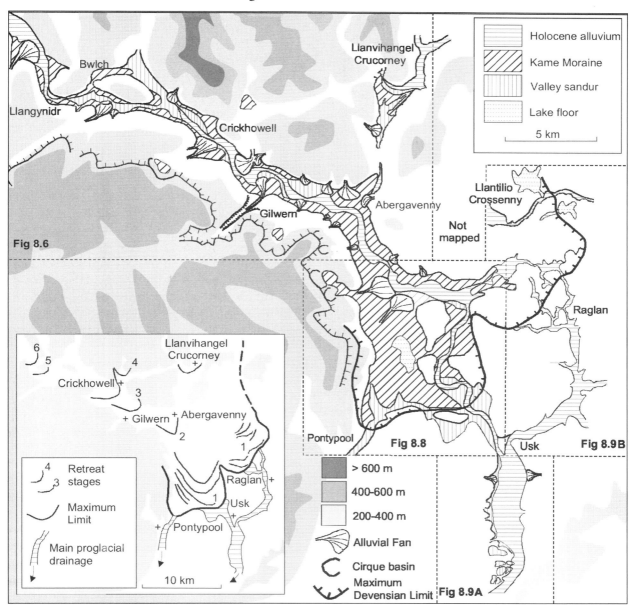

Fig. 8.4 Outline geomorphological map of the middle and lower Usk showing distribution of sediment-landform assemblages, maximum of the Late Devensian glaciation and location of geomorphological maps in Figs 8.6, 8.8 and 8.9. Inset shows major moraine systems and distribution of main proglacial drainage

i) Kame moraine assemblage
This assemblage consists of extensive tracts of hummocky topography, up to 20 m in height, comprising ridges, mounds, basins and intervening channel systems. Individual tracts occur either as arcuate bands running across valley or as irregular borders to the valley floor. They are composed of sand and gravel overlying basal diamict and bedrock. The hummocky topography was formed at, or immediately in front of an ice-margin during temporary retreat, still-stand or minor readvance, by deposition of ice-marginal sedimentation on top of dead and decaying ice. When the glacier retreated and the buried ice melted, the resultant surface was left as a series of irregular ridges, mounds and basins marking the former arcuate ice-margin. Much of this topography was later removed by meltwater stream erosion as the glacier retreated, or was buried by subsequent meltwater sedimentation and post-glacial alluviation.

Kame moraine assemblages are preserved in the middle Usk valley, especially west of Talybont, west of Llangynidr, opposite Crickhowell, at Gilwern, to the east of Abergavenny and in a major tract between Abergavenny, Pontypool and Usk. Each occurrence represents a major still-stand position during retreat of the Usk valley glacier from its maximum limit and at least six major episodes can be recognised (inset, Fig. 8.4). They also exist in the upper Usk valley, as will be described later.

ii) Valley sandur assemblage
This assemblage consists of gently sloping depositional surfaces, elevated above the modern river flood plain, that front tracts of cross-valley kame-moraine and extend down-valley. They formed as meltwater from the melting ice carried sediments down valley and deposited them on the valley floor downstream of the ice-front to form an outwash plain, or sandur. This feature was abandoned as the ice-front retreated up-valley and a new sandur was deposited. The deposits underlying the sandur surfaces consist predominantly of gravel. The older sandur deposits were at least partially eroded as the newer sandur formed. Consequently, through a succession of retreat still-stands, sandur surfaces occur as a set of stacked depositional features grading into one another down-valley. Under normal circumstances the proportion of the valley floor occupied by sandur surfaces diminishes downstream as a result of increased lateral erosion by fluvio-glacial and post-glacial stream incision. The gradients of sandur in the Usk valley are normally steeper than either that of the existing river profile or those of recent river terraces.

Valley sandur form from a single-exit meltwater stream system discharging from an ice-margin defined by the cross-valley moraine ridge on the up-valley side of the sandur. At the time of deposition outwash sediments accumulated across the whole width of the Usk valley floor, but subsequent erosion associated with the deposition of younger sandur plains and with Holocene fluvial incision has reduced the extent of the original deposits.

A typical sandur section occurs in the west bank of the Usk north-east of Gilwern (SO 262148). This displays 8 m of clast-supported, sub-rounded, pebble to cobble gravel in stacked, sub-horizontal sets indicative of high-energy, upper fan braid-bar sandur environments. The clasts are mainly of Devonian conglomerates, sandstones and mudstones derived from outcrops upstream. Apart from tuff, which may derive from Ordovician volcanic rocks outcropping in the Wye basin around Builth Wells, no rock types from outside the Usk catchment have been found.

iii) Alluvial fan assemblage
Alluvial fans occur frequently on the margins of the Usk valley. Over twenty fans, ranging in size up to more than a kilometre in area, occur between Bwlch and Usk (Fig. 8.4). These fans extend from tributary valleys out onto the floor of the main valley. They are mostly gently sloping and either rectilinear or gently concave upwards. They consist of coarse gravel that is often little rounded, intercalated with debris flow deposits.

The alluvial fans were probably initiated as deglaciation occurred. Water and debris flows transported sediment down the valley sides and from tributary valleys to form the alluvial fans on the floor of the main Usk valley. By analogy with contemporary valley glacier systems it is likely that in their early stages they were deposited partially upon the lateral margins of the glacial ice. As glacier retreat continued water and debris flows transported sediment downslope from parts of the declining ice-sheet that still occupied the upland flanks of the Usk, or, as in the case of the two large fans around Llanover and similar fans opposite Abergavenny and Crickhowell, from cirque glaciers occupying large hollows along the rim of the Carboniferous Limestone escarpment.

Some of the alluvial fans at least partially overlie and bury kame moraine and sandur deposits formed during earlier stages of valley glacier retreat. Consequently they form a diachronous series, younging up-valley.

iv) River terrace assemblage
River terraces border the Usk throughout most of its length. In general terms the horizontal and vertical extent of terracing markedly increases down-valley. Two main terraces occur, though intermediate terraces exist in some areas.

The Upper Terrace occurs at a height of approximately 4-5 m above the modern flood plain while the Lower Terrace is at a height of 2-3 m. Abandoned meander scars and cut-offs, some of the latter being still filled with water, frequently exist on the Lower Terrace.

A river bank cut on the outside of a large meander bend east of The Bryn (SO 335099), between Abergavenny and Usk, exposes details of both terraces (Fig. 8.5). The Upper Terrace, 4 m above river level, truncates a small kame mound and consists of laminated red silts and fine sands. These pass down into grey, partially laminated organic clay, with tree trunks and scattered plant debris, and then into peat. To the east the Upper Terrace is truncated by the Lower Terrace at a height of 2 m above river level. The age of these terraces is unknown but is believed to be Holocene. They probably formed as the River Usk adjusted to changes in sediment supply, climatic fluctuation and to anthropogenic influences.

Drift distribution, thickness and type

Drift deposits in the middle Usk are largely confined to the valley bottom, either as a series of flat or gently sloping terraces across the valley floor or as narrow, hummocky strips along the lower flanks of the valley. These deposits are up to 50 m thick. South of Abergavenny the valley widens, drift distribution is thinner, (usually less than 30 m), and more diffuse, forming an irregular veneer across partially buried Silurian bedrock escarpments. Drift is sparse south of the town of Usk, but the over-deepened rock floor of the valley towards Newport is overlain by sands and gravels succeeded by thick Holocene river alluvium bordered by narrow gravel terraces (Williams, 1968b).

Analysis of borehole records and mapping of bedrock outcrops (Crimes *et al.*, 1992) suggests that the middle Usk forms a series of shallow, over-deepened rock basins separated by rock-bars at or near the surface. Each basin consists of accumulations of sand and gravel overlying diamict. Rock bars occur at Llangynidr, Crickhowell and Abergavenny (Fig. 8.4) with an approximately 5-6 km spacing. This is a pattern similar to other radial valley systems investigated elsewhere in Wales (Thomas *et al.*, 1982). Although few boreholes reach bedrock, average over-deepening is a minimum of 50–60 m.

Although drift outcrop is poor, boreholes indicate that four major sediment types can be identified. Diamict is rarely exposed but occurs extensively at depth in thicknesses of up to 10 m across the bedrock floor. It comprises a massive, matrix-supported, sandy clay crowded with well-rounded, often striated clasts of predominantly Devonian and Carboniferous rock types, but with a sprinkling of more far-travelled Silurian rocks from the Wye basin (Lewis, 1970a). By its lithological characteristics and

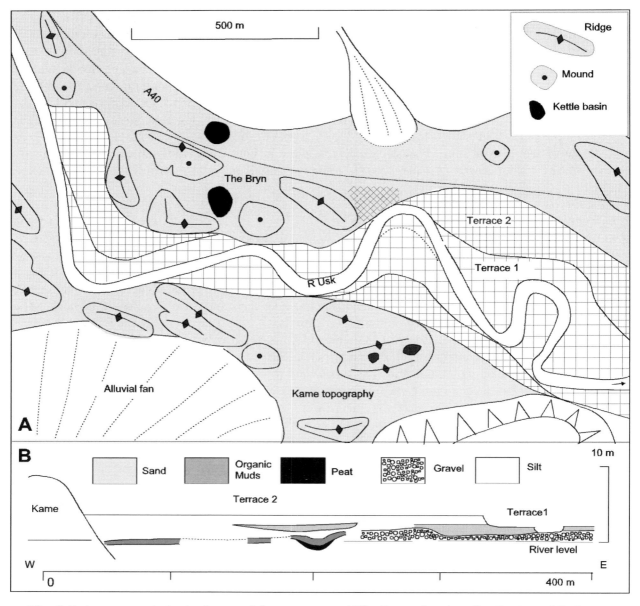

Fig. 8.5. A: geomorphological map of the area around The Bryn, showing distribution of Holocene river terraces. B: sketch section showing relationships between Devensian fluvioglacial and Holocene river terraces in section east of The Bryn

its stratigraphic position above the bedrock, the diamict is interpreted as the product of sub-glacial deposition. Sand and gravel ranges from coarse, sometimes boulder gravel, to fine sand deposited in a number of different depositional environments including ice-front alluvial fans, ice-marginal sandur and valley-sandur. Finer gravel and sand also comprises much of the alluvial flood plain. Mud includes laminated or thinly bedded brown or reddish clay, silt and fine sand occurring locally within the succession but also as extensive surface outcrop in flat-floored basins east of Llanover and in the upper headwaters of the River Trothy. These mud deposits are interpreted as glacio-lacustrine sediments.

Regional descriptions

Introduction

The underlying geology plays a significant role in the morphological development of the Usk basin. North of Abergavenny the middle Usk valley floor is narrow and developed along the strike of the mudstones of the Lower Devonian St Maughans Formation (Barclay, 1989). It is bordered on its western side (Fig. 8.4) by a prominent escarpment formed by the more resistant Upper Devonian Quartz Conglomerate Group and the Lower Carboniferous Castell Coch Limestone. This escarpment is broken north of Abergavenny by the deep entrenchment of the Clydach gorge. South of Abergavenny the valley widens along the axial trend of the Silurian Usk anticlinal inlier and some strong local relief occurs in the form of minor escarpments across the valley floor.

The middle Usk: Talybont to Abergavenny

The salient features of the geomorphology of the Usk valley between Talybont and Abergavenny are shown on Fig. 8.6. At Talybont a large alluvial fan drains north into the margin of the Usk valley and is flanked to its west by a strip of kame-moraine topography diversified by a number of arcuate cross-valley moraine ridges and marginal drainage channels. To the west of Bwlch a further area of kame-moraine topography is preserved on both margins of the Usk and displays a complex sequence of arcuate cross-valley moraine ridges, small channels running parallel to the ridges, and small kettle holes (Lewis, 1970b). Together these cross-valley moraines represent the final ice-marginal retreat stage of the Usk glacier in the middle Usk valley (Stage 6) although further ice-marginal stages occur in the upper Usk region. Further east the valley narrows abruptly into a shallow gorge below Buckland Hill and bedrock is exposed in the river bed. This marks the position of a shallow rock bar separating overdeepened rock basins upstream and downstream.

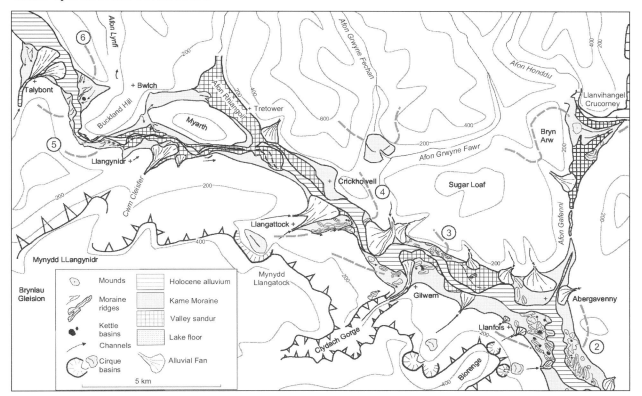

Fig. 8.6 Geomorphological map of the area between Talybont and Abergavenny. For location, see Fig.8.4

Downstream of the gorge the valley floor widens and extensive kame-moraine topography, accompanied by short arcuate moraine ridge segments and associated channels, occurs on the southern valley margin west of Llangynidr, marking an earlier stage in ice-marginal retreat (Stage 5). On the opposite side of the valley a large area of subdued kame-moraine topography extends eastwards below the slopes of Myarth and passes through the shallow col between Buckland Hill and Myarth to extend into the open valley to the east of Bwlch. The kame moraine topography on both sides of the Usk around Llangynidr is fronted by a large valley sandur (Williams, 1968a) that extends east as far as Crickhowell and is joined by an extensive sandur emerging from the Rhiangoll valley, south of Tretower. At the confluence of the Usk and Rhiangoll the extensive sandur surface on the southern side of the Usk forms a flight of terraces. Immediately to the east of Llangynidr a large alluvial fan, draining north from the Carboniferous Limestone escarpment fronting Mynydd Llangynidr and Mynydd Llangattock, via Cwm Claisfer, extends onto the floor of the Usk valley and partly buries the kame-moraine topography and sandur surface on its southern margin.

The town of Crickhowell is built on an extensive area of kame-moraine topography bordering the northern margin of the Usk. Opposite the town, two coalescing alluvial fans drain north from a large cirque basin on the rim of the Carboniferous Limestone escarpment below Mynydd Llangattock. They are bordered on their eastern side by a strip of kame-moraine topography that displays a major set of arcuate cross-valley moraines, separated by parallel channels that define a further ice-marginal retreat position (Stage 4). This stage may correlate with a prominent moraine at the confluence of the Afon Grwyne Fechan with the Afon Grwyne Fawr to the north of Crickhowell.

The geomorphology around Gilwern (Fig. 8.7 A) characterises the relationships between the various sediment-landform assemblages identified throughout the middle Usk. East and west of the town a wide strip of kame-moraine topography flanks the southern side of the valley and carries arcuate moraines that loop across the valley to further moraine fragments on a similar strip of kame-moraine topography bordering the northern side of the river. These mark another ice-marginal retreat position (Stage 3; Fig. 8.6). To the south, the kame-moraine topography is cut through by a large alluvial fan draining from the Clydach gorge, which is a major entrenchment into the face of the Carboniferous Limestone escarpment. A similar fan descends from the mouth of the Grwyne Fawr on the north side of the Usk valley and buries much of the kame-moraine sediment.

East of Gilwern extensive sandur surfaces front the cross-valley moraine and extend towards Abergavenny where they occupy much of the valley floor. The kame-moraine topography is truncated by two separate sandur surfaces that, in turn, are entrenched by the alluvial floor of the modern river. The highest sandur occurs only to the east of the cross-valley moraine, off the down-ice flank of which it grades. A matching, but more extensive sandur, exists at the same height on the northern bank of the Usk and extends downstream towards Abergavenny. The lower sandur has eroded through the cross-valley moraine and is entrenched into the upper sandur. The upper and lower sandur deposits relate to two separate outwash systems: the upper sandur formed in front of the cross-valley moraine and the lower sandur originated from an ice-margin further upstream.

Fig. 8.7 B depicts a down-valley section through the sediment-landform assemblages in the vicinity of Gilwern. None of the boreholes shown on Fig. 8.7 B reached bedrock but two penetrate a stony, red diamict that appears to underlie most of the area. This diamict is overlain by thick red sand that coarsens upwards into pebble and cobble gravel. Upstream of the moraine ridges these gravels are overlain by a stony red diamict that terminates against the outermost moraine ridge and is replaced on the down-ice side by thickening gravels. The upward coarsening signature suggests that the cross-valley moraine was built by a small-scale, localised readvance in which basal diamict was emplaced across a floor of former valley-sandur sediment. At the readvance maximum the melting of dead-ice created irregular kettle basins, which are now occupied by organic sediments in the water-filled basin of 'The Swamp'. Sediment-laden meltwater also deposited a coarse apron of sandur material immediately beyond the ice-margin.

Abergavenny is located on a large area of kame-moraine topography that extends along the north bank of the Usk and into the valley of the Afon Gafenni to the north of the town. Part of this area is buried by large alluvial fans issuing from steep valleys on the south-east flank of the Sugar Loaf. Although much topographic detail is obscured by urban development, a series of arcuate moraine features on the western

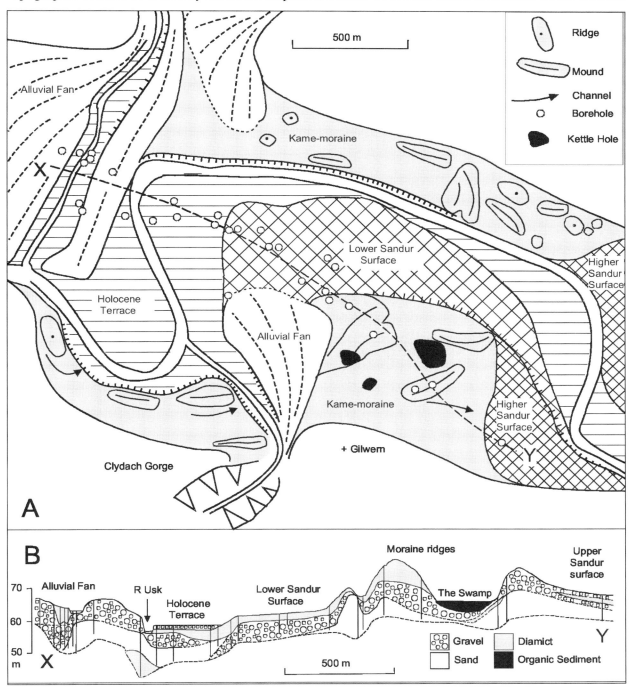

Fig. 8.7. A: the geomorphology of the area around Gilwern showing location of boreholes and section line X-Y. B: cross-section through the cross-valley moraine system at Gilwern along the line of boreholes shown on A

edge of the town correlate with a very complex area of arcuate moraine ridges, intervening channels and large kettle basins in an extensive area of kame-moraine topography on the southern margin of the valley near Llanfoist. These arcuate moraines represent another ice-marginal retreat position (Fig. 8.6; Stage 2). The outer part of this area is overlain by the toes of two small alluvial fans draining from cirque basins located on the rim of the Carboniferous Limestone escarpment below Blorenge Mountain.

The valley of the Afon Gafenni, which is a tributary of the Usk draining south towards Abergavenny (Fig. 8.6), currently carries a diminutive stream. The upper part of the valley is blocked by a very large, compound arcuate moraine at Llanvihangel Crucorney (Lewis, 1970a, Fig. 7.2), the outer face of which grades south into a large ice-front alluvial fan and sandur. To the rear of the moraine the Afon Honddu, following the arc of the moraine ridge, completes an almost perfect U-turn in flow direction from south to north. This suggests that, prior to the creation of this moraine, drainage from the upper part of the Afon Honddu flowed south, via what is now the Afon Gafenni, towards the Usk at Abergavenny.

The Black Mountains
The Honddu valley, upstream of the Llanvihangel Crucorney moraine (Fig. 8.6), flows south-eastwards down the dip slope of the Black Mountains in a trough shaped valley, the lower part of which is known as the Vale of Ewyas. The sides of the valley have been markedly oversteepened, while extensive glacial deposits exist downstream of Capel-y-ffin (BGS, 2002), indicating that glacier ice flowed down the valley and its main Nant y Bwch tributary from the uplands of the Black Mountains. Ice from mid-Wales may also have escaped south from a mid-Wales ice-sheet and breached the escarpment of the Black Mountains at Bwlch yr Efengyl (the Gospel Pass, which is the col between the Honddu and the Wye valleys south of Hay-on-Wye; Fig. 8.1) to flow down the Honddu. Erratics from mid-Wales, which would confirm this possibility, have (as has already been stated in this chapter) not yet been identified in the Honddu valley.

Extensive landslide deposits exist on the west side of the Honddu valley beneath the cliffs of Tarren yr Esgob ('Cliff of the Bishop') in the vicinity of Capel-y-ffin (SO 240310). Continuing slope adjustment occurs on the east side of the Vale of Ewyas under the spectacular Darren immediately north of Cwmyoy (SO 296245). The church at Cwmyoy was built on an unstable landslide slope and has been adversely affected by slope movements: the nave tilts at different angles to those of the rest of the building! The marked oversteepening of the sides of the Vale are evidence of erosion associated with the movement of a powerful glacier down valley.

Glaciers also flowed southwards, down the parallel valleys of the Grwyne Fawr and the Grwyne Fechan. Diamict containing occasional striated clasts, all of which were derived from the Devonian rocks of which the surrounding areas of the Black Mountains are composed, was exposed in 2002 in a road cutting some 50 m upvalley of the gate that leads into the forest plantation near Hermitage (SO 228252) in the valley of the Grwyne Fechan. These deposits indicate that glacial ice formerly flowed down the valley from adjacent higher ground. Cross valley moraines exist lower down the valley, in the vicinity of Llanbedr, where kame topography exists.

An area of kame-moraine exists in the lower portion of the Grwyne Fawr valley, upstream of the Grwyne Fawr/Cwm Coed-y-cerrig trough. Ice previously spilled across that trough to deposit spectacular kame-moraine south of Bettws (SO 300185), in the valley west of Bryn Arw (Fig. 8.6) that is the continuation of the line of the Grwyne Fawr south of the fault guided trough that cuts across the original drainage line. The kame topography, which marks the outermost morphologically identifiable limit in this area of ice originating in the Black Mountains, is evident between Llwyn-gwyn (SO 303187) and the road east of Gott (SO 299185). A sandur/alluvial fan grades towards the floor of the Gafenni valley from the kame region.

In order for a glacier from the Grwyne Fawr to cross the Grwyne Fawr/Cwm Coed-y-cerrig trough, and spill into the Bettws area, the ice must have been in excess of 40 m thick on the floor of the trough.

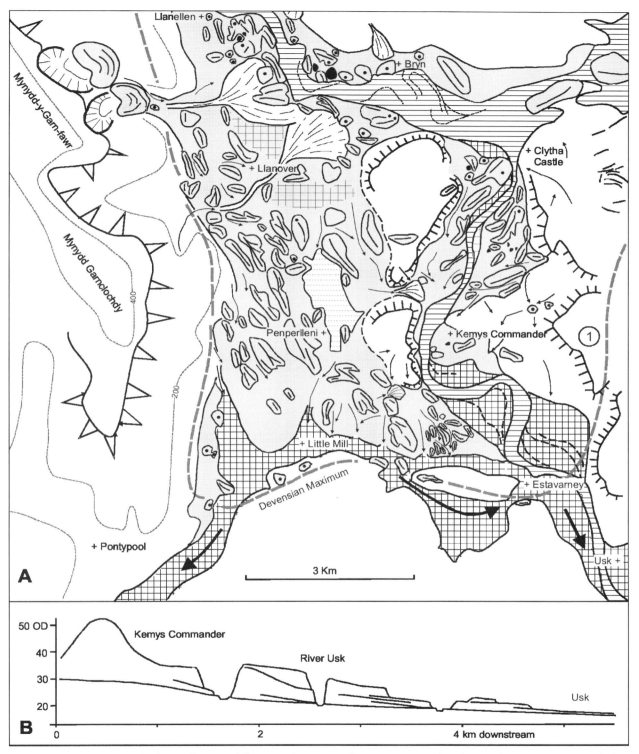

Fig. 8.8. A: geomorphological map of the area between Abergavenny and Usk. For location see Fig. 8.4. For Key see Fig. 8.6. The area around Clytha Castle is not mapped in detail and shows moraine crests only. B: section between Kemys Commander and Usk (see A) showing the relationship between sandur surfaces. (Redrawn after Williams, 1968a)

Whether diffluent ice from the Usk valley forced its way into the trough from the Crickhowell area, as Lewis (1970a) suggested, is uncertain. The glacial deposits within the trough may owe their origin entirely to glaciers flowing off the high ground of the Black Mountains and down the Vale of Ewyas.

A cirque moraine exists on the western side of the Honddu valley, as will be described later when similar features in the Brecon Beacons and adjacent areas are discussed.

The middle Usk valley: Abergavenny to Usk
The Usk valley south of Abergavenny widens rapidly along the axis of the Silurian Usk inlier, especially south of Llanellen (Fig. 8.8). Characteristically, when a large valley glacier system passes from a restricted valley to a more open region, the ice-front spreads out to become a piedmont glacier and kame-moraine topography becomes widespread. Subsequent incision by sandur and post-glacial alluviation processes occurs only at limited points along the former ice-margin, so that much of the kame-moraine is preserved.

Downstream of Llanellen the floor of the Usk valley is occupied by a very complex arcuate moraine system composed of numerous, closely-spaced but discontinuous moraine ridges and intervening channel systems in a belt five kilometres wide and six kilometres deep. The continuity of many of the moraines is interrupted by rock ridges that rise along the flanks of the Usk anticline in the centre of the valley and by a steep escarpment to the east, but the arcuate form is clearly seen on both flanks of the valley between Llanover and Clytha Castle (Fig. 8.8 A).

South of Llanellen, where the valley first widens, there is a complex area of ice-disintegration topography, including a number of large kettle basins. The morainic topography is partially buried, further south, by two large alluvial fans draining from cirque basins on the Carboniferous escarpment west of the valley.

East of Penperlleni there is an extensive linear strip of flat ground, underlaid by thick laminated silts and bounded by moraine ridges, that was probably the site of a temporary ice-marginal lake trapped between the retreating ice-front on one side and older moraines on the other.

The outermost moraine ridge terminates a little above the town of Usk and marks the maximum limit reached by Late Devensian ice advance down the valley (Fig. 8.8 A; Stage 1). Meltwater drainage from the ice-margin at its maximum was complex and the outermost moraine is fronted by a flat-floored trough, 500 m wide by 4 km long, running east from Little Mill towards Great Estavarney. The trough is underlain by more than 18 m of coarse gravel and marks the position of a major ice-marginal sandur channel that collected drainage from exit tunnels in the ice and directed it eastwards to join a similar, but more direct, sandur channel draining the eastern side of the ice-margin north of Estavarney. On the west a similar sandur trough passes south from Little Mill towards Pontypool. Neither of these major outwash conduits carry more than diminutive drainage at present and both were active only at the maximum stage of glaciation. As the ice-margin retreated from its maximum the ice-marginal sandur trough was abandoned and meltwater drainage was concentrated on the eastern side of the valley.

South of Kemys Commander the modern flood plain is flanked by a flight of sandur terraces, each graded to a particular retreat ice-marginal moraine, and subsequently incised and abandoned on further retreat (Williams, 1968b; Fig. 8.8 B).

The lower Usk and Nant Olwy
Drift deposits south of the town of Usk are limited, as the area is outside the Late Devensian ice-margin. In the lower Usk (Fig. 8.9 A) drift is restricted to extensive Holocene alluvium across the valley floor, small alluvial fans on the valley margins, some residual pre-Devensian diamict on hill slopes and a narrow sandur terrace on the western margin of the valley at heights of up to 10 m above the modern flood plain. The sandur probably formerly occupied the entire valley floor but has been reduced in extent by subsequent incision. Around Newbridge-on-Usk a fragmentary series of higher terraces occur, up to 30 m above the flood plain.

Very limited subsurface information is available for the region, but to the south, near Newport, Williams (1968b) recorded thick Holocene alluvium overlying sands and gravels of probable pro-glacial origin. Depths to bedrock in the Newport area indicate that the unconsolidated sediments partly occupy buried channel systems, which were probably cut during the Devensian cold stage, when sea levels were lower than at present. The lower Usk valley thus seems to have acted as a major outwash distributary from the Late Devensian ice-margin to the north.

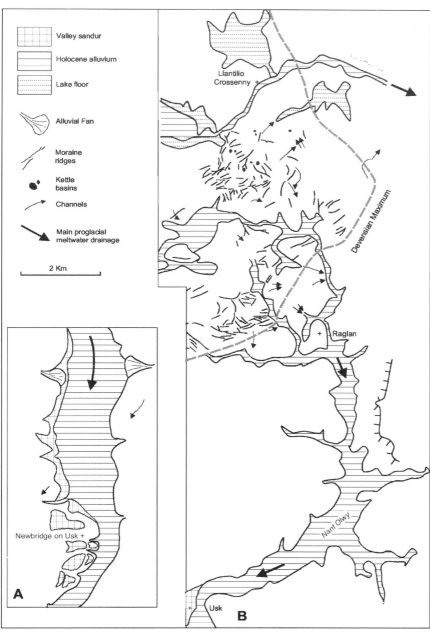

The area east of Abergavenny, between Usk and Llantilio Crossenny, lacks drift maps and has not been mapped geomorphologically in detail. A map based on the work of Williams (1968a) is shown in Fig. 8.9 B and identifies ridge crests, kettle basins and channels. These are sufficient to establish that the Devensian maximum limit south of Abergavenny continues north-east towards Raglan and then north towards the valleys draining the southern margin of the Black Mountains. It also establishes that the main marginal drainage in this area fed south into the headwaters of the Nant Olwy, an otherwise diminutive stream, the valley of which is floored with extensive Holocene alluvium, that joins the Usk at the town of Usk. Haslett (2003) indicates that a grey-brown sandy clay deposit, that may be head, underlies the alluvium and is sometimes separated from it by a palaeosol and an overlying fenwood peat. Drainage also passed eastwards along the Afon Troddi (River Trothy), a tributary of the Wye, to the east. Extensive areas of laminated clay near Llantilio Crossenny indicate the existence of large ice-marginal lake systems during early stages of retreat from this maximum ice limit.

Fig. 8.9. A: geomorphological map of the area south of Usk. B: geomorphological map of the area around Raglan. Note that moraine crests only are shown. For locations see Fig. 8.4. (Redrawn after Williams, 1968a)

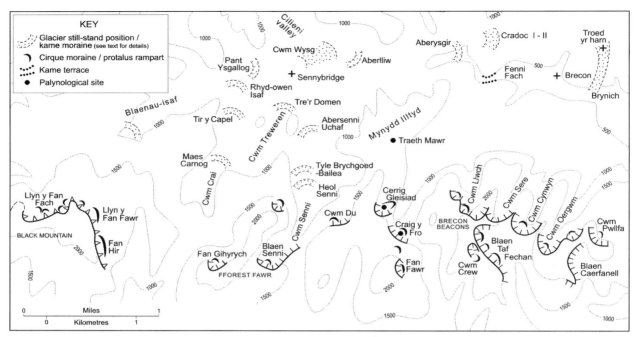

Fig. 8.10 Ice-marginal positions in the upper Usk and its right-bank tributary valleys, cirque moraines/protalus ramparts in the Black Mountain/ Fforest Fawr/Brecon Beacons area and palynological sites at Traeth Mawr, Cerrig Gleisiad and Craig y Fro. Contours in feet. (Redrawn after Ellis-Gruffydd (1972) with additions and emendations)

The upper Usk valley
This is defined as the Usk valley upstream of Talybont. Parts of the valley have been mapped by Williams (1968a), Lewis (1970a, b) and by Ellis Gruffydd (1972, 1977). The drift deposits of the area were in the process of being remapped by the British Geological Survey in 2002/3.

Both Williams (1968a) and Lewis (1970b) identified a morainic system in the Groesffordd region (SO 076280), north of Llanfrynach and some 3 km east of Brecon, where kame-moraine topography is evident on the northern side of the valley. Melt water channels south of Groesffordd, aligned towards the south-east, apparently carried drainage from the main kame-moraine area downstream, to the main Usk valley.

South of the Usk, on the flanks of the tributary valleys of the Cynrig and Menascin, which join the Usk west and east of Llanfrynach respectively, lie other meltwater channels. The channels near Tynllwyn (SO 063240) and opposite Cantref church (SO 058254), start and end abruptly and probably formed subglacially, under ice from the uplands of the Brecon Beacons that extended northwards into the Usk valley. These channels probably formed when the Usk glacier ended down-valley of Llanfrynach.

Excavations during the building of the Brecon by-pass, in the vicinity of the present road roundabout near Brynich (SO 068278; Fig. 8.10), at the junction of the A40 and the A470, exposed matrix supported diamict containing Old Red Sandstone clasts, some of which were more than three metres in length. The diamict overlay bedrock and probably formed through deposition beneath a glacier fed from source areas south and west of Brecon. No erratics from areas of Silurian and Ordovician rocks north of Brecon were identified in the diamict.

A spectacular area of kame topography occurs upslope of Brynich, immediately west of Troed yr harn (SO 066298; Fig. 8.10) where it is best seen in the fields downslope of the B4602 road. Exposures adjacent to that road when it was being improved revealed only Old Red Sandstone clasts, suggesting that

Fig. 8.11 The Tyle-crwn meltwater channel. (Photo: C. A. Lewis)

the ice responsible for the kame topography came essentially from the Brecon Beacons/Fforest Fawr uplands, rather than from northerly sources where other rocks occur.

Unconsolidated deposits are widespread north-east of Troed yr harn, on the plateau below the uplands of Yr Allt (SO 085307). Impressive meltwater channels lead from this plateau into the Dulas valley, which leads into the Llynfi and thence into the Wye (Fig. 8.1). They include the channel in which the house and buildings of Tyle-crwn farm (SO 096327) is situated (Figs. 8.1, 8.11). The Tyle-crwn channel starts suddenly on the slope above the farm buildings and ends above the level of the floor of the Dulas valley. This suggests that it formed subglacially, carrying meltwater from the plateau ice-mass downslope and into ice that occupied the Dulas valley. The meltwater then presumably drained into the Llynfi valley and thence into the Wye.

The Brecon Beacons/Fforest Fawr ice formerly flowed eastwards down the Dulas valley, leaving lateral moraines on the southern side of that valley between Llanfilo (SO 120332) and the vicinity of Tregunter (SO 135339). The glacier terminated in a ridge some 14 m high, at an altitude of 160 m, located south and upslope of the Tregunter farm dams (around SO 130335, where the ridge is partially wooded; Figs. 8.2, 8.12).

In 2002 a gravel pit near the summit of the ridge exposed 4 m of faceted and striated Old Red Sandstone clasts in a sandy and finer matrix. No clasts that might have come from the upper Wye valley were identified. The slopes of the ridge are steep, exceeding 30°, and appear to have formed as ice-contact slopes. Portions of the ridge that are lobate in plan probably formed as a result of minor ice advances in which the ice-front pushed against, and steepened, the ridge. They indicate that the ridge formed in association with ice that was not entirely dead (i.e. decaying and motionless). The ridge itself was formed by ice-marginal deposition of sediments that were mainly carried out of the ice by meltwater. The inner side of the ridge terminates in a grass-covered depression that was formerly occupied by the tongue of ice that was responsible for ridge formation.

Fig. 8.12 The ice-contact ridge south of Tregunter Farm. Glacial ice occupied the ill-drained flat area to the left of the ridge, pushing against its inner (tree covered) face to form a steep ice-contact slope. Sediment from the melting glacier, much of which was transported by meltwater, was deposited at and adjacent to the ice front to form this spectacular depositional ridge. This landform deserves protection as a site of major scientific interest. (Photo: C. A. Lewis)

On the north side of the Dulas valley, near Felin-newydd (SO 117359) and downslope of the A470, lies an area of kame and kettle topography at a similar altitude to the dead-ice ridge at Tregunter. Meltwater channels lead eastwards from this area towards Bronllys (SO 142350) and the Llynfi-Wye rivers (Fig. 8.2). Williams (1968a) and Lewis (1970a) considered that the Dulas valley had been invaded by ice from the Wye valley, but the lithology of the deposits exposed in the Tregunter ridge suggests that they interpreted the morphological evidence incorrectly. The Tregunter dead-ice ridge and the kame and kettle deposits near Felin-newydd probably mark the terminus of an ice advance from the Brecon (Usk valley) area.

Fossiliferous Silurian erratics in diamict, exposed in 2002 in excavations for the erection of a farm building at Porthamel (SO 159350), near Talgarth in the Llynfi valley (Fig. 8.2), (to which the Dulas is a tributary), indicate that ice from mid-Wales invaded the Llynfi downstream of Bronllys, although the correlation between that ice and the ice responsible for the Tregunter/Felin-newydd deposits has not been established. Neither have the latter deposits been correlated with any of the ice-marginal stages in the Usk valley.

Ellis-Gruffydd (1972) mapped seven 'moraines and associated pro-glacial and ice-contact forms' in the Usk valley upstream of Brecon (Fig. 8.10). He believed that they relate to still-stands superimposed upon the recession upvalley of the Usk glacier, or possibly to 'slight readvances superimposed upon a general retreat.'

The nearest of these features to Brecon is the area of kame topography on which the golf course was built at Cradoc (the Cradoc Moraine, I and II). Excavations associated with construction of the golf course showed that the glacial deposits at that site contained Old Red Sandstone clasts, such as might have come from the upper Usk valley and from the southerly tributaries of that valley that drain the high ground of the Fforest Fawr and the Bannau, on the border with Dyfed. No Silurian or other clasts from the Eppynt or other areas north of Brecon were identified. This indicates that the kame topography formed in association with the Usk valley glacier as ice from the upper Usk pushed into the Cradoc-Penoyre col north of Brecon, where it stagnated (Lewis, 1970b). The kame topography is at an altitude of some 220 m and lies about 70 m above the present floor of the Usk valley at Aberysgir, indicating that the Usk glacier was at least 70 m thick at the time that the kame sediments were deposited at Cradoc.

Ellis-Gruffydd (1972) considered that a terrace at Fenni-Fâch (SO 020289) also related to the main Usk valley glacier at the time that the Cradoc kames were forming. Subsequent ice-marginal positions in the upper Usk valley, he believed (Fig. 8.10), were at Aberysgir (SO 000297), Aberlliw (SN 954295), Cwm wysg-ganol, Cwm wysg-uchaf and Pant Ysgallog, at all of which (except for Cwm wysg-ganol) kame and kame terrace topography exists. Lewis (1970b) showed the Cwm wysg-ganol and -uchaf features as one and the same moraine and stated that lateral deposits of this moraine were traceable to both the Senni and Cilieni valleys, on either flank of the Usk. Ellis-Gruffydd (1972) maintained that subsequent to the formation of the Pant Ysgallog moraine glacial ice disappeared from the Usk valley west of the confluence of the Crai valley with that of the Usk. The implication of his statement is that the Usk glacier upvalley of Brecon was fed mainly by ice from the Fforest Fawr, which was limited to the Senni, Treweren and Crai valleys when the glaciers in those valleys were no longer able to reach the Usk.

Ellis-Gruffydd (1972) correlated cross-valley kame topography in the Senni, Treweren and Crai valleys, believing that those features marked the same retreat stages in each valley. In the Senni valley he identified former glacier still-stand positions at Abersenni uchaf (SN 932262); Tyle Brychgoed-Bailea (where kame topography does not exist; SN 925245); and at Heol Senni (SN 925234), where the hamlet is situated on the kame-moraine (Fig. 8.13). Only one still-stand position is identifiable in the adjacent valley of Cwm Treweren: at Tre'r Domen (SN 916267) where kame topography exists on the floor and lower sides of the valley. Road works on the eastern side of the valley formerly exposed sections in this sediment, which is of local provenance and derived from areas upvalley of the moraine. Ellis-Gruffydd (1972) identified three glacier still-stand positions in Cwm Crai: at Rhyd-Owen-isaf (SN 895281); Tir-y-capel and Maes Carnog. At all three sites kame and kame terrace topography exists. Lewis (1970a) also recorded a still-stand position near Blaenau-isaf (SN 845258) in the Hydfer valley, west of Cwm Crai, although Ellis-Gruffydd (1972) wrote that this does 'not exist.'

Both Lewis (1970a,b) and Ellis Gruffydd (1972,1977) agreed that the Usk glacier above Brecon was fed by ice from the Fforest Fawr. Lewis (1970b) also remarked on the morainic topography at Pentre'r-felin (SN 915304) in the lower Cilieni valley, which is a left bank tributary that joins the Usk immediately downstream of Sennybridge, and on deposits at Pentre-bach (SN 906330) 'and odd patches of drift further upstream' in the Cilieni valley. He suggested that they might have been formed by mid-Wales ice

Fig. 8.13 The low ridge on the valley floor in the middle distance is a kame moraine, on which the hamlet of Heol Senni is located. This landform developed at a former ice front of the glacier that occupied the upper portion of Cwm Senni towards the end of the Late Devensian. The view looks down valley, towards the inner side of the kame moraine. The flattish area grazed by sheep in the foreground was covered by the glacier when the ice front lay at Heol Senni, as was the rest of the headward portion of Cwm Senni and the adjacent uplands. (Photo: C. A. Lewis)

from the Vale of Irfon breaching the col at the head of the Cilieni valley and thence passing into the Cilieni and possibly entering the Usk. Other morainic features in the valleys of the Eppynt indicate that they were formerly occupied by ice moving southwards towards the Usk, but it is not certain whether this northern ice contributed to the Usk glacier when it was restricted to the upper Usk valley.

During the decay of ice from the Senni valley, that had extended as a piedmont glacier onto the Mynydd Illtyd plateau on the northern side of the Fforest Fawr, hummocky kame-like topography was formed on that plateau. This suggests that much of the piedmont ice decayed in situ, as dead (motionless) ice. Depressions within the hummocky terrain still contain shallow pools of water and extensive organic deposits, such as that of Traeth Mawr (SN 968255) which has been subjected to palynological analyses (J. J. Moore, 1970; Walker, 1982) and radio-carbon dating (Walker, 1980).

Deglaciation and subsequent cirque glaciation
Deglaciation

Evidence from four sites within the Wye and Usk valleys, and their immediate surroundings, indicates when those regions were deglaciated. North of Hay-on-Wye, at Rhos goch common (SO 195484; Fig.8.1) in the Bach Howey valley, an ill-drained area contains organic deposits that began to form in the Late Glacial (Bartley,1960). These deposits rest on what Bartley (1960) considered to be glacial till. They include pollen indicative of a Late Glacial interstadial vegetation. Although no absolute dates have been obtained from Rhos goch, glacial ice must have melted prior to a Late Glacial interstadial at that site.

Llyn Mire (Cors y Llyn, SO 015553) is located about 2 km south of Newbridge-on-Wye and contains organic deposits that P. D. Moore (1978) believed accumulated in 'a kettle hole in the till.' They lie about 1 km north-east of the kame-like mounds and the esker-like feature already described from near Penmincae, and are probably part of the same ice-marginal topography, marking a still-stand or minor advance in the overall retreat of the Wye glacier. The lower organic sediments at Llyn Mire evidence a typical Late Glacial vegetational sequence that includes an interstadial. Although no absolute dates have been obtained, the site must have been free of glacier ice before that Late Glacial interstadial.

On the southern side of the Usk valley, at Traeth Mawr (SN 968255; Fig. 8.10) on Mynydd Illtyd (as already described), a thickness of over 627 cm of organic deposits exists within a dead-ice depression. J. J. Moore (1970) showed that they evidence a Late Glacial vegetation sequence, passing up into Flandrian (Holocene) organics. Walker (1980, 1982) re-examined the organic deposits. He concluded that the basal organic deposits are 'likely to represent a pioneer stage in vegetational succession on to bare and unstable substrata that had recently been vacated by Late Devensian ice.' Above them, stratigraphically, lies palynological evidence of a Late Glacial interstadial that was followed by stadial conditions.

Walker (1980) obtained a radiocarbon date that calibrates to 13,822 BP (within a one sigma range of 14,006–13,637 BP; Fuls, 2003) for the base of the Late Glacial interstadial deposits, indicating that the site had been deglaciated before that date. (The remaining dates shown in this chapter have been calibrated, unless stated otherwise). Walker considered that the uncalibrated date (11,160 +/- 140 BP) was 'over 1,000 yr younger than age determinations from comparable Late-Glacial Interstadial deposits in Scotland, the Lake District, North Wales and Cornwall', possibly due to the incorporation of younger carbon into the sediment that was dated.

During the 1960s Lewis and J. J. Moore cored a basin at Bryniau Gleision (SO 086155; Fig. 8.6), at an altitude of 475 m on the plateau above the head of Dyffryn Crawnon, which is a right bank tributary of the Usk. Moore identified palynological evidence of at least one Late Glacial interstadial complex in the deposits (Lewis, 1970a), but never published his findings. The site was re-investigated by Robertson (1988), who confirmed that a Late Glacial interstadial is represented at that site.

Robertson (1988) obtained a radiocarbon date that calibrates to 20,023 BP (within a one sigma range of 20,598-19,448 BP; Fuls, 2003) for the base of the interstadial, which suggests that the Bryniau Gleision plateau was ice free by that time. He considered the date to be too old, due to the organic sediments from which it was derived being from a limestone area and consequently subject to the effects of 'hard-water' due to uptake by plants of carbonate-rich water, resulting in errors in radiocarbon dating (Lowe and Walker, 1997). Robertson suggested, on the basis of age determinations from similar pollen horizons elsewhere in Britain, that a date of 13,000–14,000 would be more probable. Whether the radiocarbon date or Robertson's suggested age-range is correct is unknown, but the uplands of Bryniau Gleision were certainly deglaciated before the Late Glacial Interstadial evidenced there, probably by 14,000 BP and possibly as early as 20,000 BP.

The evidence from the four sites so far investigated in the Wye/Usk region indicates that the landscape was deglaciated before 14,000 BP. Deglaciation may even have occurred as early as 20,000 BP on the limestone plateaux south of the middle Usk, if the suspect evidence from Bryniau Gleision is correct. The morphologically 'fresh' looking glacial landforms in the region suggest that the last ice-sheet/valley glacier glaciation of the area was Late Devensian in age. No sediments have yet been found within the area covered by this chapter to indicate the maximum age of that glaciation or of any earlier glaciation of the region.

Subsequent cirque glaciation

Mounds of unconsolidated material near the heads of valleys such as Cwm Crew and Cwm Oergwm in the Brecon Beacons, and essentially linear ridges such as those under Fan Fawr in the Fforest Fawr and Fan Hir near the borders of the Fforest Fawr and Black Mountain, bear witness to cirque glaciation subsequent to ice-sheet decay in those uplands (Fig. 8.10). Evidence of cirque glaciation north of the Old Red Sandstone escarpment, but within the catchments of the Usk and Wye, has not yet been presented.

Reade (1894) was the first to draw attention to the cirque moraines of the area, in a paper on Cwm Llwch in the Brecon Beacons (Fig. 8.14). Subsequently Robertson (1933) suggested that some of the

Fig. 8.14 Llyn Cwm Llwch, Brecon Beacons. Notice the moraine, which encloses the lake, and that the lake lies in the most shaded (and hence the coldest) portion of the valley-head, where there is least incoming solar radiation. The cirque glacier responsible for deposition of the moraine, and for cutting the over- deepened basin occupied by the lake, lay between the moraine and the shadowed area of the backwall. Reade (1894-5) wrote that '... the sun has traced out this moraine, and settled its alignment and position in the larger Cwm or valley'. (Photo: C. A. Lewis)

debris accumulations in the mountains north of Merthyr Tydfil were glacial moraines while others had accumulated at the foot of large snow patches (as protalus ramparts).

The cirque moraines and protalus ramparts of the Brecon Beacons were mapped by Lewis (1966, 1970a,b). Ellis Gruffydd (1972, 1977) remapped the area and extended coverage westwards to the Fforest Fawr and Mynydd Du. Robertson (1988) examined part of the area already covered but also added new information from the Black Mountains. Shakesby (1992, 2002) presented an overall synthesis of the information amassed by previous workers. He also published, with Matthews, two detailed studies of restricted areas in the Brecon Beacons/Fforest Fawr (1993, 1996). In 2001 Carr presented a glaciological approach for the discrimination of small ridge systems in the Brecon Beacons.

Fig. 8.10 depicts the location of depositional ridges and mounds in the Brecon Beacons/Fforest Fawr/ Black Mountain uplands south of the Usk valley. In such valley heads as those of Cwm Cynwyn and Cwm Oergwm there are a number of arcuate ridges on the valley floors that have been interpreted as glacial moraines deposited by cirque glaciers. There is also a small depositional ridge on the southwestern side of the valley head at Cwm Oergwm that may well have formed as a protalus rampart.

The Late Glacial, as evidenced by deposits within the Traeth Mawr basin (Walker, 1980, 1982), consisted of an early phase in which the ground was exposed by deglaciation, an interstadial beginning by about 14,000 BP and ending around 12,450 BP in which there were a number of climatic oscillations, and then a stadial in which mainly inorganic sediments were deposited. The stadial was terminated by Post-Glacial, Holocene, conditions, that began at about 11,300 BP. This Younger Dryas Stadial is well-known in Britain and Shakesby and Matthews (1993), in relation to their work within the Brecon Beacons National Park, refer to it as the Loch Lomond Stadial.

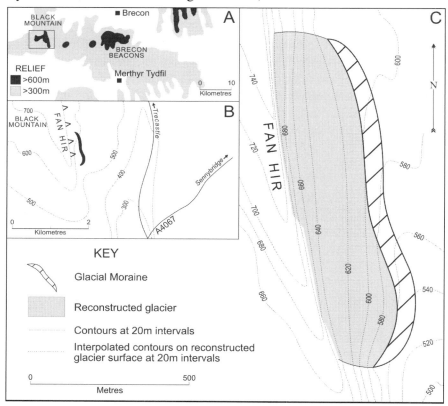

Fig. 8.15 Reconstruction of the former cirque glacier at Fan Hir, Black Mountain. Notice how the glacier was located in the area shaded from the afternoon sun by the uplands of Fan Hir, off which snow was probably blown to accumulate on the glacier surface. (Redrawn after Shakesby and Matthews, 1993)

Within the mountains, south of Traeth Mawr, Walker (1980) obtained radiocarbon dates from the lowest organic sediments within the basin that is enclosed by the cirque moraine at the foot of Craig y Fro (SN 973207), in Glyn Tarell. He also obtained radiocarbon dates for the equivalent sediments in Cwm Cerrig-gleisiad. In both cases Walker believed that organic sedimentation began in the Holocene, and dates to about 11,400 BP at Craig y Fro and to 12,769 BP at Cwm Cerrig-gleisiad. (As already stated, these are calibrated dates, according to the calculations of Fuls, 2003). Walker therefore concluded that both basins were occupied by glaciers during the Loch Lomond Stadial, as Lewis (1966, 1970a, b) had suggested. The calibrated date from Cwm Cerrig gleisiad is puzzlingly old and apparently pre-dates the Younger Dryas/Loch Lomond Stadial.

Lewis (1966, 1970a, b) suggested that some of the cirques, such as Cwm Cerrig-gleisiad, where there is an outer arc of indistinct ridges, might have harboured cirque glaciers at an earlier stage of the Late Glacial as well as during the Loch Lomond Stadial. Shakesby and Matthews (1996), however, showed that the outer ridges at Cwm Cerrig-gleisiad owe their origin to landsliding associated with collapse of part of the backwall of the cirque. They believed that the steep-sloped ridges close to the southern backwall of the cirque, and within the limits of the less obvious outer ridges, were of glacial origin and formed during the Loch Lomond Stadial.

At Fan Hir (SN 834198), on the extreme west of the Fforest Fawr, there is a ridge that is circa 1.2 km long and that is separated from the escarpment of the Black Mountain that backs it to the west by a 'gully-like hollow' (Shakesby and Matthews, 1993). The ridge is aligned essentially parallel to the escarpment and varies in height from 25 m to less than 3 m at its northern extremity. It has proximal and distal slopes of up to 32°. Towards its southern extremity the ridge curves towards the escarpment (Fig. 8.15). Some of the clasts within the ridge are striated, indicating that they have been transported in a glacier system. Shakesby and Mathews (1993) therefore believe that the ridge formed as a moraine in association with a small glacier that was located beneath the escarpment during the Loch Lomond Stadial. Their reconstruction of this palaeoglacier depicts it as covering an area of 0.28 km^2 and having an equilibrium line altitude (ELA) of 623 m. (The ELA is the dividing line between the accumulation and the ablation area on a glacier).

Robertson (1988) showed that, during the Loch Lomond Stadial, one small glacier existed in the Black Mountains in addition to those that had already been identified on the Brecon Beacons, Fforest Fawr and Black Mountain. This was at Tarren yr Esgob, where there is a moraine below the east-facing cliffs on the west side of the Vale of Ewyas near Capel-y-ffin (SO 240315). The glacier covered an area of 0.07 km^2 and had an ELA of 443 m. The other Loch Lomond Stadial glaciers of the region varied in size from that of Tarren yr Esgob to that of Craig Cerrig-gleisiad, which covered 0.38 km^2, while the glacier of Llyn y Fan Fach was almost as large as that of Cerrig-gleisiad. The average glacier size was 0.17 km^2, compared with an average of 0.5 km^2 in Snowdonia (Gray, 1982) and 0.85 km^2 for the Lake District of England (Sissons, 1980). Obviously, therefore, environmental conditions were only just severe enough for the existence of small cirque and valley-side glaciers in the Usk/Wye uplands region during the Loch Lomond Stadial.

Outside the limits of the Loch Lomond Stadial glaciers there was active reworking by periglacial processes of the already deposited glacial and other unconsolidated sediments. This produced stepped slopes, with flights of solifluction terraces on them, as between Fan Gyhirych and the A4067, south of Crai (SN 895210). On steep slopes under the escarpment of the Brecon Beacons, Black Mountain and elsewhere, talus accumulated. The absence of ice-wedge casts and pingo remnants, however, suggests that permafrost was uncommon in the area during the Loch Lomond Stadial, and that the glaciers and snow beds that existed in the region may have owed their existence largely to snow-blow rather than to severely cold climatic conditions.

Conclusion

Mid-Wales was probably glaciated on at least two occasions during the Pleistocene, but evidence for the earlier glaciation is forthcoming only for areas outside the limits of this chapter. The glacial sediments and landforms discussed in this chapter appear to be of Late Devensian age. The scale of glaciation during the

Late Devensian varied from ice-sheet to cirque glacier in size. During deglaciation the mid-Wales ice-sheet shrank to become two major valley glaciers, those of the Wye and the Usk. The oscillations of those glaciers are imperfectly known, as are the limits of mid-Wales and Brecon Beacons/Fforest Fawr sourced ice. The age of deglaciation, especially in the Wye drainage basin, is a matter for debate and much absolute dating is still needed. There is therefore much scope for research in the region covered by the Usk and Wye rivers and their tributaries.

9 Herefordshire

by Andrew Richards

Introduction

Herefordshire may well have been glaciated during a large number of Cold Stages of the Quaternary. However, geomorphological and geological evidence documents the incursion of ice-masses during only two of these phases. The first of these advances occurred during the Anglian Cold Stage, Oxygen Isotope Stage 12, around 430,000 years before the present and in the Middle Pleistocene (Richards, 1998; 1999; Coope *et al.* 2002). The latter phase of glaciation occurred during the Late Devensian, Dimlington Stadial (Oxygen Isotope Stage 2), approximately 18,000 years before the present (Brandon, 1989).

Non-glacigenic superficial deposits that occur west of the Anglian limit and east of the Late Devensian limit record the development of major Cold Stage fluvial systems. Terrace deposits in the lower Lugg, Teme and Wye Valleys record at least 4 stages of gravel aggradation during post Anglian/pre Late Devensian time. Much of this gravel is likely to have been provided by frost weathering and high run-off rates that existed in periglacial environments during Oxygen Isotope Stages 10, 8, 6 and 4. However, it is likely that the northerly-derived drainage pattern was also supplemented by meltwater derived from ice-sheets and glaciers that occupied the Welsh Massif during some periods of these cold stages.

Glaciation and landscape development in Herefordshire

Prior to Middle Pleistocene glaciation the River Lugg was the major drainage route in central Herefordshire (Richards, 1994; 1998; 1999). In eastern Herefordshire another large system, the Mathon River (Barclay *et al.* 1992), flowed parallel to the Lugg and then swung southwards along the western margins of the Malvern Hills via the Cradley Brook Valley (Fig. 9.1). At this point, the River Wye was probably a tributary of the River Lugg (Hey, 1991). The gravels that were deposited by these northerly-derived river systems and their tributaries are the oldest Pleistocene deposits in Herefordshire and have been classified as the Mathon Formation (Table 9.1; Richards, 1994; 1998; Maddy, 1999). These gravels were reworked as an ice-sheet of Middle Pleistocene age extended from Wales into eastern Herefordshire. Following this glaciation, the River Lugg maintained a southerly flowing course similar to its preglacial configuration, while the Mathon River swung eastward, through a gap in the Malvern Hills to join the River Severn at Worcester, thereby forming the present course of the River Teme (Fig. 9.1b). Minor changes subsequently occurred in the organisation of tributaries of the River Lugg and Teme on the St. Maughans Plateau, between Leominster and Bromyard, largely as a result of the release of meltwater from lakes trapped at the margins of the retreating Middle Pleistocene ice-sheet.

The lithological composition of gravel terraces within the Lugg and Wye Valleys suggests that this northerly-derived drainage network persisted for around 400,000 years, until disrupted by Late Devensian glaciation. The most striking effect of Late Devensian glaciation was probably the emergence of the River Wye as the major drainage route of the region (Fig 9.1c). The catchment of the River Lugg decreased as the Late Devensian ice-sheet diverted the Rivers Teme and Onny eastwards towards the Severn (Pocock,

1925; Dwerryhouse and Miller, 1930; Cross and Hodgson, 1975), while the Wye catchment area expanded. While the upper parts of the catchment of the River Lugg was subject to glaciation, its lower course largely remained marginal to the piedmont ice-lobe occupying the Hereford Basin. Deglaciation patterns suggest that ice occupied the middle and upper reaches of the Wye Valley for some time after its retreat from the Hereford Basin. Meltwater from the retreating ice was channelled down the valley during the latter stages of the Late Devensian (Lewis and Thomas, this volume).

Pre- Devensian glacial deposits in Herefordshire

The stratigraphy of the glacial deposits of Herefordshire and their relation to the river terraces of the Lugg and Wye valleys can be seen in Table 9.1. Older drift deposits occur east of the Late Devensian limit in the Hereford Basin. These deposits, formerly mapped by Luckman (1970) and Brandon (1989), consist of two lithostratigraphic units (Richards, 1998; 1999). The older units, the Humber Formation and Mathon Formation, represent the remnants of gravels laid down by a Cold Stage fluvial network that existed in eastern Herefordshire during the Middle

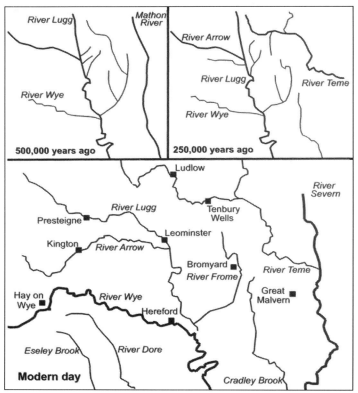

Fig. 9.1 Middle to Late Pleistocene River system development and glacial diversion in Herefordshire.
a) Principal drainage routes prior to Anglian glaciation.
b) Drainage routes following Anglian glaciation.
c) Drainage pattern subsequent to Late Devensian glaciation

Pleistocene before Anglian Stage glaciation (Richards, 1998). The composition of these units suggests that the river network flowed from north to south and included a precursor to the modern River Lugg.

The Humber and Mathon Formations are overlain by the Risbury Formation. This unit is composed of a series of geographically distinct remnants of glacial deposits that record the advance and retreat of a Middle Pleistocene ice-sheet across the Hereford Basin as far as the Malvern Hills (Richards, 1997; 1998; 1999; Coope *et al.* 2002). There has been debate regarding the provenance of the ice-sheet that deposited the Risbury Formation. Brandon (1989) believes, on the basis that the deposits contain erratics derived from the English Midlands (such as Bunter quartzite), that the ice-sheet advanced into Herefordshire from the north. However, Richards (1997; 1998; 1999) considers that the rounded and sub-rounded, northerly derived clasts that are present in the glacial deposits were reworked from the Humber Formation. Further lines to this argument lie in the geographical distribution and structural characteristics of the sediments. Both are more plausibly explained by ice advance from the west and south-west.

The glacial deposits of the Risbury Formation represent phases during the retreat of the Middle Pleistocene ice-sheet from its easterly extension in the Cradley Brook valley, immediately west of the Malvern Hills (Fig. 9.2). The deposits have been extensively eroded by later events but the sedimentology and structural geology of the remnants have allowed the reconstruction of glacial environments associated with the retreat of the ice-sheet. The Members of the Risbury Formation are shown on Table 9.1.

The Coddington and Whitehouse Members

The most easterly of these glacial deposits are the most informative in terms of relating the deposits to the Pleistocene history of the area. In the Cradley Brook Valley, pre-existing Cold Stage fluvial deposits are overlain by the Risbury Formation (Barclay *et al.*, 1992; Coope *et al.*, 2002; Richards, 1994; 1998, 1999). While the Mathon Formation represent the development of a large, braided, Cold Stage river system, a single thread fluvial channel that underlies the coarse gravels was exposed in a gravel pit at Mathon in 1992. Silts deposited by this river contain plant micro- and macro-fossils and beetle remains that suggest a relatively warm, Middle Pleistocene environment, probably part of the Cromerian Complex (Coope *et al.*, 2002). In addition, the glacial deposits of the Risbury Formation are locally overlain by organic deposits that are believed to be of late Anglian- early Hoxnian age (Bonny, 1992). Thus, the Risbury Formation appears to relate to glaciation during a Middle Pleistocene Cold Stage, probably the Anglian (Coope et al., 2002; Richards, 1998; 1999).

The northerly-derived river system in the Cradley Brook valley became increasingly influenced by the Middle Pleistocene ice-sheet

Western Herefordshire		Eastern Herefordshire
	Alluvium	
	Peat	
	Sub-alluvial gravel	
Herefordshire Formation		**Cradley Valley Formation**
Wye Valley Formation	**Lugg Valley Formation**	Ham Green Member
1st Terrace	Marden Member	
2nd Terrace	Moreton on Lugg member	
3rd Terrace	Kingsfield Member	Colwall Member
4th Terrace	Sutton Walls Member	Cradley Silt Member
	Starpit Formation	
	Risbury Formation	
Portway Member		
Franklands Gate Member		
Stoke Prior Member		Limbury Member
Stoke Lacy Member		Whitehouse Member
Newton Farm Member	Kyre Brook Valley Member	Coddington Member
Humber Formation		**Mathon Formation**

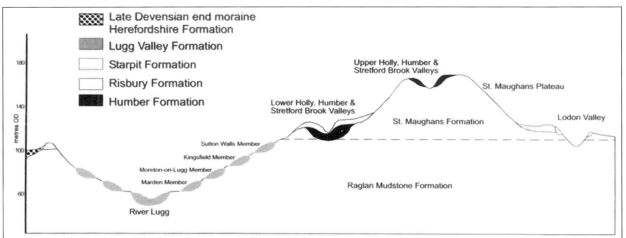

Table 9.1 *The Middle to Late Pleistocene stratigraphy of Herefordshire including schematic representation of the distribution of the units discussed in this chapter. The table includes the relative position of the units, without reference to inferred dates. Suggestions regarding the probable age of the units are given elsewhere in the chapter*

as the ice extended from the west. Meltwater and debris flows emanating from the ice-sheet flowed into the river and eventually, the river system became dammed by the ice-sheet. Subsequently, a glacial lake accumulated in the lower Cradley Brook Valley as evidenced by the laminated silts and clays of the Whitehouse Member at Mathon Pit (SO 734449). This lake drained as the Middle Pleistocene ice-sheet retreated from the area.

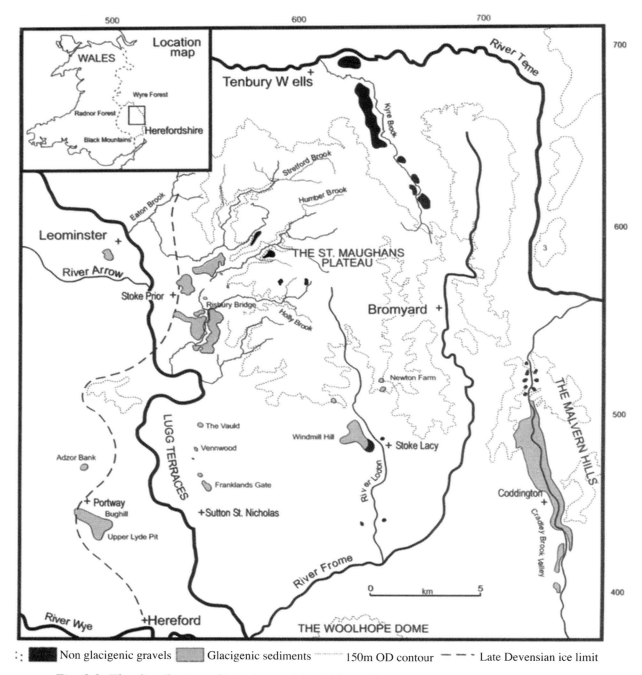

Fig. 9.2 *The distribution of Members of the Risbury Formation. No remnants occur on the St.Maughans plateau and structural characteristics suggest the depositing ice-mass moved east, from the Hereford Basin*

Newton Farm and Kyre Brook Valley Members

At its maximum, the Middle Pleistocene ice-sheet extended into eastern Herefordshire and was bordered at its northern margins by a plateau formed by the St Maughans Formation of the Lower Old Red Sandstone Formation. In the Bromyard area, glacial deposits laid down by the ice-sheet record the development, and subsequent catastrophic drainage of Glacial Lake Bromyard (Richards, 1999). South-west of Bromyard, at Newton Farm (SO630519), a now disused gravel pit exposes large delta foreset bedding. Immediately west of the gravel remnant, there is an ice-contact ridge, that represents the position of the ice-sheet, and a meltwater conduit that fed the glacio-lacustrine water body (Fig. 9.3). The coarse gravel foresets at Newton Farm are underlain by a lower, horizontally bedded, passively-deformed sand unit. These sands accumulated at the bottom of the glacial

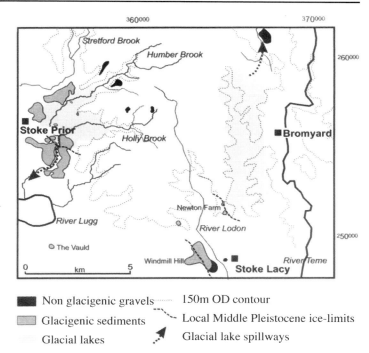

Fig. 9.3 Ice-marginal environments associated with Glacial Lake Bromyard, and Glacial Lake Humber

lake. Diamicton pods that are found within the sands are likely to represent debris carried into the lake by small ice-bergs that had calved from the melting ice-sheet. Glacial Lake Bromyard appears to have filled to a level of 165 m, before overspilling, via the col at Collington, into the Kyre Brook Valley (Fig. 9.3). Before this event, the valley contained gravels with lithologies: Millstone Grit from Clee Hill and low grade metamorphic clasts from around Church Stretton; that were carried by an ancient river system that flowed from the north (probably as part of the Humber Formation drainage network). These gravels were reworked into large fluvial bedforms, known as 'mega-ripples' by the catastrophic drainage of Glacial Lake Bromyard as the torrent flowed northwards (Richards, 1997; 1999).

The Stoke Lacy and Stoke Prior Members

Following a further phase of retreat, the Middle Pleistocene ice-sheet readvanced against a bedrock ridge at Stoke Lacy, deforming meltwater gravels laid down at the margins of the ice-sheet into an ice-pushed ridge. Trial pits at Crossways (SO 610500), and exposures seen by Luckman (1970) in a now infilled gravel pit, revealed coarse gravels, contorted into large 'open' folds by pressure from the ice-sheet. This event appears to have been short-lived and further retreat occurred as the ice-margin stagnated, depositing melt-out and flow tills that are exposed in a small pit at Windmill Hill (SO 506487), 500 metres west of the Stoke Lacy ice-pushed ridge.

During deglaciation, meltwaters also became dammed in the westernmost parts of the St. Maughans Plateau. Glacial Lake Humber was dammed by the Middle Pleistocene ice-sheet to the west and south, and by the high ground rising to the east. The main source of meltwater was a conduit on the margins of the Lugg Valley at Stoke Prior. Two small gravel pits, near Stoke Prior School (SO 527567) and at Blackwardine (SO 530564), expose large gravel foresets that formed as a coarse gravel, fan-delta advanced, or prograded, into the rapidly filling glacial lake. Higher areas to the east are overlain by subglacial tills that represent the maximum extent of the ice-sheet in the area prior to minor retreat and the development of the ice-dammed lake. Meltwater submerged small river valleys formed by precursors to the Stretford, Humber and Holly Brooks until the water level reached a critical level and the damming ice-margin became unstable. Ice immediately overlying the Holly Brook Valley at Risbury Bridge (SO 539549) then became buoyant and lost contact with underlying rock and glacial deposits. As a result, the impounded water drained back underneath the ice-sheet, eroding a steeply incised gorge through glacial deposits, pre-glacial river gravels of the Humber Formation and a further 3 metres into the underlying bedrock of the St. Maughans Formation. Today, the landscape still bears the impressions of this event. Below Risbury Bridge, the modern Holly Brook gently trickles through an impressive gorge cut by the meltwater torrent. Above Risbury Bridge a basin structure marks the position of a scour pool that was eroded by huge eddies that developed before Glacial Lake Humber drained rapidly through the subglacial tunnel (Fig. 9.3).

The Franklands Gate Member

The lower Lugg Valley, above its confluence with the River Wye near Hereford, contains an impressive incised meander near Bodenham and four terrace levels that are underlain by fluvial gravel. These gravels accumulated during at least four phases of aggradation, mainly during Cold Stages. The terrace gravels contain far-travelled rock types that suggest that, during the Middle Pleistocene, the River Lugg catchment was larger than at present. The Lugg system is probably one of the oldest in the whole of the Welsh Borderland. The terraces of the River Lugg represent fluvial development in the period between Middle Pleistocene and Late Pleistocene glaciation.

Just north of Sutton St. Nicholas (Fig. 9.2), a series of hills mark the position of remnants of the Risbury Formation. Exposure in these deposits at The Vauld (SO 534494), Vennwood (SO 543488) and Franklands Gate (SO 548464) reveals very coarse, often poorly-sorted gravels contained within a bedrock channel. The size of gravel clasts and sedimentary structures suggest that they were transported by very high flow rates. The bedrock channel has a basal height of ~110 m at Norton Court (SO 538495), falling to ~85 m at Franklands Gate. The main channel runs north-north-east to south-south-west with a gradient of ~1:100. Subsidiary, small channels run towards the west and south-west at steeper gradients (1:10—1:33; Fig. 9.4). At Franklands Gate, old gravel pit faces expose beds of very coarse gravel, sand and silt that have been deformed. Gravel units are inclined at 90° in the west with two, large-scale asymmetric

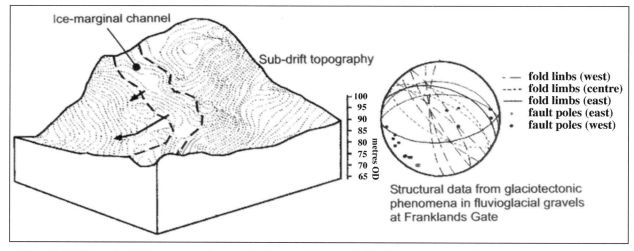

Fig. 9.4 *The sub-drift topography and characteristics of the glaciotectonic forms in the glaciofluvial gravels at Franklands Gate. The Franklands Gate Member represents an ice-marginal channel, incised into the Raglan Mudstone, that was partially supported by the Middle Pleistocene ice-margin. Subsequent oscillation deformed the glaciofluvial gravels into large folds*

folds evident in the northern section of the pit. Fold amplitude decreases and wavelength increases towards the eastern part of the pit where gravels are horizontally bedded, conforming to the underlying bedrock channel. Sand and silt beds within the pit show a variety of ductile and brittle deformation structures; small scale normal faulting, and larger scale thrust faulting associated with recumbent folds and thrust nappes.

A number of lines of evidence suggest that the channelised gravels in the Sutton St Nicholas area formed at the margins of the Middle Pleistocene ice-sheet. The main channel is joined by steeply inclined channels that probably represent chutes where ice-marginal meltwater became deflected westward, and into confinement beneath the margins of the ice-sheet. Eastern portions of the main channel are supported by a bench in the Raglan Mudstone while the western margin is likely to have been supported by ice. The evidence suggests that the oscillating ice-margin did not override the ridge and ice-movements associated with the post-depositional deformation exhibited in deposits at Franklands Gate must have been small and short-lived.

The Portway Member

The final phase of Middle Pleistocene glacial deposition recorded by units of the Risbury Formation occurs near Portway on the western margins of the Lugg Valley. At Burghill Pit (SO 48204511), Adzor Bank (SO 48234782), Burlton Court (SO 48844478) and Upper Lyde Pit (SO 49304470), two coarse gravel units and an intervening sand, silt and clay unit are exposed (Brandon and Haines, 1981; Brandon, 1989).

The lower gravels exposed at Upper Lyde (Fig. 9.2) represent braided glaciofluvial deposition with increasing proximity to an ice-sheet source. The gravels represent a sand∂ur environment with features illustrative of minor channel and bar growth and bar migration during flood stages, before increasingly ice-proximal deposition, alternation between low-water and high-water deposition under a higher flow regime and the deposition of large boulders during periods of ice-rafting. Retreat of the ice-front, or a decrease in discharge rates, resulted in the deposition of overlying finer gravel and sand beds.

The intervening sand, silt and clay units contain primary sedimentary structures and dropstones that suggest a glaciolacustrine origin. Exposures in a sand pit at Burghill (SO 48204511) suggest that initial ponding at the ice-margin was followed by readvance and the excavation and deformation of the glaciolacustrine silts and clays (Fig. 9.5). Diamicton lenses containing primary sedimentary structures and dropstones were excavated and redeposited as glacio-erosional rafts (*cf* Aber and Ruszczynska-Szenajch, 1997) while the silts and clays were deformed into large sheath-folds, similar to those recorded in exposures within the Dammer-Berge Push Moraine (Van der Wateren, 1987; 1992). Large-

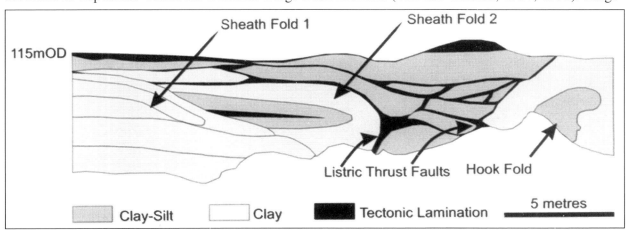

Fig. 9.5 Schematic diagram of the deposits exposed during a landfill operation at Burghill. Glaciolacustrine silts and clays have been deformed into a series of large sheath folds

scale, horizontal to subhorizontal joint sets cut through the folds, suggesting that the folds either formed subglacially, or were deformed by renewed oscillation of the ice-front soon after formation. The fact that the deformed glaciolacustrine silts and clays contain primary sedimentary structures suggests that the sediments were frozen throughout deformation (Richards, 1999).

Risbury Formation deposits at Portway (SO 485454) document a further oscillation in the Middle Pleistocene ice-margin. Following the deposition of a broad, braided outwash system which included debris transported within ice-rafts, the ice-sheet readvanced, depositing till at Adzor Bank, damming the northerly flowing meltwater to form a proglacial lake. The period of stagnation was shortlived and the fine grained glaciolacustrine sediments in the Burghill area were deformed by an oscillation of the ice-front. Sediment gravity flows were released and high volumes of meltwater from beneath the ice-sheet fed an outwash sandur that flowed in a southerly direction at the margins of the ice-sheet.

Devensian glacial deposits of the Hereford Basin
The youngest group of glacial deposits in the region has been formally designated as the Herefordshire Formation (Maddy, in Bowen, 1989). The Herefordshire Formation consists of a complex array of glacigenic deposits. The limits of this phase of glaciation are well marked, and it appears that a large piedmont ice-lobe, fed by a plateau ice-cap and a number of outlet glaciers in the Welsh uplands, expanded into the Hereford Basin as far as the current Lugg Valley (Fig. 9.6). The northern extent of this ice-mass terminated along the Silurian dipslope that runs from Kington to Orleton, while to the south, its limit can be traced from Whitfield to Hay-on-Wye, at the base of the Dittonian (Lower Old Red Sandstone) escarpment.

Subglacial tills
Tills derived from the Late Devensian ice-sheet are found throughout the Hereford Basin, in the lowlands between the Silurian and Dittonian escarpments. It is likely that much of this area is underlain by subglacial tills, probably formed by deformation of pre-existing Quaternary deposits, outwash gravels laid down in front of the advancing ice-mass and detritus from the easily eroded Raglan Mudstone bedrock. Till derived from subglacial deformation probably formed an extensive plateau throughout the Hereford Basin as the ice reached its limit near the present Lugg Valley (Fig. 9.6).

The depth of till varies markedly throughout the Hereford Basin, probably as a result of variations in the thickness of the overlying ice-mass, the configuration of the preglacial topography of the basin, variations in subglacial hydrological conditions and the occurrence of bands of more resistant bedrock.

Exposures reveal marked variability in till texture. The till often contains rotated, well-sorted gravel pods, sand lenses and beds and a variety of tectonic structures that suggest that it has been subjected to high subglacial shear stresses. Since deposition, the till sheet has been modified by incision and deposition by post-glacial river development. However, extensive till sheets still occur in some locations, frequently overlying, and often deforming, outwash gravels that were laid down as the ice-sheet advanced.

Pebbles that are found within the till include sandstones, siltstones and cornstones derived from the bedrock of the Hereford Basin. Further travelled clasts include Silurian siltstones, sandstones and shales, turbiditic greywackes, Ordovician volcanic tuffs, granodiorite, dolerite, gabbro and a variety of vein minerals. These far-travelled components outcrop at a number of locations throughout the Welsh Massif, suggesting that the ice-sheet moved from the west.

Clast fabrics recorded from subglacial tills and the probable configuration of the Late Devensian ice-sheet in the Hereford Basin can be seen in Fig. 9.8. Small patches of till occur at heights of up to 220 m on Dinmore Hill and Westhope Hill immediately west of the Lugg Valley. On the southern margins of the Hereford Basin, till and ice-marginal fluvioglacial deposits (see following section) occur at heights ranging between 210–225 m (SO 332432, SO 320442). The ice-sheet appears to have been thinner in the northern parts of the Hereford Basin as glacial deposits that overlie the Silurian escarpment do not occur above 160 m.

Morainic forms and deposits
Ice-marginal moraine ridges

Fig. 9.6 illustrates the distribution of a variety of morainic forms that occur in Herefordshire. The most continuous of these features occurs just west of Hereford and is believed to be a terminal moraine (Luckman, 1970; Brandon, 1989). The feature can be traced northward as irregular patches of till to the well-defined, arcuate moraine at Orleton (SO 500665) (Brandon, 1989). Post-depositional denudation has

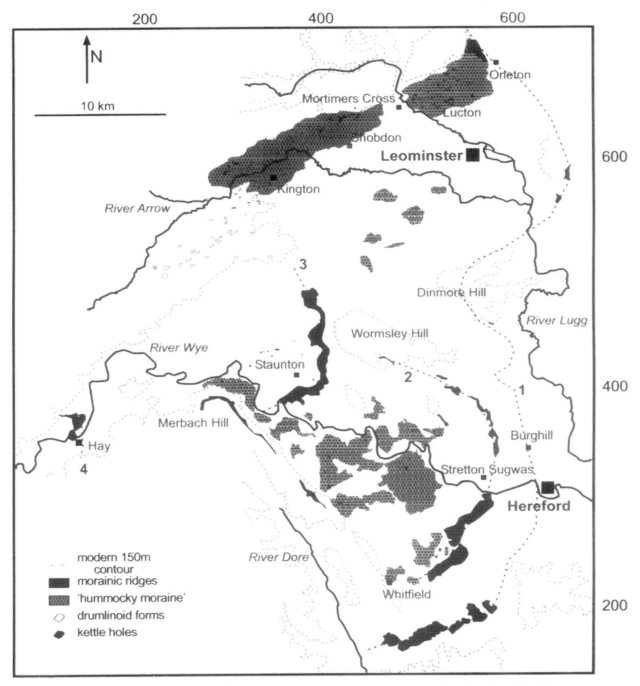

Fig. 9.6 The distribution of Late Devensian morainic forms within the Hereford Basin

had a large effect on the feature. It is most pronounced from Hereford Racecourse (SO 497410) to Burghill (SO 480449) where the moraine forms a single, irregular ridge. A group of smaller ridges near Tillington (SO 470465) and isolated patches along the flanks of Derndale Hill (SO 494490) and Stoke Prior (SO 526580) are also believed to be correlatives of this terminal moraine (Brandon, 1989). Very few exposures have been reported in this feature. However, Brandon and Hains (1981) report very poorly sorted till containing cobbles exposed in a pipeline trench (SO 48554463), and till with similar characteristics is exposed in a disused gravel pit at Burlton Court (SO 48504416). Here tills of the terminal moraine directly overlie older glacial deposits of the Risbury Formation (Anglian). Probable ice-marginal tills are found at 190–225 m along the Dittonian excarpment from Timberline Wood (SO 385375) to Merbach Hill (SO 305445). While these deposits have been extensively reworked by post-depositional periglacial activity, they probably represent remnants of lateral moraine deposited when the ice-sheet was at its maximum extent and are therefore coeval with the terminal moraine west of Hereford, which Lewis and Thomas (this volume) refer to as the Stretton Sugwas moraine.

The end-moraine west of Hereford marks a still-stand before retreat of the ice sheet. Two similar features occur near Staunton-on-Wye and near Hay-on-Wye/Clyro, which present further evidence of oscillation of the ice-margin on retreat from the Hereford Basin.

The Staunton-on-Wye moraine occurs as an irregular ridge that extends from Staunton (SO 365455), through Norton Canon (SO 380475) to Hyatt Sarnesfield (SO 380500). While the topography of the ridge is more obvious around Norton Canon, the southern parts of the ridge are overlain by 2–3 metres of hummocky, locally-kettled, deposits. The ridge is exposed at Brobury Scar (SO 353444) where two distinct units occur. The lower deposits consist of a deformed core of matrix-supported tills, silts, sands and gravel with localised hook and recumbent folds. The overlying unit consists of cobble-rich tills, coarse, crudely bedded gravels and interbedded sands that are dissected by normal faults.

An excavation into the Hay/Clyro moraine in 1989–90 exposed the upper 4 metres of part of the moraine (SO 219434). Sections revealed a lower, crudely-bedded diamicton containing lenses of sand and gravel. The water-lain deposits are traversed by a network of normal faults and are often draped, conformably, over large cobbles contained within the till. This possible melt-out unit is overlain by a complex of clast-rich diamicton units that interbed with undeformed gravel beds and lenses.

Hummocky Kettle-kame moraine
Between the defined ridges of the Hereford, Staunton and Hay moraines the glacial deposits of the Hereford Basin are characterised by extensive areas of hummocky, irregular topography with kettle holes. Augering at Bridge Sollers (SO 406422) and Turners Boat (SO 313459) in 1998 proved that many of the kettle holes contain silts and organic deposits to a depth of over 12 metres. An extensive area of hummocky topography also occurs below the Silurian escarpment west of the Orleton moraine, which Luckman (1970) named the 'Kington – Orleton Kettle Moraine'. Brandon (1989) has classified all of these deposits as kettle-kame moraine. This inference has marked implications regarding both the genesis of the deposits and the style of Late Devensian deglaciation in the Hereford Basin.

The morphology of the hummocky kettle-kame moraine is extremely variable. To the north of the region, in the areas north of Wormsley Hill (SO 424476) and Dinmore Hill (SO 589504) and south of the Kington-Orleton kettle moraine, moraines often occur as flat topped features. South of Wormsley Hill, the moraines are characterised by irregular mounds with intervening kettle-holes and a distinctive deranged drainage pattern (*e.g.* immediately south-west of Madley at SO 420392). The landforms also show marked compositional variability.

Kettle-kame moraine in the lower Arrow Valley appears to be dominated by well sorted, stratified and largely undeformed sand and gravel (*e.g.* Dilwyn, SO 421551; Bishopstone, SO 415443). These deposits are up to 25 metres in thickness and occur at heights between 85–87 m (Jackson and Sumbler, 1983). The kettled moraine that occurs between recessional moraines in the Wye Valley appears to consist of variable thicknesses of till and deformed silts, sands and gravels. At Bridge Sollers, undercutting by the River Wye (from SO 40654220 to SO 40774259) reveals kettled till directly overlying up to 7 metres of subglacial till, which in turn overlies sands and gravels that are dissected by normal, reverse and thrust faults. A similar sequence is represented downstream at Weir Cliff (SO 44594166 to 44634133).

Sections observed from 1989 to 1993 at Stretton Sugwas gravel pit (SO 447424) have revealed a consistent stratigraphic sequence. Up to 8 metres of subglacial till directly overlies the Raglan Mudstone bedrock. This till is overlain by approximately 6 metres of sand and gravel that has been deformed into a range of large simple and recumbent folds. Similar features are illustrated by Luckman (1970, plate 23). The deformed gravel unit is overlain by a till complex that is locally over 8 metres in thickness. The lower till beds within the unit are often densely consolidated and, in their lower portions, contain gravel pods and rafts that have been excavated from the underlying gravels. The till sequence rises into diamicton beds with increasing clast content and clast-support. These upper tills have been locally deformed and also contain locally deformed laminated silts, sands and gravels (Luckman, 1970; Brandon and Hains, 1981; Brandon, 1989). The sequence is completed by 6.5 metres of undeformed sand and gravel that extends to underlie the ridge immediately east of the gravel pit (SO 464523).

A further extensive section occurs upstream at Turners Boat (SO 313459). Here incision by the River Wye has exposed hummocky kettle moraine deposits that occur west of the Staunton recessional moraine. Up to 8 metres of well-sorted sands and gravels interbed with thin (0.2–0.6m) diamicton deposits. The diamict units are crudely bedded, contain subrounded to subangular clasts and often interdigitate with waterlain gravel units. Clast fabrics in the upper diamicton units suggest that the diamictons may represent gravitational reworking of pre-existing glacial deposits. While disturbed by soil formation and vegetation, the sequence appears to be capped by a deformed diamict, silt and sand complex, similar to the upper units exposed at Old Weir (SO 434419) and Brobury Scar (SO 345442).

While classified by Luckman (1970) as a single morphological unit, the Kington- Orleton kettle moraine is a feature of appreciable morphological and sedimentological variability. The moraine can be broken down into three regions; the first extending from the Upper Arrow Valley to Kington, an intermediate region from Kington to Mortimers Cross, and a final region from the River Lugg to the Orleton Moraine.

The moraine that extends from the Upper Arrow Valley to Kington comprises of drumlinoid features overlain by 2–3 metres of subglacial till. In many places, augering has proved that these features are partly rock-cored, while the smallest of the features are composed wholly of till, often containing very large boulders. These features are roughly aligned from south-west to north-east. The intermediate portion of the ridge is defined by an irregular, hummocky topography with common kettle holes. As in the Hereford region, although small, many of these kettle holes contain appreciable thicknesses of sediment. Augering at Stansbach (SO 365618) proved over 13 metres of silt, sand and organic sediments.

The intermediate region of the 'Kington-Orleton Moraine' appears to be dominated by a complex of tills, deformed sands and silt. Well-sorted sands and gravels are also found underlying the till complex in some areas. A gravel pit at Shobdon (SO 405615), since infilled, exposed up to 5.2 metres of undeformed sand and gravel underlying 3.5 metres of kettled till with interbedded, deformed, sand and silt.

The final part of the Kington-Orleton Moraine extends from Mortimers Cross to Orleton. As with the region in the Upper Arrow Valley, beyond Lucton (Fig. 9.7a), the ridge is defined by drumlinoid forms that are aligned roughly from south-west to north-east. However, here the drumlinoid hills appear to be entirel composed of gravel with minor medium to coarse sand beds. Till is rare in this portion of the ridge, occurring as a small kettled remnant at Oaker Wood (SO 465635). Near Mortimers Cross (SO 440633), the alignment of the features is slightly different, from south-east to north-west. The deposits are exposed by the River Lugg below Mortimers Cross (SO 442626 to 435628) and in a small pit at Lucton (SO 433638). The latter exposes 6 metres of clast-supported gravel with rare coarse sand beds and lenses. Bedding planes dip gently to the west at 8–10°, while imbricated clasts document flow towards the northeast (Fig. 9.7b).

Meltwater deposits and features

Deposits and features associated with glacial meltwater produced by the Late Devensian ice-sheet are found within and beyond the lateral and terminal moraines that mark the limits of glaciation during this stage. As discussed above, some of these deposits are closely associated with morainic features and/or have been deformed during the advance of the ice-sheet.

Fig. 9.7 a) The drumlinoid topography near Lucton. b) Exposure in fluvioglacial sediments in a drumlinoid feature at Lucton c) Fluvioglacial, possible esker, sediments exposed at Stretford. Line represents 2.4 metres

Within the Late Devensian limit, major zones of proglacial and subglacial meltwater activity are marked by the positions of current drainage routes in the Hereford Basin (Fig. 9.6). In the area north of Wormsley Hill extensive glaciofluvial gravel units underlie the alluvium of the River Arrow, Tippets Brook and Stretford Brook. South of Wormsley Hill, Brandon and Hains (1981) have mapped a channel within the underlying Raglan Mudstone that was incised and subsequently infilled with subglacial meltwater deposits. At Stretford a sequence of sub-parallel ridges, trending from south-east to north-west, are underlain by glaciofluvial deposits. A small pit (SO 439562) reveals 6 metres of well-bedded gravels that define a crude, coarsening-upward sequence. The gravel beds are inclined at angles of 35–40° from both sides into the centre of the ridge, and are traversed by a sequence of normal faults. The gravel sequence is overlain by a core of deformed gravel, sand, silt and coarse-grained diamict (Fig. 9.7c).

Glaciofluvial deposits also occur in gaps in the Silurian escarpment, north of the Kington Orelton Moraine at Byton (SO 370630) and Aymestrey (SO 430660). An extensive pit exploited the sand and gravel at Aymestrey from the mid 1960s to the late 1980s. The deposits here recorded a classic, tripartite sequence of a Gilbert Type delta, marking an input of water through the gap at Amestrey into a lake that occupied the Wigmore Basin, north of Aymestrey. The lower parts of the sequence are dominated by horizontally bedded, locally deformed sands, silts and clays that extend throughout the basin. In the Aymestrey pit, these deposits were overlain by coarse gravel foresets dipping to the north and north-east, which in turn were overlain by horizontally bedded silt, sand and gravel beds.

Glaciofluvial gravels derived from the ice-sheet have been mapped as river terraces beyond the Late Devensian limit at the margins of the River Lugg from Dinmore to the river's confluence with the Wye immediately south-east of Hereford City. In this area, glaciofluvial gravel occurs as a spread that underlies much of Hereford City. Similar terraces occur in the Wellington Brook and Tillington Brook Valleys. Meltwater from the ice-sheet is also thought to be associated with a variety of erosional features. Many of the gaps and transverse lines of drainage within the Silurian and Dittonian escarpments are thought to

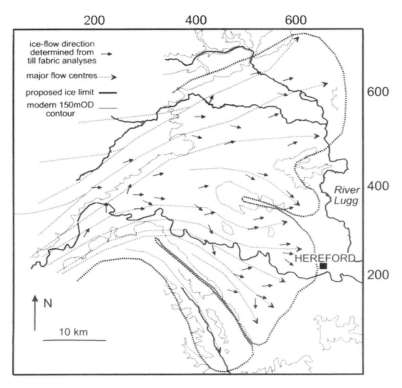

Fig. 9.8 The probable configuration of the Late Devensian ice-sheet and associated flow directions derived from till fabric analysis

have been formed, or at least have been radically modified by, meltwater activity (Luckman, 1970; Brandon, 1989). Glaciofluvial meltwater activity therefore played a prominent role in landscape evolution.

The nature of Late Devensian glaciation in Herefordshire

The Late Devensian ice-sheet spread across the Hereford Basin as a large piedmont outlet glacier, fed by icefields covering the Welsh Massif. Ice-flow within the Basin, while exhibiting local deviations, generally moved from west to east, reaching its maximum near the current Lugg Valley. No dateable deposits have yet been recovered underlying these glacigenic deposits. However, the stratigraphic relationships of these deposits to pre-existing river terrace deposits of the Rivers Lugg and Wye and older glacial deposits that occur in eastern Herefordshire suggest that they formed during the last, major cold snap of the Pleistocene. Recent work on sediments recovered from kettle holes found within the ice-limit provide the only other means of dating this phase of glaciation. Organic detritus within basal kettlehole sediment at Bridge Sollers (SO 406422) and Stansbach (SO 365618) yielded radiocarbon dates of 15,411–13,841 and 15,196–13,499 cal. yr BP respectively, providing the youngest possible date for deglaciation of the Hereford Basin (Stokes, 2003).

The configuration of the ice-sheet at its maximum extent can be seen in Fig. 9.8. It appears that the ice-lobe was thinner in the areas north of Wormsley Hill, and most of the evidence for stages in the withdrawal of the ice-sheet occurs south of Wormsley Hill, within the Wye Valley. In this area the following stages can be recognised:

1. When the ice was at its maximum extent, meltwater deposited a sequence of outwash gravels, forming a platform on which much of the city of Hereford is built. It is likely that much of this gravel spread derived from a large subglacial conduit in the Wye valley. The ice-margin then oscillated, deforming the proglacial outwash deposits, and stagnated, resulting in the deposition of an extensive band of hummocky, kettled, moraine.

2. A similar series of events occurred after the ice-sheet had retreated to the area west of Wormsley Hill. The Staunton moraine marks a renewed phase of ice-advance, the deformation of previously deposited gravels and the formation of a terminal moraine that became partially covered by hummocky moraine as the ice-front melted. The height and characteristics of the terrace of gravels and interbedded diamicts running from Turners Boat to Bredwardine suggests that a kame terrace formed below the Dittonian escarpment at this stage.

3. Following this event, the ice became constrained by the topography of the Wye Valley below Hay-on-Wye, with a further phase of readvance and subsequent stagnation forming the moraine that traverses the Wye valley from Hay to Clyro.

The stratigraphy of the sediments associated with each of these stages appears to show the same pattern. A basal till is overlain by proglacial gravel (often intensely deformed near the position of the ice-terminus at each stage) in turn overlain by a sequence of flow and melt-out tills that interbed with a complex of deformed glaciofluvial and glaciolacustrine sediments (Fig. 9.9).

Representatives of these phases of ice-retreat appear to be absent from the areas surrounding the River Arrow, north of Wormsley Hill. In that region the ice-sheet appears to have been thinner than further south, with more evidence of widespread glaciofluvial activity. The current course of the River Arrow is underlain by appreciable thicknesses of fluvioglacial gravel, as are many other tributaries of the River Lugg in the area (Jackson and Sumbler, 1983). In addition, the sedimentology and structural geology of sediments exposed in ridges at Stretford is characteristic of esker sedimentation (Fig. 9.7c). The sequence represents a coarsening-upward, flood event, followed by the collapse of what must have been an englacial or supraglacial channel, before the emplacement of diamictons into the depression formed as the ice-sheet stagnated.

The origin of the Kington-Orleton moraine is probably very complex, but again appears to owe much to glaciofluvial action. The 'moraine' is a composite feature, formed by a number of processes that occurred in discrete stages. The western, 'drumlinised' component of the feature is likely to be of purely subglacial origin, formed as ice spilled into the Hereford Basin from the Welsh Massif. The section that runs from Kington to Mortimers Cross consists of hummocky moraine deposited as the ice stagnated, covering pre-existing subglacial and proglacial gravel. The final section of the feature, running from Mortimers Cross to the terminal moraine at Orleton, is dominated by drumlinoid features. Here there is no hummocky moraine, till appears to be absent and exposures reveal that the features are entirely composed of fluvioglacial gravel.

Drumlins and drumlinoid forms are commonly accepted to be of glacial origin, but they are equifinal forms. Inferences regarding their mode of formation cannot be made with reference merely to their form. The drumlinoid features between Mortimers Cross and Orelton may have formed subglacially, as a consequence of a subglacial outburst flood when meltwater could not be catered for by the subglacial drainage network that incorporated precursors to the modern drainage network of the region. This would account for the 'mega-ripple' or dune- bedding characteristics exhibited by deposits at Lucton. A subglacial, confined, linked cavity system would provide the flow rates for the formation of such large primary bedforms. Alternatively, a proglacial meltwater network may have moulded fluvioglacial gravels into drumlinoid forms after the ice-sheet retreated from the area. The latter may account for the lack of overlying glacigenic deposits. A further possibility is that the features may be associated with delta deposits that occur north of a gap in the Silurian escarpment at Aymestrey. The delta is believed to have formed as meltwater from the ice-sheet within the Hereford Basin spilled through the gap and into the Wigmore Basin. Topsets in the glaciodeltaic sequence suggest that the level of Glacial Lake Wigmore reached a height of 128–131 m before the lake overflowed, cutting the present gorge of the Teme to Downton (Luckman, 1970). The orientation of drumlinoid features immediately south of Mortimers Cross suggest that they may have been modified by an outburst from the lake.

Both parts of the ice-sheet, north and south of Wormsley Hill, may have reacted in concert to climatic and/or glaciological forcing. It may be that any evidence of still-stands or oscillations in the ice-mass on retreat have been removed by the action of pro-glacial meltwater in the northern parts of the Hereford Basin. However, as the northerly section of the ice-lobe was thinner, it may have been more sensitive to climatic amelioration and as a consequence became disassociated from the main ice-sheet.

Conclusions

The Pleistocene deposits of Herefordshire represent discrete phases during c. 500,000 years of landscape evolution. Despite the paucity of absolute biostratigraphic control there is no doubt that Herefordshire has been subject

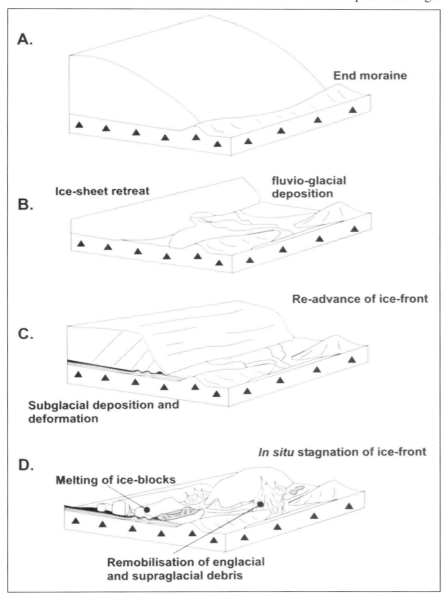

Fig. 9.9 Schematic representation of moraine formation in the Hereford Basin.
a) Ice advance and the deposition of the terminal moraine at Stretton Sugwas.
b) Retreat of the ice-front and fluvioglacial deposition.
c) Readvance and deformation of subglacial deposits.
d) Stagnation of the ice-front and formation of 'Hummocky moraine'

to both Middle and Late Pleistocene glaciation. Both Luckman (1970) and Brandon (1989) highlighted the lack of reliable records of Pleistocene fauna and flora recorded from Herefordshire. Before the discovery of organics in the Cradley Brook Valley (Barclay *et al.*, 1992; Bonny, 1992; Coope *et al.* 2002) and the recovery of organic sediments from kettle holes within the Late Devensian limit in the Hereford Basin (Stokes *et al.* In prep), records of both vertebrate remains in the terraces of the Rivers Lugg (*i.e.* Symonds, 1872; Dawkins, 1869) and Wye (Symonds,1872) and benthic foraminifera (bottom dwelling, marine, protozoa) from lower terraces of the River Wye (Wright, 1905) were the only evidence of Herefordshire's Pleistocene biology.Unfortunately the stratigraphic position of their discoveries remains unclear, as is the classification of at least the finds of Wright (1905). Glacial deposits rarely incorporate organic materials (and when they do, the fossils have often been reworked from pre-existing sediments) so that additional stratigraphic evidence is most likely to be recovered from the fluvial terraces of the Teme, Lugg and Wye valleys.

The dearth of exposure in the terraces of the Wye, Lugg and Teme means that it is difficult to obtain biostratigraphic evidence. The same problem prevents a more detailed understanding of the nature of the Late Devensian glaciation of the Hereford Basin. In the case of Middle Pleistocene glaciation, while the evidence is open to debate, it is unlikely that the small remnants present will yield any more clues. However, the landscape of Herefordshire presents a fine example of the effects of glaciation on the lowland margins of a major ice-centre during the Cold Stages of the Pleistocene.

10 South Wales

by D. Q. Bowen

Introduction

Evidence for glaciations and interglacials is examined in Pembrokeshire, Carmarthenshire, Glamorgan, southernmost Powys and western Gwent (Fig.10.1). Estimating the age of these events is the basis for understanding their place in the evolution of the climate system. This is facilitated through their correlation with oxygen isotope stages ($\delta^{18}O$ stage) and oxygen isotope events ($\delta^{18}O$ events) (Bassinot *et al.*, 1994), as well as millennial events from ice cores and other records (Alley, *et al.*, 1999). All heights are in metres above Ordnance Datum. All ages are in thousands of years (ka) before present. Unless otherwise stated, radiocarbon ages are in calibrated (calendar) years.

Overall, the 'pre-glacial' landform and its drainage system were probably not greatly unlike that of today, with a high degree of correlation between rocks and relief. Modification by glacial erosion was

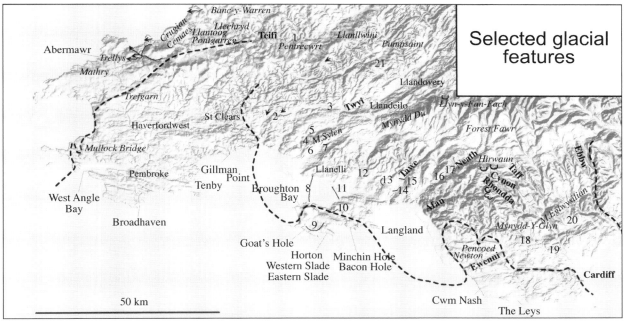

Fig. 10.1 Last Glacial Maximum shown by broken line; selected meltwater channels by small arrows; cirques by 'c' symbol; end moraines numbered thus: (1) Pont-tyweli, (2) Sarnau, (3) Llanarthne, (4) Llandyfaelog (Gwendraeth Fach), (5) Pontanwn (Gwendraeth Fach), (6) Pontnewydd (Gwendraeth Fawr), (7) Ponthenri (Gwendraeth Fawr), (8) Whitford-Pembrey, (9) Salthouse-Machynys, (10) Oldwalls, (11) Paviland, (12) Waun Gron, (Pontarddulais), (13) Tirdonkin (and younger Pontlassau moraines), (14) Glandwr (Landore), (15) Glais, (16) Tonna (Aberdulais), (17) Clyne, (18) Talbot Green, (19) Radyr, (20) Llanbradach, (21) Talley

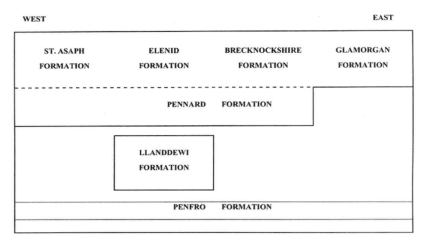

Table 10.1 *Pleistocene Lithostratigraphical Formations in South Wales*

generally unspectacular except in some coalfield valleys. Thus many 'preglacial' elements persist, including the overall surface geomorphic profile of a southerly tilted Oligocene surface (Brown, 1960b; Walsh *et al.*, 2002), modified as upland plateaux (Brown, 1960a) and, below ~200 m, as extensive coastal plateaux, that T. N. George (1938) ascribed to 'an infinity of shorelines' of possible Pliocene age. Off-shore, the 'preglacial shore platform', of possible late Miocene-Pliocene age (Wood, 1974; van Vliet-Lanoë, *et al.*, 2000), has been modified by interglacial marine transgressions, although it 'forms only an insignificant nick in the coastal cliffs' (Wood, 1974) where raised beaches are preserved (George, 1932).

An unambiguous lithostratigraphical classification is necessary to organize, interpret and understand evidence of Pleistocene events. The fundamental unit is the formation, a body of sediment that is mappable, and which may be subdivided into members and beds (see Richards, this volume). Rapid lateral and vertical change is characteristic of many Pleistocene deposits so that diversity may be a form of unity provided it is mappable, as in the case of the Pennard Formation. Lithostratigraphical units represent actuality. They exist. But interpretation about their precise origin may change or be controversial. So a distinction between actuality and inference (indicated by the references cited) should be clear, and genetic names for lithostratigraphical units, especially on maps, should be avoided. Seven formations are used as unifying concepts as well as a context for geomorphological interpretation (Table 10.1, Fig. 10.2).

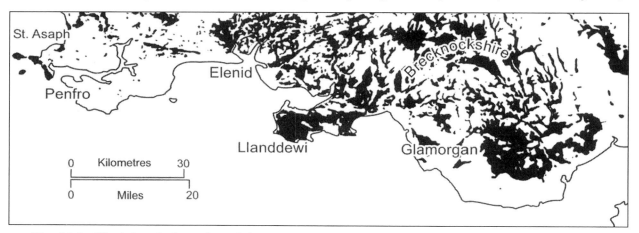

Fig. 10.2 *Glacial and glaciofluvial deposits in South Wales assembled from 1:50,000 maps of the British Geological Survey with generalised outcrops of the lithostratigraphical formations (reproduced from Bowen, 1970)*

The Penfro Formation

The Penfro Formation consists of glacial and glaciofluvial deposits of an Irish Sea ice-sheet. Its erratics show that it penetrated the lower Teifi Valley (Griffiths, 1940; Jones, 1965), crossed Pembrokeshire, west Carmarthenshire (Strahan et al.,1914; Griffiths, 1940), Carmarthen Bay, west Gower (Strahan and Cantrill, 1907b; George, 1933) and, more controversially, east of Swansea Bay (Fig. 10.3). At Pencoed, Strahan and Cantrill (1904) described 9.5 m of stiff 'deep red and purple clay' with distinctive igneous rocks of western provenance whose identification by Howard was confirmed nearly 90 years later (Bevins and Donnelly, 1992). Farther east erratic boulders occur at Pentre, near Llantrithyd (tuff), Newton (quartz felsite), St Athan (quartz felsite) (Strahan and Cantrill, 1904), and Maendy Pool, Cardiff (pyroxenic keratophyre—possibly from New Inn, Pembrokeshire (Griffiths, 1940), although this was not confirmed (Waters and Lawrence, 1987). Glaciation may have extended into Somerset, where glacial deposits with striated Carboniferous Limestone boulders and other erratics occur at Kenn (Gilbertson and Hawkins, 1978).

At West Angle Bay, the Penfro Formation, described as 'scratched stones, including igneous rocks, in stiff purplish clay' is overlain by raised beach deposits (Dixon, 1921). The Penfro Formation underlies the Glamorgan Formation at Pencoed (Griffiths, 1940), and its heavy minerals and erratics were respectively reworked into the Elenid Formation in south-east Carmarthenshire (Griffiths, 1940) and the Llanddewi Formation in Gower (Bowen, 1973b). The extent of any Welsh ice coeval with the Penfro Formation is unknown.

The Irish Sea ice crossed Mynydd Preseli (537 m) where it quarried 'spotted bluestones' from the dolerite sills at Carn Meini and transported them 20 km farther south (Fig. 10.3) (Griffiths, 1940). Kellaway (1971) suggested that the ice carried large 'bluestones' to Stonehenge, but this is not supported by any evidence of glaciation in Wiltshire where over 50,000 pebbles examined from Pleistocene river gravels are all local (Green, 1973). In 1991 fragments of 'bluestone' from the 'heel-stone' at Stonehenge (provided by Claire Conybeare of the Salisbury and South Wiltshire Museum) gave a ^{36}Cl cosmogenic age of 14 ± 1.9 ka (Purdue University). This shows that the 'bluestone' samples were first exposed on

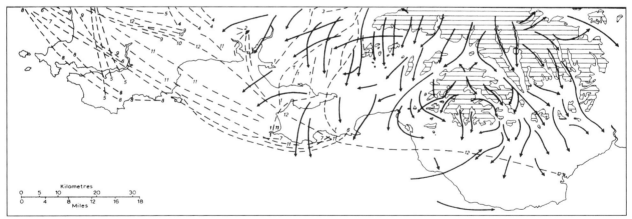

*Fig. 10.3 Ice movement in South Wales (after Griffiths, 1940; Pringle and George, 1937; Woodland and Evans, 1964; Bowen, 1970). Erratics (after Griffiths, 1940): (1) Llangadog rhyolite; (2) Llangynog rhyolite, agglomerate and diabase; (3) Builth Wells olivine dolerite; (4) Preseli dolerite ('bluestone'), rhyolite and slate; (5) Roch-Haycastle andesite; (6) St David's granite; (7) Ramsey Island quart albite porphyry; (8) St David's Head gabbro; (9) Clegyr agglomerate; (10) Llandeloy porphyrite; (11) Harlech (Barmouth) grit; (12) New Inn pyroxenic keratophyre; T: trilobites (*Asaphus *and* Ogygia *from the Ffairfach Grit near Llandeilo (reproduced from Bowen, 1970)*

Mynydd Preseli to the atmosphere and cosmogenic rays at that date, long after any glaciation could have transported them to Wiltshire.

Llanddewi Formation

The Llanddewi Formation of west Gower consists of glacial deposits of an ice-sheet from mid-Wales (Strahan and Cantrill, 1907b; George, 1933). Millstone Grit quartzite erratics are common and George (1933) believed that occasional Irish Sea erratics showed a 'commingling' of Irish Sea and Welsh ice sheets. But such is the contrast between the greatly dissected outcrop of Irish Sea glacial deposits west of Carmarthen Bay and the extensive outcrop of the Llanddewi Formation (Fig. 10.2), that it was suggested two glaciations occurred, and the younger Llanddewi glaciation reworked pre-existing Irish Sea erratics into its deposits (Bowen, 1969). A defining characteristic of the Llanddewi Formation is its weathered condition, unlike the Elenid and Brecknockshire Formations to the north and east (Davies, 1986). Such weathered deposits are exposed at Western Slade and Eastern Slade (Slade Member) as unsorted fan deposits that were reworked from the Llanddewi Formation (Bowen, 1971, 1977).

The Paviland moraine (Fig. 10.4) is so degraded that it escaped recognition for over 75 years. It may mark the extent of the Llanddewi glaciation because cliff top plateaux and coastal valleys (slades) between Port Eynon and Rhosili are free of glacial or glaciofluvial deposits. The Hills borehole (SS 45258615) was drilled on the crest of the Paviland Moraine and proved 21.5 m of deeply weathered sands and gravel overlying 1.5 m of red clay (Bowen and Jenkins, in Bowen *et al.*, 1985; Bowen *et al.*, *unpublished* 1992). The Hangman's Cross borehole (SS 483867) drilled through one of the shallow depressions mapped as 'sink holes' on the 1907 Geological Survey field slips, proved 10 m of gravel, and the Western

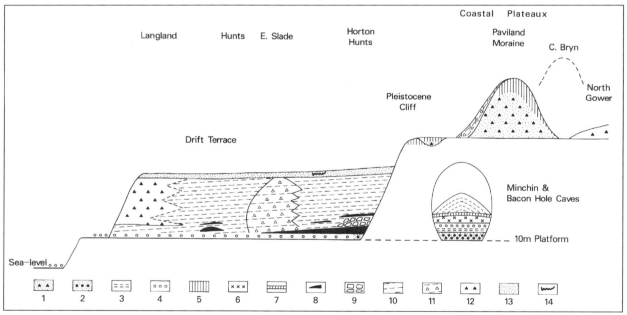

Fig. 10.4 Diagrammatic and reconstructed figure of the Pennard Formation in South Gower.
Key: (1) Llanddewi Formation; (2) Minchin Hole Bed 1 (inner beach); (3) Minchin Hole Bed 2 ('red cave earth'); (4) Minchin Hole Bed 3 (Patella beach); (5) weathered deposits; (6) Minchin Hole Beds 4 to 6, Bacon Hole Beds 4-7; (7) stalagmite; (8) red beds (colluvium); (9) blocky head; (10) head (scree); (11) reworked Llanddewi Formation deposits (Slade Member); (12) Elenid Formation in North Gower, Brecknockshire Formation in Southeast Gower; (13) Horton Member (loess);
(14) cryoturbation structures

Slade Farm borehole (SS 48208605) proved 7.6 m of dark blue clay on limestone. The Bay View borehole (SS 46808585) proved 17 m of weathered and finely bedded small gravels in a sand matrix and, unlike Hills and Western Slade Farm, lacked calcite and pyrite that may have been weathered out (Catt, pers. comm). The Bay View deposits are interpreted as slope deposits on the distal margin of the moraine (Fig. 10.5) (Bowen *et al.*, 1999). Borehole control assisted the interpretation of a resistivity survey that revealed the dimensions and extent of the moraine (Reid, 1985; Smith *et al.*, 2002).

The Llanddewi Formation *in situ* does not overlie raised beaches. But its reworked deposits have become constituents of younger deposits such as the western slope of Oxwich Bay, as individual erratics and erratic lenses in head, or fan deposits at Western Slade and Eastern Slade (Bowen, 1970, 1971). The relatively undissected Llandewi glacial deposits may suggest a relatively recent age (Bowen, 1970), but the weathered state of many of its erratics and the greatly degraded nature of the Paviland Moraine (Fig. 10.4) point to earlier rather than later glaciation. Underlying Dinantian limestone has probably ensured a state of 'arrested dissection' in an area of modest contemporary precipitation (<900 mm).

Fig. 10.5 Exposure of head in the lee of Gilman Point, Carmarthen Bay (SN 227075). This shows blocks of Dinantian limestone, some of which are joint and bedding plane bounded (blocky head). Many limestone blocks are coated with stalagmite. The coarse head is overlain by finer calibre scree set in a variable matrix of finer slope wash material. Characteristically clasts in the head have a preferred orientation and clast long axes lie perpendicular to the slope (old cliff), and lie at low angles parallel to the crude bedding. The head is overlain by loess (Horton Member of the Pennard Formation) and the entire exposure represents all of the last (ice age) glacial cycle (Photo: D. Bowen)

The Pennard Formation

The Pennard Formation includes a diversity of deposits, but its unity as a mappable unit is provided by its co-extensive coastal terrace that overlies the 'pre-glacial shore platform' (Fig 10.4). While its principal lithology is head (Fig. 10.5), generally angular fragments of local bedrock produced by frost action (George, 1933; Bowen, 1970; John, 1970), it also includes raised beaches of different ages. The *Patella* beach (Hunts Bay Member) is exposed at: Poppit, Porth Clais and Druidston Haven (John, 1970); throughout Milford Haven (Cantrill *et al.*, 1916) and at Freshwater West Bay (Dixon, 1921); Broadhaven (John, 1970), Manorbier, and Swanlake (Dixon, 1921); Marros and Ragwen Point (Bowen, 1970), throughout Gower (Fig. 10.6), and at Goldcliff, Gwent (Allen, 2000). At West Angle Bay beach gravels are overlain by 'loam'—fine grained fossiliferous estuarine deposits that include a peat unit (Dixon, 1921; John, 1970; Bowen, 1974). Pollen analysis of the loam showed an interglacial vegetational spectrum (Stevenson and Moore, 1982), and radiocarbon dating of wood from *Alnus* tree stumps in growth position on the surface of the loam provided an age of >35,500 years (Birmingham 327) (Bowen, 1974). Many raised beaches contain erratic pebbles from the South Wales Coalfield, north Pembrokeshire, North Wales and Scotland, that were reworked from older glacial deposits. The Pennard Formation overlies the Penfro Formation at West Angle Bay and is overlain by the St Asaph, Elenid, and Brecknockshire Formations (Fig. 10.4).

Two fossiliferous raised beaches discovered at Minchin Hole Cave (Fig. 10.7) are separated by head (Sutcliffe and Bowen, 1973) (Table 10.2). Bed 1 (inner beach) overlies the marine abraded floor of the cave between 9.9 and 11.8 m and consists of gravel and stratified sand that yielded a single erratic tuff. Bed 2 (red cave earth) is angular limestone material in a red clay matrix, that yielded a large form of the cold climate *Microtus oeconomus* (northern vole). Bed 3 (outer beach) is the *Patella* beach (George, 1932) that lies between 9.4 and 11.0 m and overlies, with unconformity, Beds 1 and 2 (Sutcliffe and Bowen, 1973). Bed 1 (stratified sand) is interpreted as typical of low tide level and Bed 3 (*Patella* beach) as typical of high tide level (Sutcliffe and Bowen, 1973).

The ratio of the amino acids, D-alloisoleucine to L-isoleucine (D-aile/L-Ile), in the protein residue of marine gastropod fossils is a means for estimating the ages of the raised beaches. During life only L-isoleucine exists but on fossilisation it undergoes progressive interconversion (epimerization, a form of racemization) to D-alloisoleucine. The D-aile/L-Ile ratio is the basis for recognizing aminozones with different D/L ratios and is also a means of correlation (aminostratigraphy). In Bed 1 at Minchin Hole the

Fig 10.6 The fossiliferous Patella *beach and colluvial silts at Hunts Bay (East Cove) Gower (SS 566866). The cemented* Patella *beach lies on the 'pre-glacial' shore-platform between present low water mark to just above high tide level and consists of poorly to moderately sorted deposits of cobbles, shingle, sand and shell debris and is stratified into coarser and finer laminae. Clast shape ranges from well-rounded to angular. Clasts are dominated by Dinantian limestones, with a variable proportion of Devonian sandstones and conglomerates and Upper Carboniferous sandstones, siltstones, gritstones and quartzites. There are numerous whole or fragmented gastropod fossils of* Patella vulgata, Nucella lapillus, Littorina littorea *and* Littorina littoralis. *The raised beach also contains shallow water temperate marine foraminifera and ostracods. Overlying the beach are colluvial silts (see text). (Photo D. Bowen)*

Fig. 10.7 Minchin Hole Cave at low tide (SS 5628680). The cave runs along the line of a cross fault seen along the foreshore, entrance and top of the cave. The abandoned and degraded cliff line meets the plateau surface at approximately 70 m OD (see also surveyor standing on the raised platform remnant to the left of cave entrance for scale). Inside the cave, only some of the beds described in the text are now exposed. (Photo D. Bowen)

D-aile/L-lle ratio for *Patella vulgata* is 0.17 ± 0.1 (aminozone 1), and in Bed 3 it is 0.1 ± 0.02 (aminozone 2) (Table 10.2; Fig. 10.8) (Bowen *et al.*, 1985). Aminozone 1 is calibrated by a thermoluminescence age on sand grains from the beach of 191 ± 32 ka (Southgate, 1985). Aminozone 2 is calibrated by $^{234}U/^{230}Th$ (Uranium-series) ages between 130 and 101 ka from a stalagmite block that was embedded on the surface of the *Patella* beach—Bed 3 (Table 10.2) (Sutcliffe and Currant, 1984). Thus, aminozone 1 is correlated with $\delta^{18}O$ event 7.1 at 194 ka (Bassinot *et al.*, 1994), and aminozone 2 is correlated with $\delta^{18}O$ event 5.5 at 122 ka (Bowen *et al.*, 1985).

A rich fauna of terrestrial gastropods that indicate a temperate climate was embedded in the stalagmite block (Appendix 10.1). Such a climate is also indicated by mammalian fossils from interglacial deposits at Minchin Hole and the adjacent caves of Bacon Hole and Ravenscliff including Hippopotamus, Woodland Elephant, slender nosed Rhinoceros, Fallow Deer and Hyena (Sutcliffe *et al.*, 1987). The *Neritoides* beach

MINCHIN HOLE CAVE			BACON HOLE CAVE		
Sutcliffe 1981	Henry 1984	Age (ka)	Age (ka)	Henry 1984	Stringer *et al.*, 1986
Inner & Outer Talus cones	travertine Bed 7 Inner travertine		13 ± 13 [U] 12.8 ± 1.7 [U] 81 ± 18 [U]	travertine Bed 9 Lower travertine	stalagmite breccia stalagmite
Sandrock	Bed 6			Bed 8	Upper cave earth
Earthy	Bed 5			Beds 6 & 7	Upper sands & Brown sands
Breccia	Bed 4c			Bed 5	Clays silts sands
'Series' (cave earths)	Bed 4b	101 ± 16 [U] 107 ± 10 [U] 127 ± 21 [U] 130 ± 26 [U] Pv 0.11 ± 0.01	116 ± 18 [U] 122 ± 11 [U] 125 ± 26 [U] 128 ± 20 [U] 129 ± 16 [U] 129 ± 30 [U] Cn 0.12 ± 0.02	Bed 4	Shelly sand
Neritoides Beach	Bed 4a	Llrs 0.1 ± 0.01 Lsax 0.11 ± 0.01	Pv 0.1 ± 0.12 Llra 0.12 Llrs 0.1 ± 0.01 Lsax 0.1 ± 0.01		Sandy cave earth
Patella beach	Bed 3	Llra 0.12 ± 0.018 Nl 0.136 ± 0.01	Pv 0.1± 0.01	Bed 3	Sandy breccia conglomerate
Lower red cave earth [head]	Bed 2	*M. oeconomus*	*M. oeconomus*	Bed 2	Coarse grey sands Coarse orange sands
Inner beach	Bed 1	191 ± 32 (TL) Pv 0.17 ± 0.01 Llra 0.18 ± 0.017		Bed 1	Basal pebbles

Table 10.2 *Correlation of the Pleistocene beds at Minchin Hole Cave and Bacon Hole Cave Gower. To simplify existing stratigraphical nomenclature (Bowen, 1999b) a Minchin Hole Member and a Bacon Hole Member of the Pennard Formation have been subdivided into beds*

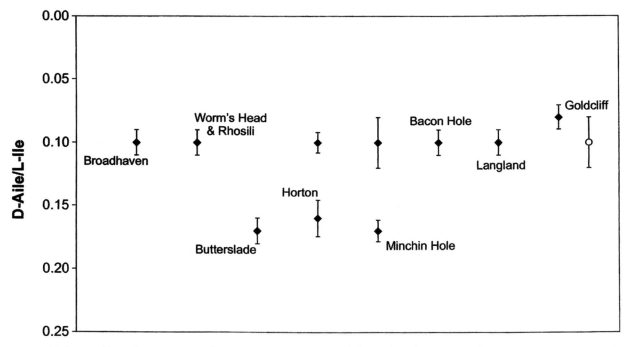

*Fig. 10.8 D-alloisoleucine/L-isoleucine ratios measured from fossil gastropods and bivalves in South Wales. The mean ratio and one standard deviation are shown. Minchin Hole Cave is the type site where two aminozones are defined. All ratios are from Patella vulgata except, at Horton (*Littorina littorea *and* Nucella lapillus*) and Goldcliff (*Littorina sp.*, and* Macoma balthica*). Data from: Bowen* et al.*, 1985 (with additional ratios since 1985 for Minchin Hole Bed 3) and Allen (2000). Aminostratigraphy is a powerful means of detecting reworked shells (Wehmiller* et al.*, 1995), a feature characteristic of most raised beaches as illustrated at Horton. An earlier analysis used a different sample preparation method (Miller, 1985) that produced higher and more variable ratios*

(George, 1932) is transitional to cave earths (Table 10.2) that contain terrestrial gastropods, which help to show the transition from marine to terrestrial conditions at the end of the interglacial (Fig. 10.12).

The age of aminozone 2 is confirmed at Bacon Hole Cave where Bed 4 contains *Patella vulgata* with D-aile/L-Ile ratios of 0.1 ± 0.1 and broken stalagmite with U-series ages between 129 and 101 ka (Table 10.2) (Currant *et al.*, 1984). Bed 4 also yielded the terrestrial gastropod *Cepaea nemoralis* with D-aile/L-Ile ratios of 0.12 ± 0.02. This allows further correlation with blown sand overlying the *Patella* beach east of Langland Bay (0.13 ± 0.01 for *Cepaea* and 0.12 ± 0.01 for *Helicella sp.*), that is probably the same age as cemented blown sand below head at Three Cliffs Bay that contains abundant *Pupilla muscorum* and *Trichia hispida* and sparse *Vitrina pellucida* (J. G. Evans, *unpublished*).

Red beds overlying the raised beaches (Fig. 10.6) were described as 'fluvioglacial' (George, 1933) or *in situ* fossil soils (Ball, 1960), but were reinterpreted as colluvial deposits that included reworked fossil soil material (Bowen, 1970). They contain glacial erratics (George, 1933) and at Pwll Du Head contain the terrestrial gastropods *Pupilla muscorum* (3 specimens), *Discus rotundatus* (1), *Oxychilus cellarius* (3), *Clausilia bidentata* (3), *Helicella itala* (8) and *Cepaea nemoralis* (2) (J. G. Evans, *unpublished*). At Horton red colluvial material forms a matrix for the raised beach.

The Slade Member (Fig. 10.4) consists of unsorted pebble and boulder gravel in a variable matrix with deeply weathered clasts (Bowen, 1977b). Its geometry ranges from small lenses in limestone head to the fan-like units that infill small valleys at Horton, Western Slade and Eastern Slade (Bowen, 1970a).

It was suggested that the Slade Member was deposited by small valley glaciers (Wilkinson, in Strahan, 1970b) or glaciofluvial fans (Rijsdik, 2000). But its intrinsic characteristics of fabric and bedding, and lithostratigraphical and geomorphological field relationships, were interpreted as a cold climate fan deposit that consists largely of redeposited constituents from the Llandewi Formation (Bowen, 1969, 1970, 1977a, 1977b). The Horton Member (Fig. 10.4) consists predominantly of silt, interpreted as loess (Case, 1977, 1983, 1984), while head at Nash Point is overlain by Holocene tufa (Evans *et al.*, 1978).

Hiatuses indicative of breaks in head deposition and climatic amelioration are indicated by pedogenesis at Hunts Bay (Case, 1993), stalagmite formation with U-series ages of 36.8 ± 4 and 85 ± 10 ka at Goat's Hole Cave (Aldhouse-Green, 2000), and of 81 ± 18, 12 ± 10, 12.8 ± 1.7, 13 ± 3 at Bacon Hole Cave (Currant *et al.*, 1984). Pollen from two horizons at Long Hole Cave was correlated with the Chelford and 'Upton Warren' interstadials (Campbell, 1977).

Palaeolithic human occupation in South Wales is evident from artifacts at: Hoyles Mouth and Little Hoyle, near Tenby; Coygan Cave near Laugharne; Cat Hole, Bacon Hole, Long Hole and Goat's Hole in Gower. At the latter Buckland (1823) discovered the 'Red Lady' of Paviland (Aldhouse-Green, 2000). AMS Radiocarbon ages on bones of $25,840 \pm 280$ and $26,350 \pm 550$ when calibrated to sidereal years gave respective ages of 29,650 and 29,900, close to interstadial 4 in Greenland, a time of climate amelioration (Bowen, 2002). More recent radiocarbon age calibration (Bard, 2004; Hughen *et al.*, 2004) supports the correlation with interstadial 4.

Elenid Formation

Named after the country around Pumlumon and Drygarn Fawr in central Wales, the Elenid Formation unifies glacial and related deposits over a wide area (Bowen, 1999b). Its glacial and glaciofluvial deposits, coeval with the Brecknockshire and St Asaph Formations (Fig. 10.2) have been described in the middle Teifi Valley (Price, 1976), lower Teifi Valley (Lear, 1985; Davies *et al.*, 2003), around Ammanford (Strahan *et al.*, 1907), Carmarthen (Strahan *et al.*, 1909), south-east Carmarthenshire, Gower (Strahan, 1907a, 1907b; Griffiths, 1940), and around Cross Hands and north-west Gower (Campbell, 1984).

Glacial and glaciofluvial deposits of the Elenid Formation contain Lower Palaeozoic erratics, with indicator erratics from igneous rocks at Builth, Llangadog and Llangynog (Fig. 10.3). A major component of the Elenid ice was the 'Towy Glacier' (Pringle and George, 1937), also nourished by ice from north of Mynydd Epynt and from Mynydd Du. It spilled south through old wind gaps on the south bank of the Tywi between Llandeilo and Carmarthen. The ice flowed predominantly to the south-west, but at maximum glaciation its upper layers also flowed from north to south (Fig. 10.3). This 'basal ice shed' (Hollingworth, 1931) was described on Mynydd Sylen (284 m) and on Mynydd-y-Bettws (300 m) (Strahan *et al.*, 1907c; Bowen, 1970).

North Carmarthenshire has not been mapped systematically so that the upper limit of the Elenid Formation between the Tywi and Teifi valleys is uncertain. The country around St Clears marks the south-western extent of the Elenid Formation (Bowen, 1970). In the Teifi Valley Price (1976) suggested the morainic ridge with kettle holes near Pont-tyweli, Llandysul marked its limit (Fig. 10.1), although Griffiths (1940) believed it extended downstream to Newcastle Emlyn.

End-moraines occur in most valleys (Fig. 10.1). Meltwater channels are ubiquitous, notably north of a line between Carmarthen and Llandeilo (George, 1942; Bowen, 1970). The highest channel at Pant-y-rhyg (SN 515339) lies some 100 m below local summits. Lake clays show glacial lakes that ranged in size from Lake Rhydodyn in the Cothi Valley where 5 m of laminated clays were proved in the Edwinsford (Rhydodyn) Bridge borehole, north of the Talley moraine (M. S. Parry, *unpublished*); to Lake Teifi indicated by widespread lake clays (Jones, 1965; Bradley, 1980; Lear, 1986; Fletcher and Siddle, 1998).

J. K. Charlesworth proposed that a glacial Lake Teifi existed during 'the period of maximum advance of the Newer Drift' (Charlesworth, 1929, page 340), a view supported by others (Bowen, 1967;

Fig. 10.9 Exposure of lacustrine clays at Lake Teifi at the disused clay pit near Llechryd, Cardigan (SN 211436). The clays overlie medium to coarse gravel with well rounded pebbles, many of which are erratics. Sands, which are also exposed, are stained with iron and manganese and are false-bedded. (Photo: D. Bowen)

Lear, 1985). O. T. Jones first described lake clays in the Teifi valley (Fig. 10.9), followed by Bradley (1980), Lear (1985), Fletcher and Siddle (1998), and Davies *et al.* (2003). Lear (1985) mapped them upstream to Pentrecwrt, and lake clays up to 104 m were proved in the Llandudoch boreholes (Fletcher and Siddle, 1998; Davies *et al.*, 2003). Jones (1965) also described 'steep foreset bedding' east of Pantydwr at 125 m, as well as shorelines cut into the Pentrecwrt delta, with the highest at ~130 m. Unlike Charlesworth, Jones believed that Lake Teifi came into existence 'during the retreat of the ice' (Jones, 1965, page 247).

Between Llandysul and Llanllwni, 'morainic' deposits (Griffiths, 1940) were mapped as: till, glaciofluvial sands and gravels, kame terraces and kettle holes by Price (1976). She suggested they showed ice between Lake Teifi and a separate lake upstream of Llanllwni—here named Lake Llanllwni. The margins of Lake Llanllwni are marked by deltas at Llanllwni (Price, 1976) and others at Rhuddlan, Llanybydder, Llanwnnen, Pencarreg and Lampeter (M. S. Parry, *unpublished*). Those at Llanybydder and Pencarreg have kettle holes and all delta surfaces lie between 125 and 130 m. Price (1976) proposed that Lake Llanllwni was drained subglacially into Lake Teifi. Downstream of Llanllwni, deeply incised river gorges run sub-parallel to abandoned drift-plugged former courses of an earlier Teifi. Davies *et al.*, (2003) suggested they were postglacial, unlike the glacial meltwater flood or subglacial stream erosion hypothesis of Jones (1965) and Bowen (1967) respectively.

Charlesworth (1929) and Jones (1965) suggested stages in the draining of Lake Teifi were indicated by the heights of overflow channels. They both suggested overflow through the Pont-gareg (Pontgarreg) channel at ~140 m, then later through the Cipin channel at ~107 m. Jones (1965), however, misidentified the Pont-garreg channel and called it the 'Llantood' (Llantwyd) channel that lies 0.65 km north of Charlesworth's channel at Pontgarreg. But both channels slope in the reverse direction to that required by an overflow hypothesis and were classified as subglacial (Gregory and Bowen, 1966). Earlier and higher levels for Lake Teifi were proposed from the heights of cols between Cynwil Elfed and Pentrecwrt at ~245 m, Crymych at 180 m (Griffiths, 1940), and Pedran at 198 m (Jones, 1965; Fletcher and Siddle, 1998). Palaeogeographical reconstructions of the lake were proposed along the 200 m (Pedran) and 130 m (Llantood) contour lines by Fletcher and Siddle (1998).

The Elenid Formation overlies a raised beach (Hunts Bay Member) at Broughton Bay in north-west Gower from which it contains reworked gastropods with D-aile/L-lle ratios characteristic of aminozone 2. It also contains bivalves of *Macoma calcarea* with characteristic Late Devensian D-aile/L-lle ratios (Davies, 1988; Bowen, *in preparation*), as well as wood fragments with radiocarbon ages older than 42 ka (Campbell and Bowen, 1989). A ^{36}Cl cosmogenic rock exposure age of 23.2 ± 2 ka (calculated previously as 22.8 ka) from Arthur's Stone in Gower, a large erratic boulder composed of Millstone Grit conglomerate (Owen, 1973), provides an age for the maximum extent of the Elenid glaciation (Phillips *et al.*, 1994; Bowen *et al.*, 2002).

Brecknockshire Formation

Edgeworth David (1883) recognised a Brecknockshire Drift defined by distinctive Old Red Sandstone erratics (ORS) (Fig. 10.2). Its glacial and glaciofluvial deposits have been described around Merthyr Tydfil (Barclay *et al.*, 1988), Cardiff (Waters and Lawrence, 1987), Newport (Squirrell and Downing, 1969), and Swansea (Strahan, 1907a). The ice grew on the dip slopes of Fforest Fawr but in its movement south was unable to surmount the Pennant Measures scarp at Craig-y-Llyn (600 m), below which it bifurcated south-east into the Cynon and Taff glaciers and south-west to the Neath Valley glacier. Drumlin orientation near Hirwaun shows these separate ice flows. In the Afan Valley, ORS erratics show overspill of the Neath glacier (Woodland and Evans, 1964). Rock basins were excavated by the ice, notably in the Tawe and Neath Valleys (Jones, 1942; Anderson, 1974; Culver and Bull, 1979; Al-Saadi and Brooks, 1973). Some are infilled by lacustrine clays and silts (Anderson and Owen, 1979).

The combined Neath and Tawe glaciers expanded into Swansea Bay and ORS erratics around Porthcawl, and lake deposits in the Ewenni valley (Wilson *et al.*, 1990), suggest the ice-margin may have dammed Lake Ewenni (Fig. 10.1). The eastern extent of the Brecknockshire Formation lies approximately along the Ebbw Valley (Squirrell and Downing, 1969). A basal ice-shed was located on Mynydd Eglwysilian (Bowen, 1970). In the Cardiff area the ice-margin extended offshore (Bowen, 1970), north of which numerous kettle holes occur in 'morainic drift' below the great scarp from the Rhymni valley westwards (Squirell and Downing, 1969; Bowen, 1970; Griffiths, 1995). End moraines occur in the major valleys (Fig. 10.1).

The Brecknockshire Formation is coeval with the Elenid and Glamorgan Formations. Stratigraphically, it overlies the Hunts Bay Member raised beach of the Pennard Formation in east Gower, notably east of Langland Bay (Fig. 10.4), and is overlain by Devensian Late Glacial organic deposits in kettle holes at Swansea (Trotman, 1963) and Cardiff (Griffiths, 1995).

Glamorgan Formation

Edgeworth David (1883) also recognised a Glamorgan Drift that consists of glacial and glaciofluvial deposits containing erratics exclusively from the South Wales Coalfield. It was described around Maesteg and Pontypridd (Woodland and Evans, 1964), Pencoed (Wilson *et al.*, 1990), at the former Llanilid open-cast (Harris and Donnelly, 1991) and near Pontypridd where it was redeposited by periglacial slope processes (Harris and Wright, 1980). Ice grew at steep valley headwalls of the coalfield valleys, such as the Afan, Llynfi and Ogwr. Ice from the Graig Fawr and Cwm Saerbren fed the Rhondda Fawr glacier. Glacial oversteepening of steep valley slopes, notably at the base of the sandstone above the No. 2 Rhondda seam, later caused extensive slope failures (landslips) (Woodland and Evans, 1964). From the distribution of glacial and glaciofluvial deposits it is clear that large upland areas were ice free (Woodland and Evans, 1964), but it is difficult to fix upper limits to the ice.

South of Mynydd-y-Glyn, the Brecknockshire and Glamorgan Formations merge (Woodland and Evans, 1964) and terminate in the Vale of Glamorgan (Strahan and Cantrill, 1904; Charlesworth, 1929). During deglaciation small temporary lakes developed (Driscoll, 1953). Small eskers at Creigiau and Nelson show stagnant ice (Squirrel and Downing, 1969; Bowen, 1970). End moraines occur at Talbot Green and upstream at three other locations (Fig. 10.1) in the Ely Valley (Anderson, 1977).

Other than for being coeval with the Brecknockshire Formation there is no stratigraphical evidence for the age of the Glamorgan Formation. A calibrated ^{14}C age of 15.8 ka (Walker *et al.*, 2003) from the Llanilid kettle holes near Pencoed (Bowen, 1970) shows the ice had disappeared long before that time.

St Asaph Formation

Defined from the St Asaph Drift of McKenny-Hughes (1887), this formation unifies 'Irish Sea' glacial and glaciofluvial deposits of Late Devensian age (Bowen, 1999b). As well as erratics of widely varying ages from the Irish Sea Basin, Scotland and North Wales (Fig. 10.3), it contains marine shell fragments and, occasionally, entire shells. Characteristically, the St Asaph Formation consists of red, purple or blue-black clay, often calcareous, and sands and gravels. At Abermawr, it has been interpreted as till (John, 1970) or glaciomarine mud (Eyles and McCabe, 1989; Edwards, 1997). In the Teifi Valley Jones (1965) reported Irish Sea erratics and Lear (1985) described calcareous till at Cenarth. South of the Teifi estuary on the high ground of Crugiau Cemaes, the Trewyddel Member (Charlesworth, 1929; Bowen, 1999b; Hambrey *et al.*, 2001) consists of a variety of glacial and glaciofluvial deposits (Gregory and Bowen, 1966; Davies *et al.*, 2003). Charlesworth (1929) believed these were part of the 'South Wales End Moraine', but others suggested the ice was marginally more extensive and reached the northern slopes of Mynydd Preseli (Gregory and Bowen, 1966) and the Roch-Trefgarn ridge (Griffiths, 1940), thence along the west coast to St Ann's Head. Nearby the Mullock Bridge sands and gravels were interpreted as partly deltaic (George, 1982), but there is no evidence for a proglacial lake within Milford Haven (Cantrill *et al.*, 1916).

At maximum glaciation, Charlesworth (1929) suggested that Irish Sea ice dammed up Lake Teifi (*see Elenid Formation*) which drained via ice-marginal drainage channels between Cardigan and Fishguard, but these were thought to be subglacial and submarginal glacial drainage channels (Bowen and Gregory, 1965; John, 1970).

Stratigraphically the St Asaph Formation overlies part of the Pennard Formation (Fig. 10.10). Radiocarbon age estimates on shell fragments in the Trewyddel Member yielded ages between 31,800 and 37,960, >36,300, >40,300 and >54,300 ka (John, 1970). D-aile/L-Ile ratios on bivalves from the St Asaph

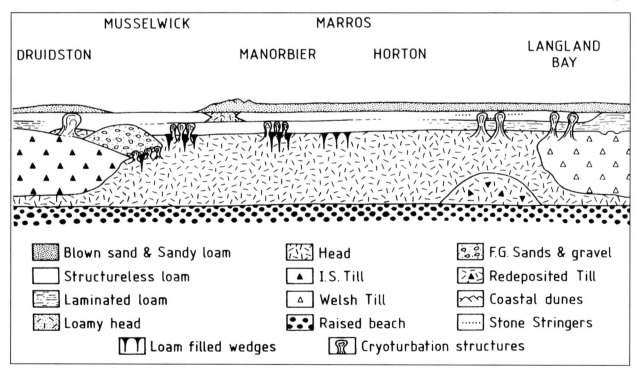

Fig. 10.10 Diagrammatic and reconstructed figure of the Pennard Formation from St Bride's Bay, Pembrokeshire to east Gower. It includes glacial deposits of the Brecknockshire Formation at Langland; and glacial deposits of the St Asaph Formation at Druidston Haven (reproduced from Case, 1984)

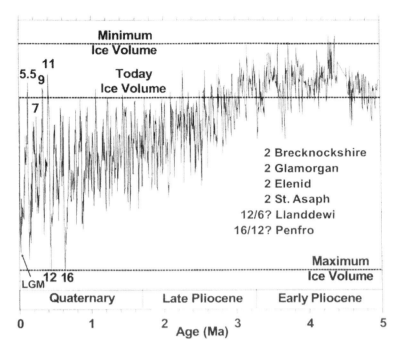

Fig. 10.11 Signal of ice ages and interglacials based on benthic $\delta^{18}O$ record from odp Site 849 East Equatorial Pacific (Mix et al., 1995) as reconstructed by Zeigler et al., (2003). Although the International Union of Geological Sciences age for the base of the Pleistocene is 1.81 Ma, it is likely that it will be revised shortly to 2.5 Ma close to the first time when northern hemisphere ice sheets reached the tidewater as shown by the IRD record. The large glaciations of mid-latitudes started about 800 ka and were dominated by the 100,000 eccentricity pacing. An interpretation of the South Wales record is shown. Interglacial sea levels at 7.1 and 5.5 are only the ones for which evidence is known

Formation show that they may range between early to late Pleistocene (Davies, 1987), but it is axiomatic that the lowest ratio indicates the maximum age of the deposit (Andrews et al., 1983; Miller, 1985). Minimum D-aile/L-lle ratios of ~0.07 on *Macoma* and *Arctica* fragments at Abermawr (Bowen et al., 2002) and from the Mullock Bridge delta gravels (Davies, 1987), suggest a Last Glacial Maximum (LGM) age for the fauna.

The sequence of events

Enhanced understanding of the Pleistocene events in the context of climate system evolution is possible by correlations with $\delta^{18}O$ and millennial events (Fig. 10.11; Table 10.3). But because only a few ages are available before $\delta^{18}O$ stage 5, earlier events should be viewed in a 'floating timescale' (Table 10.3), and more and improved age estimates are needed.

A trend towards increasing ice volumes over the last million years (Fig. 10.11) is not invariably reflected in the extent of successive glaciations in South Wales (Bowen, 1999a) and elsewhere (Roy, et al., 2004). The explanation may be that earlier, relatively low gradient, ice-sheets were able to advance farther because of their deformable beds on softer Oligocene-Miocene sediments, the former extent of which is shown by numerous outliers in Britain and Ireland (Bowen, 1999a). Once older and harder rocks were exposed, ice-sheet profiles became steeper and their extent more limited (Clark and Pollard, 1998).

The Penfro glaciation

Irish Sea ice crossed South Wales from north-north-west to east-south-east (Fig.10.3) because local glaciers were insufficiently powerful to exclude it. Stratigraphically, the glaciation is older than raised beach deposits ascribed to $\delta^{18}O$ event 5.5 at West Angle Bay. A view that the entire sequence at West Angle Bay is glacial in origin and Late Devensian in age (Rijsdik, 2000) is contradicted by the interglacial pollen spectrum from loam and peat units (Stevenson and Moore, 1982), as well as the radiocarbon age of >35.5 ka on *Alnus* tree stumps in growth position on the surface of the loam (Bowen, 1974).

If the Penfro glaciation is the same age as the Kenn Formation in Somerset, it may be older than 400,000 years (Andrews et al., 1984), and as such was correlated with $\delta^{18}O$ stage 16 (Bowen, 1999a), the first major glaciation of the hundred thousand 'eccentricity world' (Raymo, 1997). Alternatively, because the age of the Kenn Formation is uncertain (Bowen *et al.*, *in preparation*), the Penfro Formation may be the same age as the Wolston Formation in the Severn Valley of Anglian ($\delta^{18}O$ stage 12) age (Maddy *et al.*, 1995; Maddy and Lewis, this volume). Both are highly dissected (Fig. 10.2 and Wills, 1938) and both represent the oldest known glaciations of their respective provinces.

Lithostratigraphy	Events/landforms	Correlation	$\delta^{18}O$	Age (ka)
ELENID	Cirque moraines	Younger Dryas/ Bølling-Allerød		~ 14.5 to 11.6
GLAMORGAN	Clyne Moraine	Heinrich Event 1?		~ 16.5
BRECKNOCKSHIRE & ST. ASAPH	glacier readvances valley end moraines	'precursor' Heinrich events	2	19 to 17.5
FORMATIONS	Maximum extent of the glacial formations	Heinrich Event 2 Last Glacial Maximum		~23
Goat's Hole Member	Human burial site	interstadial 3	3	29
PENNARD FORMATION	Head deposits		4	71
Minchin Hole cave earths			5	
Minchin Hole Bed 3	*Patella* beach	SW England beaches	5.5	122
Minchin Hole Bed 2	red cave earth		6	
Minchin Hole Bed 1	Raised beach	SW England beaches	7.1	194
LLANDDEWI FORMATION	Welsh Glaciation Paviland Moraine	Ridgacre/Anglian?	6/12	160/430
PENFRO FORMATION	Irish Sea Glaciation	Anglian/?	12/16	430/630
	'preglacial' shore platform	Late Miocene - Pliocene		~ 2.8 – > 5 Ma

Table 10.3 *Correlation of the sequence of events. Because of uncertainty about the ages of the Llanddewi and Penfro Formations, they appear in a 'floating timescale' with alternatives indicated in columns 3–5. The range of the Pennard Formation, indicated by arrows and bars, is from 194 ka to the end of the Pleistocene at 11.6 ka. Geochronology and oxygen isotope ($\delta^{18}O$) stages and events are from Bassinot et al., (1994). See text for explanation of hypotheses and uncertainty*

The Llanddewi Glaciation

Glaciation from mid-Wales reworked Irish Sea erratics of the earlier Penfro glaciation into the Llanddewi Formation of west Gower. Its possible age is adduced from its intrinsic characteristics and field relationships because no unequivocal lithostratigraphical relationships are known, nor are there any geochronological ages available. Critically, however, Llanddewi glacial deposits *in situ* do not overlie raised beaches of the $\delta^{18}O$ 5.5 or 7.1 events, into which Llanddewi erratics were reworked at Horton and Butterslade. Erratics from the Llanddewi Formation within post-$\delta^{18}O$ 5.5 colluvial red beds (George, 1933; Bowen, 1970) show that the Llanddewi Formation was being reworked and redeposited early in the last glacial cycle, or even during the interglacial/glacial transition, as indicated by temperate terrestrial gastropods in the red beds at Pwll Du and the colluvial matrix of the raised beach at Horton. Such early deposition, together with the lithostratigraphical and field relationships of the Western and Eastern Slade deposits, and the relative age implications from clast weathering, invalidate any notion that the Western and Eastern Slade fans are outwash from a Late Devensian glaciation (Rijsdik, 2000).

The field relationships suggest that the Llanddewi Formation is older than $\delta^{18}O$ stage 5, and possibly $\delta^{18}O$ stage 7, and that its deposits were finally reworked into the Pennard Formation. Ice volume indicators (Fig.10.11) show that stages 10 and 8 were times of relatively low ice volume (Mix *et al.*, 1995), but that stages 12 and 6 corresponded with enhanced ice volume. Because the Llanddewi glaciation extended farther south than any previous or subsequent known ice advances from mid-Wales, it is not unreasonable to correlate it with either the extensive Anglian or Ridgacre glaciations of the Severn Valley (Maddy *et al.*, 1995; Maddy & Lewis this volume): that is, with either $\delta^{18}O$ stage 12 or $\delta^{18}O$ stage 6.

Minchin Hole Cave Bed 1 interglacial

Bed 1 (aminozone 1) at Minchin Hole Cave represents the high sea level of $\delta^{18}O$ event 7.1 about 190 ka, which is also recorded at Horton, Butterslade and in south-west England (Bowen *et al.*, 1985; Vliet-Lanöe *et al.*, 2000). Its upper height at 11.8 m at Minchin Hole and about 10 m elsewhere probably indicates subsequent uplift because the estimated global sea level at that time is well below that of the present (Schellmann and Radtke, 2004).

Minchin Hole Cave Bed 3 interglacial

Greater accuracy and precision for the Minchin Hole and Bacon Hole U-series ages using Thermal Ionization Mass Spectrometry (TIMS) is needed. But the high reproducibility of the marine D-aile/L-lle ratios shows their robust nature (Fig. 10.8) and terrestrial ratios from *Cepaea* and *Helicella* enable correlation with the 'last interglacial' in English valley systems (Bowen *et al.*, 1989; Maddy *et al.*, 2003; Maddy and Lewis, this volume).

The temperate mammalian fauna from Gower caves shows an interglacial probably warmer than present. Chemical weathering of cliff slopes and the glacial deposits of the Llanddewi Formation occurred, and soils that require warmer conditions that at present developed (Clayden, 1977; Crampton, 1966). The transition from marine to terrestrial conditions, based on the replacement of marine by terrestrial gastropods at Minchin Hole, precludes any subsequent high sea level during $\delta^{18}O$ stage 5. Orbital tuning suggests a duration of 128 to 120 ka for the $\delta^{18}O$ 5.5 event, but U-series ages on corals suggest between 135 to 114 ka (Muhs *et al.*, 2002), as does U-series terrestrial geochronology (Winograd, 1997). These suggest the importance of orbital forcing at 65 degrees South rather than the traditional 65 degrees North (*e.g.* Henderson and Slowey, 2000). The range of amino acid ratios from the 5.5 raised beaches (Fig. 10.8) could support a longer rather than a shorter interglacial. But would this be an appropriate analogue for the present interglacial or is $\delta^{18}O$ stage 11 better?

In a review of the last interglacial shoreline around the Bristol Channel, Allen (2002) suggested that its high tide coast was between 5 and 10 m. But he assumed that the $\delta^{18}O$ 5.5 sea level was 6 m above the present, whereas other estimates are as low as 2 m. Most 5.5 raised beaches in Gower lie on platforms at about 10 m and are similar to present shingle beaches at present high water. Mean high water at Mumbles is ~3 m thus, even with allowing for a higher 5.5 sea level, some uplift has occurred over the last 122,000 years. It may also explain platforms at the Leys, between Porthcawl and Rest Bay, and the earlier 30 m cliff line at Newton (Driscoll, 1958). This is consistent with uplift of river terraces in southern Britain including the Severn Valley (Maddy, 1997, 2002), as well as raised beaches in southern England (Bowen, 1994, 2003).

The last ice age (Devensian)

Insufficient geochronological ages are available for specific correlation with events in the eccentricity, obliquity and precessional and millennial bands, but general correlations are feasible. The anomalously numbered $\delta^{18}O$ stage 3 (an 'interglacial'!) is explained because Emiliani (1955) originally thought that interglacials were paced by the 41,000 obliquity cycle. The transition between the 'last interglacial' and the last ice age is seen in the Minchin Hole succession (Table 10.2) where the fall of sea level is tracked by marine and terrestrial gastropods (appendix 10.1). Together with a stalagmite age of 81 ka from Bacon Hole Cave, it makes it unlikely that sea level returned to near its present position after $\delta^{18}O$ event 5.5.

The traditional view of Devensian glaciation in Great Britain and Ireland is that ice-sheets grew after 26,000 radiocarbon years ago, reaching their maximum extent (LGM) between ~20 and 18,000 radiocarbon years ago. This contrasts with Scandinavia, Canada and New Zealand, where extensive ice-sheets were present before the LGM. Cosmogenic ^{36}Cl rock exposure ages indicate that Ireland was extensively glaciated before ~36 ka that compares with a calibrated maximum age of ~35 ka from Four Ashes in the west Midlands. It is possible that the LGM glaciation may have overrun earlier Devensian glacial deposits because it has not been possible to separate similar deposits of different ages from the same ice sources (Griffiths, 1940; Squirrell and Downing, 1969). This may account for several LGM reconstructions by different workers (Bowen, 1981). Basal ice-sheds in Carmarthenshire and Glamorgan may also represent earlier glaciation. Regardless of glacial geology uncertainty, the record of ice-rafted fractured quartz in the offshore Barra Fan shows that British glaciations pulsed at millennial timescales in the Devensian before the LGM (Knutz, 2000; Bowen et al., 2002).

South Wales may be divided into two areas: one glaciated at the LGM, the other ice free during the entire Devensian. In the ice free area the deposition of the Pennard Formation commenced with a phase of sheet washing when fine grained sediments, especially red interglacial soils, were reworked and deposited at the base of cliffs. At Western Slade these contain erratics from the Llanddewi Formation and they grade upwards into unsorted fan deposits of reworked sediments from that formation. Periglacial erosion of cliff faces first removed chemically weathered rock to form 'blocky' head often joint and bedding plane bounded. Then frost action on unweathered rock produced finer calibre scree. This periglacial sequence of progressive rock removal is seen, for example, at Gilman Point (Fig. 10.5), between Horton and Western Slade and at the east side of Hunts Bay (Bowen, 1970, 1971, 1977). Thin stalagmites in head, or head coated by stalagmite, indicate hiatuses in head deposition because of climatic amelioration. U-series ages should be obtained from these for millennial timescale correlation.

The LGM ice-margin, between 23 and 19 ka (Clark and Mix, 2002), is shown as a broken line (Fig. 10.1), that is guided by lithostratigraphical control where glacial deposits overlie raised beaches of the $\delta^{18}O$ 5.5 event, and otherwise guided by the extent of the St Asaph, Elenid and combined Breckonshire-Glamorgan Formation extent inland. The ^{36}Cl age in Gower corresponds with Heinrich event 2 (H2) about 23 ka when ice-sheets on either side of the North Atlantic surged into the ocean. Deglaciation followed and ^{36}Cl ages in Ireland show some retreat before 22 ka (Bowen et al., 2002). An age of 15.2 ka from Llanilid kettle holes shows that the ice disappeared before then.

Radiocarbon ages on marine shell fragments in west Wales suggest glaciation from the Irish Sea Basin sometime after 32 ka (36.5 ka cal) (John, 1970), but the shells may have incorporated younger radiocarbon, thus giving ages that are too young (Shotton, 1967). More promising is the ^{14}C calibrated LGM (23–19 ka) age of D-aile/L-Ile ratios of 0.07 on marine shell fragments of *Macoma balthica*, *Macoma calcarea* and *Arctica islandica* from the St Asaph Formation (Davies, 1988; Bowen *et al.*, 2002). These occur in fine-grained deposits of the St Asaph Formation at relatively low levels along the coast, in contrast to higher level shelly sand and gravel deposits at Trellys and Banc-y-warren that yield older D-aile/L-Ile ratios of the δ^{18}O 5.5 age (Davies, 1988; Bowen *et al.*, *in preparation*). It was previously thought that this could show an age difference: the lower deposits being ascribed to the LGM, and the higher ones to an earlier Devensian glaciation (Bowen *et al.*, 2002); but the higher deposits could equally have been emplaced during glaciation and early deglaciation, whereas the lower ones were deposited during later deglaciation, and under glaciomarine conditions (Eyles and McCabe, 1989). Glasser *et al.* (2004) have suggested that iron cemented gravels in subglacial drainage channels also point to two glaciations. Unfortunately neither proposal for two glaciations is based on the necessary criteria of lithostratigraphy or geochronology.

The age of the marine fauna found in deposits at lower elevations shows that marine conditions existed sometime between before, during and after the LGM (23–19 ka) (Bowen *et al.*, 2002), and is consistent with the views of Eyles and McCabe (1989) who proposed glacio-marine conditions at a rapidly retreating ice-magin. At maximum glaciation and greatest isostatic depression of the crust, the gravitational attraction of the large European ice-mass warped glacial sea level significantly upwards at its margin (Mitrovica, 2003). Any ensuing high marine limits close to the ice-sheet margin are not simulated by a geophysical model of the Irish Sea Basin (Lambeck and Purcell, 2001) first, because it simulates anomalously low sea level rates (Mitrociva *et al.*, 2001); second, because it does not include all of the relevant field evidence (Stephens and McCabe, 1977; Clark *et al.*, 2004) and is based only on late-glacial and Holocene ^{14}C age determinations. Polarised views of the Irish Sea glacier are misleading because neither is mutually excusive. Indeed, the legacy of glaciers in tidewater locations consists of both terrestrial and glacio-marine deposits: as in Scotland (Sissons, 1974), Ireland (Eyles and McCabe, 1989; McCabe and Clark, 1998; 2003) and the Gulf of Maine, U.S.A. (Belknap, 2002).

After the LGM, a record of retreating ice is marked by end moraines in the valleys of South Wales (Fig. 10.1). Without age estimates they cannot be correlated, but it is possible to envisage their general timing (Table 10.3). It is proposed that they represent re-advances of the ice-margin that correspond with times of increased precipitation from the North Atlantic (*cf* Alley, 2000). Flat-topped end moraines in the Neath Valley may have been planed off by glacial lakes, and boreholes show that 'sand, gravel and some large boulders' overlie lacustrine clay silt and sand (Anderson and Owen, 1979). Because glacial deposits of the region are characteristically gravelly in nature (Woodland and Evans, 1964) it is possible that this succession indicates rapid ice movement across lake beds. An example in the Neath Valley is the succession of 28 m of clay, silt and sand overlain by 10.5 m of sand and gravel with large boulders upstream from the Clyne Moraine (Fig. 10.1). Together with possible indications of fast ice flow from drumlins farther upstream, a hypothesis for testing would be to audition the Clyne Moraine as a candidate for the H1 (~17 ka) readvance.

It is doubtful if glacial lakes in the Teifi Valley were ever as extensive as has been suggested (*see Elenid Formation*). Currently available evidence suggests a maximum lake elevation of 125–130 m, and the proposal that Lake Llanllwni drained subglacially (Price, 1976) is also applicable to Lake Teifi. Such outburst floods (jökulhaups) may have provided catastrophic floods that fashioned the glacial drainage system at Fishguard (Bowen and Gregory, 1965). These and similar discharges of freshwater into the North Atlantic may correlate with δ^{18}O spikes in marine records (Knutz *et al.*, 2002).

A global marine transgression at 14.7 ka (Meltwater Pulse 1A) preceded the Bølling-Allerød interstadial events, towards the end of which ice grew in Welsh cirques. Chlorine-36 rock exposure ages show that cirque glaciation was at its maximum extent by 12.9 ka and had retreated to about half its size by 11.6 ± 1.3 ka at Cwm Idwal (Eryri) (Phillips *et al.*, 1994). Such unexpectedly early growth during the later Bølling-Allerød was probably driven by precipitation enhancement, followed by precipitation deficit during the Younger Dryas (Bowen, 1999a). This sequence of events is also proposed for Llyn y Fan Fach, Craig-y-llyn, Cwm Saerbren and Graig Fawr.

Periglacial processes operated throughout the last ice age beyond, and above, the LGM ice, and extended farther inland as the ice retreated, where head and solifluction deposits are widespread. Considerable thicknesses of head accumulated below steep slopes after deglaciation, as shown at Rhosili Bay where it overlies LGM gravels (George, 1933; Bowen, 1970). Glacial deposits on steep slopes were also reworked (Harris and Wright, 1980). Ice wedge casts, indicative of permafrost occur in Pembrokeshire (John, 1970; Case, 1984) (Fig. 10.5) and west Carmarthenshire (Bowen, 1970). Possible remains of pingos occur around Llanpumsaint and Pontarsais, and north and north-west of Pumsaint (Bowen. 1974). Loess, characteristic of dry, cold and windy conditions is ubiquitous in South Wales and is widely exposed as the Horton Member (Figs. 10.4, 10.5) along the coastline (Case, 1977, 1983, 1984); unexpectedly, its heavy minerals have a western provenance (Case, 1983).

APPENDIX 10.1

Terrestrial Gastropods at Minchin Hole Cave and Bacon Hole Cave
by J.G. Evans and C. French

Minchin Hole Cave

Figure 10.12 shows the distribution of marine and terrestrial gastropods from Bed 1 to Bed 5 in Minchin Hole Cave that was exposed towards the cave entrance. Figure 10.13 shows the succession of terrestrial gastropods from thicker deposits inside the cave from Beds 4b, 4c and 5 (analysis: Evans and Betty Race).

The marine species indicate a littoral environment trending from mid to upper shore (Fig. 10.12). The species *Littorina littorea* and *Nucella lapillus* are characteristic of mid-shore conditions, while *Littorina saxatilis* and *Littorina neritoides* are more typical of the upper shore. *Littorina neritoides* is very specific to the high-tide zone and even occurs in the splash zone above that. This suggests a declining marine influence through the sequence, an interpretation supported by land snails present only in the upper sequence. There is a suggestion of a succession with a slightly more diverse, possibly more shaded, fauna lower down, and a more open, grassland, fauna towards the top of the sequence, indicated by *Helicella itala*, *Pupilla muscorum*, *Vallonia excentrica* and *Trichia hispida* that suggest environmental change, perhaps associated with overlying Bed 6 of blown sand.

A block of stalagmite on the surface of Bed 3 (*Patella* beach) contained terrestrial gastropods. They represent an altogether more woodland fauna than those in Beds 4b to 5 despite two specific open-country species, *Pupilla muscorum* and *Helicella itala*. They are: *Carychium tridentatum*, *Vertigo angustior*, *Pupilla muscorum*, *Lauria cylindracea*, *Vallonia costata*, *Acanthinula aculeata*, *Spermodea lamellata*, *Punctum pygmaeum*, *Vitrea sp.*, *Aegopinella* cf *nitidula*, *Oxychilus* cf *cellarius*, *Deroceras sp.*, *Euconulus sp.*, *Clausilia bidentata*, *Balea perversa*, cf *Helicella itala*, *Cepaea*, *Arianta*. This rich fauna represents temperate climate conditions.

The raised beach and cave-earth sequence shows a decrease in the influence of the sea, the development of a diverse vegetation in a terrestrial maritime environment, and then an opening of this towards

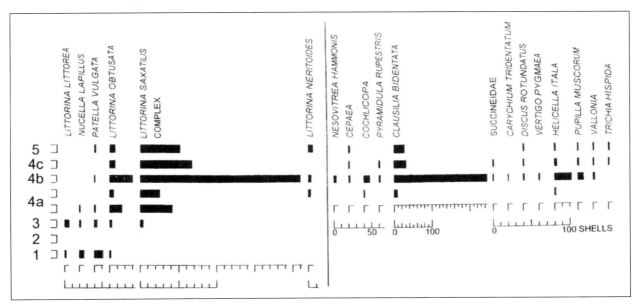

Fig. 10.12 Marine and terrestrial gastropod succession at Minchin Hole Cave, Gower towards the cave entrance. 1–5 are the Pleistocene beds as shown in Table 10.2

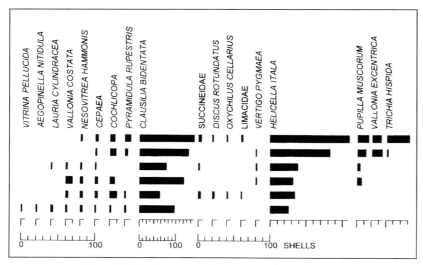

Fig. 10.13 Terrestrial gastropod succession inside Minchin Hole Cave

conditions of grassland. The climate was temperate, although almost all the species are recorded by Sparks (1964) from the later part of the Ipswichian Interglacial in the pine woodland and steppe zone. There are similarities between the cave-earth faunas at Minchin Hole, Bacon Hole, Pwll Du Head and Caswell Bay.

Bacon Hole Cave

Bacon Hole Bed 2 (coarse-sands) yielded *Cochlicopa lubricella* 4, *Pyramidula rupestris* 68, *Pupilla muscorum* 80, *Lauria cylindracea* 45, *Vitrina pellucida* 1, *Oxychilus cellarius* 1, *Oxychilus* cf *alliarius* 6 and *Cepaea sp.* 1. This fauna differs significantly from the other faunas discussed in the low diversity and high numbers of three species. It is a much more narrow fauna indicative of rock surfaces (*Pyramidula* and *Lauria*) and grassland (*Pupilla*), and the absence of *Clausilia* (rupestral) and *Helicella* (grassland), common at Minchin Hole in similar habitats suggests a different age. The *Pupilla* shells are significantly longer than those of the modern fauna, suggesting a cold-climate origin or a narrow sand-dune and rock environment. This fauna is significantly different from the cave-earth faunas at both caves.

The terrestrial gastropod fauna from Bed 4 (sandy cave-earth) at Bacon Hole is similar to that from Bed 4b in Minchin Hole in the predominance of *Clausilia bidentata* and the presence of *Pupilla muscorum*, *Vallonia sp.*, *Discus rotundatus*, *Nesovitrea hammonis*, *Helicella itala* and *Cepaea sp.*

11 South-west England

by Stephan Harrison *and* David H. Keen

Introduction

Identification of the encroachment of glacial ice on to the South-west Peninsula goes back to Maw (1864) and the topic has been revisited throughout the 20th century (Campbell *et al.*, 1998). Despite this continuing interest, much remains uncertain about the glaciation of the area. In particular the type of glaciation, the extent of the ice and the age or ages of the glacial incursions are all still subjects of debate.

Particular problems are occasioned by the difficulty of dating the glacial sediments themselves and the need to date overlying or underlying deposits to give a framework of time for the ice advances. Although modern advances in dating such as amino acid geochronology, electron-spin resonance and radiocarbon have all been used to date aspects of the South-west England successions (Bowen *et al.*, 1985; Scourse, 1991; van Vliet-Lanoë *et al.*, 2000) the small number of determinations and the uncertainties of dating in the late Middle Pleistocene have still left considerable room for debate about the age of the sequences.

The examination of the glacial deposits of South-west England is best tackled on a regional basis. The following groups have been chosen to facilitate description; the tills of the north coast (Trebetherick, Lundy, Barnstaple Bay); the Somerset Levels and Bristol area; the Scillies; the erratics of the south coast and the possible glaciation of the upland moors (Exmoor and Dartmoor) (see Fig. 11.1).

Glaciations of the North Coast

The main area of evidence for glaciation on the north coast is around Barnstaple Bay and on Lundy, although the deposits at Trebetherick on the Camel estuary have also been proposed as glacial sediments.

Trebetherick

The sequence at this site was first described by Ussher (1879). The most detailed account is that of Arkell (1943) whilst a full review of the history of research at Trebetherick can be seen in Campbell *et al.* (1998). The main units are found at the base of a raised beach, and composed of both local and erratic material including elvan, dolerite and flint. This is overlain by blown sand up to 6 m thick, and 3 m of head (a diamicton called 'the boulder bed') which sometimes forms a discrete unit, and at other times is incorporated into the head, an upper head and Holocene blown sand. The boulder bed also contains erratics including porphyry, granite, flint and mica schist.

The identification of glacial deposits at Trebetherick depends on the interpretation of the origin of the erratics in the raised beach and the boulder bed. Arkell (1943) was of the opinion that all lithologies exotic to the section could be traced to an origin within North Cornwall and were not from glacial sources. Mitchell & Orme (1967) however, referred to the boulder bed as outwash from Irish Sea ice, and Stephens (1970) identified the erratics as having a source in weathered till and outwash and being from regions outside the South-west Peninsula.

The age of the sequence has also been much discussed. Arkell (1943) suggested that the raised beach was equivalent in age to the Boyn Hill Terrace of the Thames (Hoxnian?) (Marine Isotope Stage [MIS]

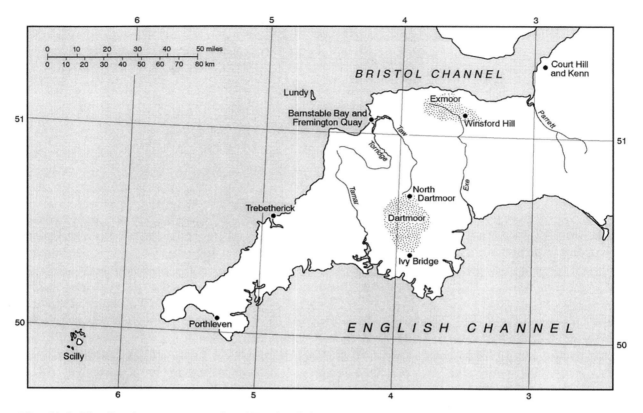

Fig. 11.1 The South-west peninsula of England the Scilly Isles showing locations discussed in the text

11 or 9) and the boulder bed as being of 'Wolvercote Interglacial' (Marine Isotope Sub-stage [MISs] 5e) age and probably deposited by the River Camel. Stephens (1970) agreed that the raised beach was of Hoxnian age, but preferred the boulder bed, which he regarded as of glacial origin, as dating to the 'Wolstonian' glacial stage (MIS 6). Amino acid determinations by Bowen *et al.* (1985) gave an age in MISs 5e for the raised beach, thus dating all of the overlying units to the Last Glaciation. Scourse (1996), although not ruling out a glaciation of the Camel estuary, thought that all of the erratics in the basal raised beach and the boulder bed could be provenanced in the Camel catchment. He, too, suggested an age in MISs 5e for the raised beach thus dating all of the overlying sediments to the Last Glaciation, whatever their depositional origin.

Although the Camel Estuary is exposed to the north-west and might be expected to provide a trap for glacial sediments and thus be the site of the formation of glacial deposits from Irish Sea ice, unequivocal evidence for a glaciation of this part of the Cornish coast is lacking. It is probable from amino acid evidence that the sequence of deposits at Trebetherick are all of MISs 5e age or younger.

Lundy
The island of Lundy lies in Barnstaple Bay *c.* 30 km west of Morte Point. The bedrock of the island is almost entirely of Tertiary granite, but the northern plateau has extensive scatters of pebbles of erratic sandstone, flint, chert and greywacke reaching 107 m O.D. The granite of the north-west coast exhibits ice-moulding and these lines of evidence were used by Mitchell (1968) to identify ice movement from west-north-west to east-south-east across the island. The age of the glaciation was uncertain, although both Anglian (MIS 12) and Wolstonian (MIS 6?) ages have been suggested by Mitchell.

In view of the uncertainty of the glacial origin of the Trebetherick deposits and those around Barnstaple (below) the strong evidence for glaciation on Lundy is important in indicating that an ice-mass did occur in the Bristol Channel at some stage of the Pleistocene.

Barnstaple Bay

The evidence for glaciation of the Barnstaple area is in the form of both erratics within the modern shore zone on the north side of Barnstaple Bay at Croyde and Saunton, and diamicton at Fremington Quay and Brannam's Pit to the south.

a) Saunton and Croyde

The Croyde and Saunton erratics are a varied suite of igneous (pink granite, spilite, quartz porphyry, dolerite), metamorphic (granulite gneiss) and sedimentary (Carboniferous Limestone, sandstone, arkose) rocks, the largest of which are up to *c*. 50 tons in weight (Maw, 1864; Stephens, 1970). Petrological work by Dewey (1910) suggested that those that could be matched to an outcrop originated in northern Scotland. The erratics occupy a narrow zone up to about 10 m above O.D. and are not found inland of the coast. This occurrence strongly suggests an origin by ice rafting, although the size and petrology of the boulders would indicate ice-berg rather than floe ice transport from a calving margin to the north or west of their current position. Despite the abundance of erratics no undoubted till has been recorded in the area, although Stephens (1970) notes diamicton which he regards as weathered till at several places in Croyde Bay.

The age of deposition of the boulders is uncertain. At Saunton they are overlain by raised beach, blown sand and head (Campbell *et al.*,1998). The beach has been dated to the Hoxnian (MIS 11 or 9) by Stephens (1970), but amino acid determinations by Bowen *et al.* (1985) suggest a possible date in MISs 5e. In the first case an ice rafting event prior to MIS 11 is indicated, in the second, the erratics are older than MISs 5e.

b) Fremington Quay

The exposures at Fremington Quay have been regarded as crucial for establishing the relationship between the diamicton at Brannam's Clay Pit (below) and the raised beaches of the coastal sections which allow some possibility of dating the inferred glacial deposits. Both the succession at Brannam's Pit and at Fremington have at their base rolled gravel which may be of raised beach origin, although some authors (see Campbell *et al.*, 1998 for a full discussion) prefer either a fluvial or glacio-fluvial origin for these sediments.

At Fremington Quay the sections extend for 500 m along the south side of the Taw estuary. They consist of 0.5 to 1.5 m of gravel overlain by 1.6 m of sandy silt and sand and 2.0 m of gravel (Campbell *et al.*, 1998). The gravel contains erratics, some of which are striated, but the silts and sands are without foreign stones. Although Stephens (1970) regarded the Fremington silts as till or soliflucted till, Kidson & Wood (1974) pointed to the lack of glacial features in the silts and denied a glacial origin for them. Suggestions by Campbell *et al.* (1998) that the basal gravels and underlying bedrock exhibit glacio-tectonic structure however, point to a possible glacial origin.

As elsewhere in Barnstaple Bay, the dating of the possible glacial deposits is closely linked with their relationship to the raised beach. Stephens (1970) regards the beach as being of Hoxnian age (MIS 11 or 9) at both Brannam's Pit and Fremington, thus the overlying sediments must be of MIS 10 age or younger. Kidson & Wood (1974) showed that the clast content of the gravels at Fremington and Brannam's Pit are different, with only the former exhibiting erratics. They conclude that the coastal sections show a raised beach which they regarded as Ipswichian (MISs 5e). Therefore the overlying deposits (which they did not regard as a till) must date from the last glaciation (MIS 4 or 2). The gravels inland, without erratics, were undateable and could give no indication of the age of the overlying material.

c) Brannam's Clay Pit
The deposits to the S of Fremington village and 2 km inland from the coastal localities noted above at Fremington Quay, have received attention from Earth Scientists since the middle of the nineteenth century (Maw, 1864; Ussher, 1878). The extent of the sediment body is far larger than any of the other possible glacial remnants in the northern part of the South-west Peninsula. The Fremington Clay occurs over an east/west spread of 4 km and is around 800 m wide at its widest point. Its thickness proved in boreholes is 27 m. The basal unit of the succession is a gravel composed of local sandstone and shale. Overlying this are clays and clayey sand with occasional pebbles of lignite, sandstone and mudstone. Rare cobbles of undoubted erratics are sometime found (Maw, 1864; Dewey, 1910, Campbell *et al.*, 1998) and lithologies recorded include andesite, spilite, quartz dolerite, olivine dolerite and quartz porphyry. Some of these erratics are striated, but the majority of the clasts in the clay show no evidence of fabric such as would be expected from ice transport. In addition to the erratics some horizons contain shell fragments, and pollen and spores reworked from early Tertiary sources have also been reported (Croot *et al.*, 1996).

Because of the undoubted, but sparse erratic content of the Fremington Clay, most authorities have accepted that it was deposited in cold conditions and with some glacial connection, although whether as till, glacio-marine or glacio-lacustrine sediment is still the focus of debate. The lack of fabric derived from ice movement and the occurrence of dropstone structures around some of the clasts has suggested to Croot *et al.*, (1996) that the clays were deposited in standing water. The lack of a marine fauna led them to suggest a glacio-lacustrine environment, although Eyles & McCabe (1991) in keeping with their general preference for the origin of many of the glacial deposits at the limit of the Irish Sea ice, suggested a glacio-marine origin for the Fremington Clay.

The age of the deposits at Fremington has also been the subject of considerable discussion. Early ideas of the continuity of the basal gravel with the raised beach of Barnstaple Bay and thus an age younger than the gravels (see above) have been challenged by Kidson & Wood (1974) who saw no connection between the gravel at Fremington Quay and at Brannam's Pit. Amino acid ratios on the shells from Brannam's Pit (Bowen, 1994) proved inconclusive, although an age in the Anglian (MIS 12) was tentatively suggested. Optically stimulated luminescence (OSL) dates by Croot *et al.* (1996) also favoured an Anglian age for the glaciation.

Bristol area and the Somerset Levels
The distribution of erratics in the Bristol area was known to nineteenth-century geologists (see review in Campbell *et al.*, 1998), but the discovery of undoubted till during the construction of the M5 motorway in the 1970s, and in drainage ditches in the northern Somerset Levels, gave clear proof of the incursion of ice into the region. The age of the glaciation, as usual in South-west England is uncertain and depends on the dating of overlying deposits.

Court Hill

The deposits at Court Hill were revealed by the cutting for the M5 at the western end of the Mendips (the Failland Ridge) as the motorway reached the summit of the ridge. The sediments are 24 m thick and consist of tills and flow-tills to the north of the section deposited by ice abutting the north-west side of the ridge and part-filling a channel cut into the Carboniferous Limestone bedrock. To the south, the tills are replaced by outwash of cobbles and boulders. The tills and gravels interdigitate in the central part of the sequence. The glacial origin of the deposits are underlined by the occurrence of Chert erratics and foraminifera from the Cretaceous, probably from Irish Sea sources. The age of the glaciation was originally assumed to be 'Wolstonian' in keeping with the then generally accepted age of the most southerly extent of ice when the Court Hill deposits were described (Gilbertson & Hawkins, 1978a). Further discussion of the age of the glaciation of the Bristol area will be found below.

Kenn

Diamictons occur beneath younger interglacial deposits at Kenn in the northern part of the Somerset Levels (Gilbertson & Hawkins, 1978b). The sequence consists of up to 4 m of cobble gravel overlain by silty and sandy diamicton with striated boulders up to 1 m in diameter. The overlying sediments consist of shelly sands and silts named by Gilbertson & Hawkins (1978b) as the Yew Tree Formation.

The Mollusca and pollen from these deposits are from a fully interglacial context and indicative of summer temperatures up to 2°C warmer than the present (see discussion in Campbell *et al.*, 1998). Their position overlying the diamicton thus gives a minimum age for the underlying glacial sediments. Shells from the Kenn deposits have been dated by amino acid analysis and gave ratios indicative of an age in MIS 15 (Andrews, Gilbertson & Hawkins, 1984) which would date the glacial sediments to MIS 16 at the youngest.

The age of the glaciation of the Bristol area

The amino acid ratios from the sites around Kenn with an age in MIS 15 (see review in Campbell *et al.*, 1998) would place the interglacial deposits in the Cromer Complex (Bowen, 1999). Sediments of such an age would be expected to produce a molluscan assemblage like that of the Cromerian sites of East Anglia (Preece & Parfitt, 2000). Significant 'Cromerian' taxa such as *Bithynia troscheli*, *Tanousia* and *Valvata goldfussiana* should be present. However, none of these species have been recorded from any of the localities at Kenn. Instead species such as *Bithynia tentaculata* and *Valvata piscinalis* are prominent members of the fauna, both of which are typical of post-Anglian assemblages (Preece & Parfitt, 2000; Keen, 2001). Also prominent in the Kenn deposits are *Belgrandia marginata* and *Corbicula fluminalis*. The former species is rare in the Cromerian, being found only in deposits regarded as being of very late Cromer Complex age from Sidestrand, Norfolk (Preece & Parfitt, 2000) and nowhere else in the Cromerian. *C. fluminalis* has never been found in the Cromerian but is common in MIS 11, 9 and 7 (Keen, 2001). Despite the amino acid evidence, the molluscan biostratigraphy indicates very strongly that the interglacial deposits around Kenn are not as old as the Cromerian. A more likely age would be in MIS 11 or 9 when both *C. fluminalis* and *B. marginata* are common in fluvial deposits in southern Britain (Keen, 2001). The suggestion by Gilberton & Hawkins (1978b) of summer temperatures of 2°C higher than the present would tend to favour an age in MIS 9 for the Kenn deposits as it is becoming clear that the MIS 9 interglacial was somewhat warmer than MIS 11 or MIS 7 (Green *et al.*, in press). An age in MIS 9 for the interglacial deposits would suggest a date no younger than MIS 10 for the underlying glacial deposits, an age that would also apply to the tills of Court Hill if their connection with Kenn is accepted.

South Coast Erratics

Along the south coasts of the South-west Peninsula numerous erratics occur either on the shore platform within modern tidal limits or within the low raised beaches (Campbell *et al.*, 1998; van Vliet-Lanoë *et al.*, 2000). These erratics occur along the south coast of England as far east as Selsey in Sussex, (Mottershead, 1977), although concentrations occur in Cornwall around Porthleven and in the coastal section between Start and Prawle Points in Devon. The most spectacular of these erratics is the 50 ton garnetiferous microcline gneiss block known as the 'Giant's Rock' at Porthleven (Flett & Hill, 1912). Despite the suggestions of Kellaway *et al.* (1975) of a general glaciation of the Channel in the Middle Pleistocene, the lack of any erratics from heights in excess of 10 m above sea level clearly indicates a drift ice origin for them rather than emplacement from glacial ice. This explanation also holds good for similar erratics found along the north Brittany coast thought by all authors to be well beyond the limit of glacial ice (van Vliet-Lanoë *et al.*, 2000).

Despite the agreement that these erratics owe their current position to ice rafting of some form rather than deposition from a glacier, the exact mechanism of their deposition remains obscure. The size of the Giant's Rock rules out movement by floe ice (van Vliet-Lanoë *et al.*, 2000) and suggests that the erratics were liberated from an Irish Sea glacier at a calving margin in the Western Approaches and were then blown up Channel to ground on the coasts of South-west England and Brittany. However, it is generally thought that glacial sea-levels were much lower than those of the present interglacial so stranding icebergs would have actually left their erratic content far down the submarine slope from the modern shore zone. The smaller erratics could have then been moved in-shore by storms, but the Giant's Rock is too big for such a process and Flett & Hill (1912) note that it is never moved even by the most violent Atlantic storms. Deposition of the erratics under conditions of isostatic depression of the Channel crust and rebound to the current level seems a most likely explanation for the position of the erratics, but the full details of this hypothesis remain to be determined.

Although the petrology of most of the erratics suggest an origin within the British Isles (see also the Saunton erratics, above) (Flett & Hill, 1912), the Giant's Rock does not fit with any known Scottish metamorphic terrain and Bowen (1994) has speculated that it could be from Greenland and brought into the Channel during a Heinrich-type event by ice-bergs from the collapsing Laurentide Ice-Sheet. A similar origin might have allowed the Icelandic basalt recorded from Brittany to reach Finisterre (van Vliet-Lanoë *et al.*, 2000).

The age of emplacement of the erratics is uncertain. It is not even clear whether they are the product of a single phase of ice-berg transport or were introduced by every ice-sheet which reached the Western Approaches. Erratics are found in the raised beaches below 10 m O.D. which are mostly dated to MIS 7 or younger (Bates, Keen & Lautridou, 2003) suggesting ice rafting events in MIS 6 and 4/2. More extensive glaciation in MIS 12 and 16 probably also gave the pre-conditions for ice-berg rafting into the Channel (van Vliet-Lanoë *et al.*, 2000), but confirmatory geochronometric dates of such events have so far not been obtained.

The Isles of Scilly

The Isles of Scilly are located some 40 km due west of Land's End and they are regarded as having been glaciated during either the Wolstonian (*sic*) (Mitchell and Orme 1967) or the Late Devensian (MIS 2) (Scourse, 1991). Their position and the nature of the glacial evidence suggests that they represent the furthest southerly extent of British glaciation at this time. As a result, correct interpretation of the sedimentological and landform evidence is crucial for understanding the nature and extent of glaciation in the south-west of England.

Scourse (1991) sub-divided the Isles of Scilly into two distinct zones: a northern glaciated part, and a southern zone which lay outside the ice limits (Fig. 11.2). This differentiation is reflected in the sedi-

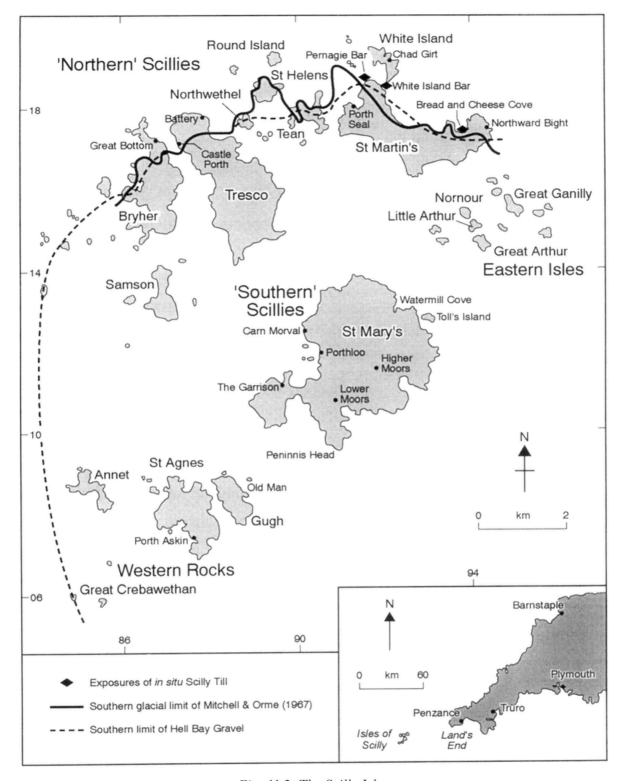

Fig. 11.2 The Scilly Isles

ments exposed in these zones and in their geomorphology. The northern zone contains a number of sedimentary units including the Scilly Till, Tregarthen Gravel and the Hell Bay Gravel. The southern occurrence of the latter unit marked the ice limits. In the southern, unglaciated, part of the Scilly Isles solifluction of sediments and aeolian deposition produced a number of distinct deposits including the Porthloo Breccia and Old Man Sandloess. Organic sediments underlie the glacial units in the northern zone and thus must predate their deposition. Radiocarbon dating of these sediments indicate deposition between 34,500+/-850 years and 21,500+/-890 years BP. In the unglaciated southern zone, the deposition of a coarse aeolian sediment (the Old Man Sandloess) was regarded by Scourse (1991) as having occurred in association with the glacial advance, and is dated between 18,600+/-3700 and 26,000+/-10,000 years BP by TL and OSL techniques. As a result, Scourse & Furze (2001) argue that this evidence suggests that glacial ice advanced to the northern Scillies during the Late Devensian (MIS 2).

There is also a strong relationship between the geomorphology of the two zones and their environmental history. In the northern zone marine bars at St Martin's and White Island are underlain by Scilly till and contain erratics on their surfaces. Scourse (1987) argues that these bars may originally have had a morainic origin, although their present form has been modified by marine processes. The shape of the Scilly tors also shows a pronounced change between the two zones. Smoothed, eroded tors occur in the northern, glaciated zone and 'castellated' tors in the southern, unglaciated zone; the inference here is that to the south of the glacial limit, periglacial activity affected tor development.

Glaciation of the South-west England moors

Most work on the glaciations of South-west England has concentrated on the evidence along the northern coast of the peninsula, but there has also been considerable research aimed at assessing the evidence for former glaciers on the upland hills and moors of the area. Much of this research dates from the 1940s and earlier and concentrates on the landforms and sediments found on Dartmoor. However, recent research showing striking evidence for small glaciers on Exmoor may serve to highlight some of the unanswered questions concerning possible glaciation of Dartmoor and provide a stimulus for future work.

Exmoor

The evidence for glaciation on Exmoor can be subdivided into two; the evidence for cirque glaciation and for a small valley glacier and an ice-cap.

Cirque Glaciation

The Punchbowl is a deep north-facing hollow cut into the slopes of Winsford Hill in central Exmoor (Fig. 11.3). The floor of the semi-circular basin lies at around 300 m OD whilst the backwall rises at an angle of about 70° to 400 m OD. The hollow is up to 350 m in width and is fronted by a subdued arcuate ridge up to 11 m in height and 40 m wide. The sediments within the ridge show characteristics of glacial deposition, including compression macro-fabrics and edge rounding. Taken together, the sedimentological and morphological evidence suggested that the Punchbowl was the site of a former glacier (Harrison *et al.*, 1998).

Valley Glacier and Ice Cap

The original work on the Punchbowl described the evidence for a small cirque glacier with an ELA of 334 m (Harrison *et al.*, 1998). Subsequent work showed that the glacier was more extensive, and had extended several hundred metres downvalley from the mouth of the hollow at some stage (Harrison *et al.*, 2001a). This small valley glacier deposited subglacial tills which are exposed in the valley-bottom. They form a matrix-rich deposit incorporating subrounded and rounded stones and the orientation of these show that the deposit underwent

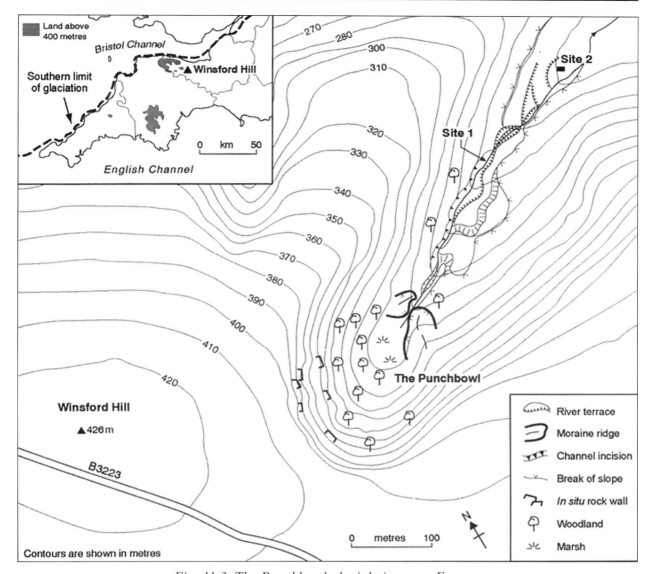

Fig. 11.3 The Punchbowl glacial cirque on Exmoor

considerable subglacial deformation. To the north of these, subdued valley-floor fluvial terraces were interpreted as glaciofluvial landforms, deposited by meltwater from the glacier. Harrison *et al.* (2001a) argued from the presence of deformation tills some 500 m north of the backwall of the Punchbowl, that the valley glacier was unlikely to have been constrained by the hollow and probably flowed from a small ice-cap located on the broad plateau surface of Winsford Hill.

While there has been much debate about the nature of the landforms and sediments developed at this site (see Prudden 2001; Harrison *et al.*, 2001b; Straw 2002; Harrison *et al.*, 2002), the evidence for glaciation at this site seems unequivocal and should not be surprising. There is considerable evidence to show that large ice-sheets reached the coasts of South-west England at times during the Quaternary (see above) and, immediately outside the limits of contemporary ice-sheets and ice-caps (and even small valley glaciers) it is usual to find high ground nurturing small glaciers and ice-masses. The distribution of these is controlled by the local equilibrium line and the nature of the topography. Thus it is surprising that glacial landforms and sediments on Exmoor have only recently been discovered. It also allows us to speculate on whether the much more extensive and higher plateau surfaces of Dartmoor also held glaciers and ice-caps during the Quaternary.

Dartmoor
The evidence for small glaciers and ice-caps at relatively low altitudes on Exmoor suggests that similar ice-masses should have developed on the high plateaus and deep valleys of Dartmoor during the Pleistocene. Whilst the present general consensus amongst geomorphologists is that Dartmoor remained ice-free during the Quaternary (*cf* Charman *et al*. 1996), there is evidence that could be taken to support the alternative view that Dartmoor was glaciated at one time (Harrison, 2001).

Discussions on the glaciation of Dartmoor have a long history (Ormerod, 1869; Somervail, 1897; Pillar, 1917). In the most detailed treatment, Pickard (1943) argued for the former existence of extensive glaciers and small ice-caps on Dartmoor during the Quaternary and discussed in some detail the evidence for this. He subdivided the landform evidence into: '(a) worn boulders, (b) moraines, (c) grooved rocks, (d) slabs, (e) the terrain, and (f) gravels' (Pickard, 1943, p. 28). He considered many of the 'worn boulders' to have been transported and modified by ice; ridges of blocky debris in many of the valleys he suggested were moraines and he argued that asymmetrically-shaped boulders and cobbles, and shallow grooves on these and on bedrock, resulted from glacial erosion. Although no *detailed* rebuttals of these arguments have been published, the modern consensus is that the landform evidence presented by Pickard did not reflect the former presence of glacial ice on the moor.

There are, however, two main lines of landform evidence which point to the former existence of ice-masses on Dartmoor. First, overdeepened valleys and valley heads exist on the northern flanks of the moor especially in the vicinity of Sourton Common; Okehampton Common and South Tawton Common (Fig. 11.1). Associated with these are linear boulder spreads and ridges which run along the east-facing hillsides above the West Okement River at the Slipper Stones and in the Taw Valley. These and other boulder ridges were interpreted as medial and lateral moraines by Pickard (1943) and it is difficult to see how such depositional and topographically distinct ridges could have been formed by periglacial or fluvial processes. They do not possess the characteristics of protalus ramparts, nor do they appear to be structurally controlled. Second, at several places around the margins of Dartmoor, thick accumulations of bouldery debris have been described from valley-bottom locations and interpreted as the product of fluvial reworking of periglacial sediments. For instance, at Ivybridge on the southern edge of Dartmoor, Gilbertson & Sims (1974) have described a suite of debris cones and alluvial fans containing poorly sorted deposits and boulders up to 2 m in length. They assumed that these were deposited by periglacial 'earthflows' and 'slushflows'. However, such accumulations are also common in glaciated areas where they reflect the importance of sediment transport by meltwater. No modern accounts of these accumulations and other deposits derived from Dartmoor (such as those infilling Plymouth Sound and in the South Hams district) exist in the literature to allow a reinterpretation of their significance. However, it is clear that there are facets of the landscape which require further investigation. It is also clear that the evidence for the glaciation of Dartmoor (if it exists at all) is equivocal.

The possible nature of the Dartmoor ice-masses and some implications for landform development
If glacier ice had existed on Dartmoor during the Quaternary, what form might it have taken and what sort of sediment accumulations could it have given rise to? The ice-masses were probably thin and, in the absence of steep slopes, subglacial shear stresses would have been low and the ice slow moving. Since

the former presence of extensive permafrost at low altitudes in southern England during the Late Devensian is reflected by ice wedge casts (*cf* Ballantyne & Harris, 1994) we would expect that at these, and earlier, times the climate on the Dartmoor uplands would have been colder. As a result, the glaciers here would probably have been cold-based on the highest ground and perhaps warm-based or polythermal only along the deeper valleys. Accumulation of ice would have been facilitated by drifting on the extensive plateaus and by the addition of rime ice on exposed bedrock surfaces. If glacial ice did develop on Dartmoor where is the unequivocal landform evidence? The observations presented by Pickard (1943) and subsequent work suggest that this evidence is subtle and open to other interpretations and this equivocality might well reflect the nature of the ice-masses. This can be explained in three ways:

1. Thin ice on low-angled slopes would only generate low shear stresses (perhaps in the order of 40 to 45 kNm^{-2}) and as a result, movement would be slow. Consequently, it is unlikely that ice velocities would have been high enough to generate lee-side cavities in hummocky bedrock terrain and, therefore, erosional bedrock features such as roches moutonées would be unlikely to have developed. The absence of small-scale erosional features such as striations and chattermarks simply reflects the fact that the coarse crystalline Dartmoor granites are not suitable for the preservation of such features. Striations are preserved, however, on the Exmoor slates (see Harrison *et al.* (1998) Fig. 8).

2. Glacier ice frozen to its bed would play a protective, rather than erosive, role allowing even delicate periglacial landforms and sediments to survive perhaps repeated glaciations (*cf* Kleman, 1994; Kleman & Borgström, 1994).

3. The absence of high bedrock cliffs above the accumulation areas of the ice-masses means that rock debris did not fall onto the glacier surfaces to be incorporated to form the material deposited as glacial moraines and till. The only material available for entrainment would have been the unconsolidated weathered granite common on Dartmoor. Low ice velocities and the prevailing thermal regime may have precluded the large-scale entrainment and transport of this material, but if this had happened locally, it might be difficult to differentiate between glacially-transported weathered granite and that moved by slope processes.

Whilst the role of glacial ice over much of the high ground of Dartmoor might be expected to be protective, it is likely that along the deeper valleys (such as in the West Okement Valley) ice accumulation would have been sufficient to increase subglacial shear stresses sufficiently to enable some glacial erosion to have occurred. It may be at these locations that evidence for glacial erosion should be sought. If found, this would demonstrate that Dartmoor was not a true periglacial landscape but one which owed much of its form to glacial protection and, perhaps, limited glacial modification.

Summary and conclusion
South-west England was peripheral to the ice-sheets of the Pleistocene, but evidence of glaciation on both large and small scale is preserved in the region. The evidence for the encroachment of land ice from Irish Sea sources on to the Peninsula is only strong in the Scillies, although Lundy, the Barnstaple Bay sites and those around Bristol also provide some evidence for the presence of ice. There is little evidence of the limits of the ice, but it seems probable that the ice-sheets responsible for the tills of the Scillies and the north of the Peninsula in general did not have the power to pass far inland even in the subdued terrain of the Somerset Levels. There is no plausible evidence for glaciation of the south coast of the Peninsula,

but the constant presence of ice-berg rafted boulders gives abundant evidence of ice-sheet calving in the Western Approaches and possible ice transport of rocks from as far away as Greenland and Iceland. With an ice front as close to the north coast as Lundy and the Scillies, it is no surprise to find evidence of small glaciers on Exmoor, although the evidence for similar glaciers on Dartmoor remains elusive.

Despite the advances in geochronometric dating, in the 35 years since the first edition of the *Glaciations of Wales and adjoining regions* was published advances in dating of the glaciations of Southwest England have been slow. Much of the data for putting the evidence for glaciation in time order is equivocal and for many sites the only evidence for relative age is superposition. The short timescale for the Middle and Late Pleistocene in use when the first edition was published no longer finds favour, and the longer timespan available for events now accepted, gives much more scope for a number of glacial events to be recognised. Despite this, only the radiocarbon-determined ages for the glaciation of the Scillies seem in any way reliable, and it is possible that glacial ice reached the Peninsula in a number of cold stages prior to the Last Glaciation.

12   The Irish Sea Basin

by Jasper Knight

Late Pleistocene glaciers in the British Isles spread generally from upland source areas and terminated on coastal lowlands and adjacent continental shelves which were eustatically dry (Bowen *et al.*, 1986). The record of glaciation in the British Isles, therefore, is imprinted in landform-sediment assemblages which are located both onshore and offshore the present land-sea boundary. Linking together the glacial evidence from these two distinct realms is not an easy task because they have been investigated using different methodologies, techniques and scales of observation. In onshore areas satellite imagery, digital elevation models (DEMs) and detailed spatial data on relief and substrate type are readily available, and field-based geomorphic mapping and lithostratigraphic recording of sedimentary exposures are commonplace. In offshore areas, relief is usually subdued, glacial landforms may be draped by Holocene marine sediments, remote-sensing imagery (*e.g.* Chirp and boomer sub-bottom profiler) is expensive to acquire and process, and ground-truthing is achieved by coring which provides only very limited spatial resolution. There are therefore important differences in spatial coverage and data quantity/quality between onshore and offshore locations, even over relatively short distances. An outcome of this disparity is that reconstruction of ice retreat stages, ice-margin positions and processes is often very detailed onshore, but may be completely lacking offshore, and land-sea correlation remains an important issue (Bowen, 1999).

The above discussion is relevant to the reconstruction of late Pleistocene glacial events around the Irish Sea Basin (ISB) which involves, by necessity, integration of data across the present land-sea boundary. The Irish Sea coasts of Ireland, Scotland, England and Wales have a long history of investigation, and ice-marginal positions are generally well known (Mitchell, 1963; Eyles and McCabe, 1989). This contrasts with the marine environment of the ISB which was only investigated systematically from the 1960s onwards (Jackson *et al.*, 1995). The disparity in knowledge between onshore and offshore areas is demonstrated in reconstructions of ice-marginal positions across the ISB. Many works purporting to define stages of ice retreat in the ISB merely draw smooth (simple) lines across the basin in order to link together evidence from disparate terrestrial locations. Evidence for submarine end-moraines across the floor of the ISB, upon which other works identifying the position of ice-margins are based (*e.g.* Mitchell, 1963), has not been demonstrated. Because of limitations of geophysical techniques, it remains difficult to link onshore and offshore locations together seamlessly, although this has been achieved in some local case studies (*e.g.* Harris *et al.*, 1997; Williams *et al.*, 2001).

Understanding of the late Pleistocene glacial history of the ISB, and evaluating its relative role in the wider glaciation of the British Isles, is therefore more poorly developed than in many adjacent locations onshore. This chapter summarises our current knowledge of the Irish Sea glacier and identifies interpretive problems which are illustrated using examples from the ISB coasts of Ireland and Wales. For further details, the reader is referred to the references cited in this chapter. Ages are cited in radiocarbon (^{14}C) or calibrated (cal.) years before present (BP).

Physical geography and geology of the Irish Sea Basin

The ISB refers to the marine area of the Irish Sea, extending from the North Channel to St George's Channel (between 55–52°N), and bounded to the west and east by Ireland and Britain respectively (Fig. 12.1). The Irish Sea can be considered a semi-enclosed macrotidal basin or epicontinental sea and is geometrically complex, varying in width from 40 km in the North Channel to 200 km south of the Isle of Man. The bathymetry of the ISB is also varied, and comprises relatively flat nearshore platforms (<60 m water depth) into which are cut north-south aligned, semi-enclosed trenches in the central and western Irish Sea which reach a maximum depth of 318 m (in the North Channel) (Jackson et al., 1995).

The pre-Pleistocene geology of the ISB mainly comprises Carboniferous to Triassic rocks which were developed in subsiding, fault-bounded sedimentary basins (Maingarm et al., 1999). Vertical movement along fault planes of different orientations has occurred throughout geological time in the ISB region (Needham and Morgan, 1997). In the northern and central ISB, bedrock types and structures are identical in both onshore and offshore areas; in the southern ISB, onshore and offshore areas are very different geologically, with Permian to Palaeogene sedimentary rocks offshore and Ordovician to Carboniferous sedimentary rocks, and igneous intrusives, onshore (including Wales and southern Ireland) (Tappin et al., 1994; Jackson et al., 1995). In the southern ISB, therefore, great contrasts exist between onshore and offshore bedrock relief, structural history, and likely responses to events such as ice loading and unloading.

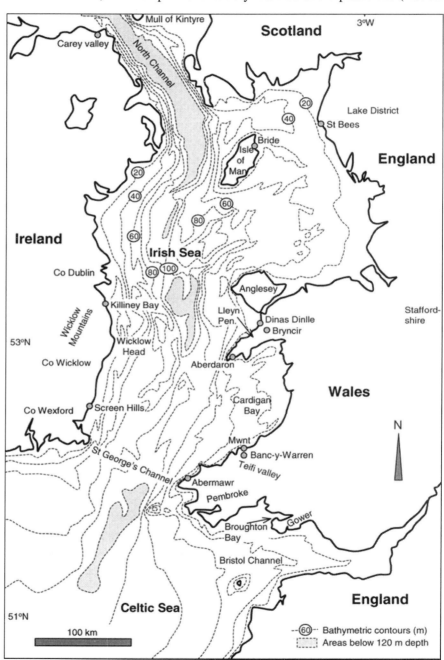

Fig. 12.1 Map showing the generalised bathymetry of the Irish Sea Basin and the location of places named in the text

Pleistocene sediments in the ISB are highly variable in distribution and thickness (Cameron and Holmes, 1999). Pleistocene sediments are mainly thin or absent in the North Channel, northwest of Anglesey, around Pembroke, and within parts of the Bristol Channel. Sediments are thickest—up to 375 m thick locally—within the deep, semi-enclosed trenches that are located in the central and southern ISB (Fig. 12.1). These trough areas are therefore considerably overdeepened and were formed prior to sediment deposition (Wingfield, 1989).

The semi-enclosed nature of the ISB also strongly influences patterns of present-day waves, tidal flow and sediment circulation and, by extension, how these patterns were different when glacial ice covered part or all of the basin or during eustatic filling of the basin. The ISB is fetch-limited and swell waves enter from the North Atlantic through both the North Channel and St George's Channel (Carter, 1988; Barne et al., 1997). The influence of such waves decreases towards the centre of the basin. Storm surges enter mainly from the south and commonly cause water levels to be elevated by over 2 m (Carter, 1988). Tides are modified by basin geometry, particularly by areas of water constriction (Ramster and Hill, 1969). Amphidromic points located off the Mull of Kintyre (north of the North Channel) and on land in southern Ireland (degraded amphidromic point) control the timing and amplitude of the main M_2 (lunar) tides. Mean spring tidal range varies from 1 m in the North Channel to over 4 m in the ISB proper, and therefore range from micro- to macrotidal (Barne et al., 1997; Brown et al., 2003). Very strong tidal flows, and therefore sediment transport velocities, occur within the North Channel (Ramster and Hill, 1969; Knight and Howarth, 1999). Tidal bores, caused by the constriction of water flow upon approach to land, also take place around the ISB, such as in the Bristol Channel. Sediment transport directions reflect the combination of these water movements, and present-day bedload partings are located in the North Channel and between the Lleyn Peninsula (Wales) and Wicklow Head (Ireland) (Stride, 1982). Despite the large literature on present patterns of surface and bottom currents, tides, role of storms and wind-forcing (e.g. Barne et al., 1997) there have been very few studies that consider the geometry of the ISB in the geological past and its influence on water circulation, tidal range or sediment dynamics (Austin and Scourse, 1997). This is an underexploited area for future research.

Late Pleistocene glaciation of the Irish Sea Basin
Many previous studies have looked at the aspects of the glacial history of all, or part, of the ISB (e.g. Mitchell, 1960, 1972; Kidson and Tooley, 1977; Eyles and McCabe, 1989; McCarroll et al., 2001). Work on ISB coasts has also appeared in a number of significant regional volumes (e.g. Lewis, 1970; Edwards and Warren, 1985; Johnson, 1985; Campbell and Bowen, 1989). Throughout, paucity of evidence from the marine environment, noted above, has limited the scope and wider significance of these studies.

The dynamics of the British ice-sheet during the Late Devensian glaciation (~22,000–13,000 cal. years BP) and, most likely, in previous glaciations also, was controlled largely by ice drainage though the ISB. This control is mainly due to the geographical position of the ISB with relation to ice source areas on adjacent land masses in the north of Ireland, southwestern Scotland and northern England (Bowen et al., 1986; Eyles and McCabe, 1989). This geographical arrangement was conducive to the development of convergent ice flow into the ISB from these surrounding source areas, and the formation of a distinctive 'Irish Sea Basin Glacier'. Evidence for the convergent nature of amphi-ISB ice comes from geomorphic evidence on land including patterns of drumlins in north-eastern Ireland, south-west Scotland and north-west England which are clearly convergent towards coastal lowlands (Eyles and McCabe, 1989). Evidence for the distributive nature of the ISB glacier comes from patterns of erratic dispersal. For example, the distinctive Ailsa Craig microgranite, from the Firth of Clyde, has been identified within ISB glacial sediments throughout the length of the Irish Sea and in locations in Wexford (Ireland), Pembroke (Wales) and Staffordshire (England) (Charlesworth, 1939; 1957). Likewise, distinctive erratic trains have been identified from source locations in western Scotland, the Lake District, north-eastern Ireland and

Anglesey. Basal sediments associated with the Irish Sea Glacier have been collectively termed 'Irish Sea Drift' (Wright, 1937), an umbrella term for a range of diamictons and muds containing, commonly, marine shells. The Irish Sea Drift is found throughout the ISB and in many coastal locations onshore (Eyles and McCabe, 1989) and is usually overlain by glacial diamictons and sand and gravel deposits of terrestrial origin (Mitchell *et al.*, 1973). As such, therefore, the presence of Irish Sea Drift has been used as a regional-scale lithostratigraphic marker for the activity of ISB ice (*e.g.* Bowen, 1973).

This signature of ice activity can also be identified offshore. Based on seismostratigraphic techniques and ground-truthed by coring, two main late Pleistocene-age formations are identified in the ISB, named the Cardigan Bay Formation and the Western Irish Sea Formation (Cameron and Holmes, 1999). Both these formations were developed on, and were then truncated by, erosional unconformities which are of glacial origin. Sediments comprising these formations are muds and muddy diamictons. Based on amino acid ratios, these formations are of very variable age (Oxygen Isotope stages 6–2), and microfaunal analysis shows that water masses of both formations were of varying provenance (from arctic to cool boreal and temperate). These characteristics suggest the two formations were formed time-transgressively across the region and that they were associated with changes in water mass provenance and thus circulation patterns, which may also be a reflection of water column stratification in a basin with complex bathymetry. The Irish Sea Drift may be correlated with the uppermost part of the Western Irish Sea Formation (Tappin *et al.*, 1994; Jackson *et al.*, 1995). If this correlation is correct, it may aid the stratigraphic ordering of sediment successions across the land-sea interface in the ISB. The presence of Irish Sea Drift onshore therefore likely represents only the most recent in a long series of glacial events in the ISB region.

The landward extent of ISB ice around the margins of the basin is fairly well known through geomorphic and sedimentary evidence onshore. These landward marginal positions are a reflection of the energenetics of ISB ice and the presence of barriers to flow including topography and obstructing ice from terrestrial sources. The southern-most extent of ISB ice in the St George's Channel/Celtic Sea region, however, is not clearly known. End-moraine ridges are not observed bathymetrically in this area (*cf* Mitchell, 1963). Marine cores containing 'till-like sediments' have been recovered from as far south as 49°N (Pantin and Evans, 1984; Scourse *et al.*, 1990) in association with marine rhythmites which are interpreted to reflect sub-ice shelf sedimentation. Glacigenic sediments which are texturally diverse may be difficult to distinguish from those deposited by mass wastage processes. Scourse *et al.* (1990) identified a possible grounding line limit at about 49°30'N based on the presence of subglacially-eroded materials within the glacigenic sediment. The presence of a thin glacial till (Scilly Member, St Martins Formation) in the Isles of Scilly (50°N) may record the south-eastern limit of ISB ice (Scourse and Furze, 2001). This till may be correlated with the Upper Till member of the Cardigan Bay Formation (Tappin *et al.*, 1994). Organic sediments beneath the Scilly Member till yield a minimum radiocarbon age of 21,500 +890/-800 (Scourse and Furze, 2001).

Timing of ice advance and retreat
The timing of maximum ice extent (last glacial maximum) in the ISB is not clearly known. ^{36}Cl rock-exposure ages from sites in Ireland and Wales show that terrestrial ice reached a maximum extent about 22,000 cal years BP (Bowen *et al.*, 2002), which presumably also corresponds to the period in which most ice was being fed into the ISB. This is noticeably before the timing of the global sea-level minimum at about 19,000 cal years BP (Yokoyama *et al.*, 2000), and suggests that ISB ice had retreated from its maximum extent in the Celtic Sea prior to eustatic sea-level rise (Scourse *et al.*, 1990). This phasing relationship is consistent with ISB deglaciation initiated by reduced ice flux from surrounding source areas as a result of global temperature increases from *c.* 21,000 cal years BP onwards (McCabe and Clark, 1998). AMS radiocarbon dates, largely on shells of marine microfauna incorporated within ice-marginal

sediments which were deposited during ice retreat, constrain patterns of deglaciation in the ISB. These dates, and glacial events associated with the formation of sediment piles within which the dated sediments are located, suggest that ice in the ISB retreated from stillstand positions (including where marine margins were pinned to bedrock outcrops) with the same pacing as high-frequency climate changes in the North Atlantic region (McCabe and Clark, 1998; McCabe et al., 1998; Scourse et al., 2000; Richter et al., 2001; Knight, 2003). These North Atlantic climate changes and climate drivers include the episodic Heinrich (ice-berg-rafting) events (duration of a few hundred years; Chapman et al., 2000), and the more regular Bond cycles (3000–6000 years) and Dansgaard-Oeschger cycles (1000–1500 years; Andrews and Barber, 2002). For example, dating evidence suggests that reactivation and/or readvance of ISB ice from source areas in Ireland and, likely, in Scotland took place at the time of the North Atlantic Heinrich event 1 (~14,500 ^{14}C years BP; ~17,000 cal. years BP) and was immediately followed by ice-sheet collapse and ice-margin retreat (McCabe et al., 1998; McCabe and Clark, 2003). If this pacing is correct, it suggests a near-immediate downstream response of ISB glacier source areas to external forcing through the coupled ice-ocean-atmosphere system. Correlation between hemispheric-scale climate events and glacial events in the ISB region (Knight, 2003) shows that the British ice-sheet was sensitive, and responded dynamically, to regional changes in climate throughout the last deglaciation (McCabe and Clark, 1998).

Environments of sediment deposition in the Irish Sea Basin

Different interpretations of deglacial sediment successions around the ISB have major implications for the style, patterns, processes and controls on Late Devensian ice retreat. Lack of consensus on sedimentary depositional environments has hindered development of a single deglacial model for the ISB as a whole (Knight, 2001; McCarroll, 2001). Argument focuses on the interpretation of a glacial marine environment, rather than a terrestrial environment, for the formation of certain landforms and sediments in the ISB. These differing interpretations arise mainly from the important paper by Eyles and McCabe (1989) who presented evidence from around the ISB, including the coasts of eastern Ireland and Wales, for sediment deposition in a glacial marine environment during ice retreat. Eyles and McCabe (1989) argued that ice loading of the land surface in the ISB region was sufficient to cause high relative sea-level (RSL) conditions even during a period of globally-low eustatic sea-level. Evidence for this comes from the presence of marine deltas which are today found raised to high levels (<150 m OD), arguably because of the degree of ice loading (Eyles and McCabe, 1989). Initial stages of deglaciation were linked to ice-mass thinning as a result of extension following fast ice flow (McCabe and Clark, 2003). Uncoupling of the ice terminus (from a grounded to a floating ice shelf) as RSL rose resulted in rapid retreat of the ice-margin, areal stagnation on coastal lowlands, and marine flooding across a land surface that was still isostatically-depressed (Eyles and McCabe, 1989; McCabe and O'Cofaigh, 1996; McCabe, 1997). The sedimentary signature of marine transgression is the presence of fossiliferous marine mud beds which drape subglacially-formed bedforms (including drumlins and erosional s-forms) and which are found at different elevations throughout the ISB (Haynes et al., 1995; McCabe, 1996).

In support of the glacial marine depositional model for the ISB is the interpretation of a range of geomorphic, sedimentary and biofacies evidence (Table 12.1). These interpretations focus on ice-marginal processes taking place in the marine environment, and are broadly supported by observations of modern glacial marine settings and with reference to late Pleistocene sediment successions elsewhere. The presence of seaward-going foresets within elevated, flat-topped landforms such as the Carey valley delta, Northern Ireland coast (McCabe and Eyles, 1988), suggests seaward progradation of glacial fluvial sediments to an elevated water plane, interpreted as high contemporary RSL. Sediments contained within glacial landforms are typically sorted and stratified, characteristic of deposition in water (McCabe et al., 1984), and mud beds contain Arctic marine microfaunas which are interpreted as *in situ*.

Evidence	Support for glacial marine model	Support for terrestrial ice model
Landforms	The presence of high-level flat surfaces, interpreted as Gilbert-type delta tops, were formed by seaward progradation of glacial marine sediments to an elevated water plane up to 152 m OD (McCabe and Eyles, 1988; Eyles and McCabe, 1989). Waterlain sediments are present within morainal banks that front inland drumlin swarms, indicating ice activity and drumlinisation towards marine-based ice margins (Haynes *et al.*, 1995; McCabe *et al.*, 1998). Resedimentation down unstable marine slopes into palaeo-lows following formation of push-moraine ridges such as at Bride, Isle of Man (Eyles and Eyles, 1984). Presence of raised beaches which are bounded by or terminate at moraines suggesting that glacial ice controlled the inland extent of marine waters (Stephens and McCabe, 1977; Eyles and McCabe, 1989).	Flat-surfaced landforms formed in proglacial lakes impounded between ISB glacier edge and terrestrial uplands, such as Teifi Valley (Fletcher and Siddle, 1998) and Lleyn Peninsula (McCarroll, 1995). Possible evidence for subaerial exposure and periglaciation offshore Anglesey (Wingfield, 1987), therefore a time-gap between ice retreat and eustatic-driven marine transgression.
Sediments	Presence of regional mud drapes and glacial diamictons, interpreted as transgressive marine and glacial-marine sediments respectively, across subglacially-formed drumlins and s-forms, suggesting very rapid ice retreat with coeval marine transgression (McCabe, 1996). Presence of intertidal boulder pavements underlying morainal banks and glacial marine sediments (McCabe and Haynes, 1996).	Presence of 'deforming bed' glacigenic sediments on SE Ireland indicative of grounded ice conditions during onshore ramping of ISB ice (O'Cofaigh and Evans, 2001; Evans and O'Cofaigh, 2003).
Biofacies	Presence of biocoenoses (living assemblages) of arctic marine microfaunas, including the foraminiferas *Elphidium clavatum, Haynesina orbiculare, Cibicides lobulatus* and *C. fletcheri*, within mud beds within morainal banks and mud drapes overlying subglacial landforms. Foram tests are generally intact and of all size ranges suggesting they are *in situ* and have not been reworked from elsewhere (Haynes *et al.*, 1995).	Some key sites such as Dinas Dinlle are almost barren (Haynes *et al.*, 1995) whereas other sites such as Broughton Bay (Gower) contain a high percentage of derived temperate faunas (Shakesby *et al.*, 2000). Foraminifera reflect cold-water rather than deglaciating arctic environments (McCarroll, 2001). Foraminifera at Aberdaron (Lleyn Peninsula) are derived from onshore transport into the terrestrial environment rather than deposition in the marine environment (Austin and McCarroll, 1992). Palaeowater depths reconstructed from microfauna asemblages in Celtic Sea cores support geophysical model predictions of water depths (Austin and Scourse, 1997).
RSL changes	Evidence for different postglacial rebound along pre-existing faults around the ISB as a result of ice unloading (Lloyd *et al.*, 1999; Knight, 1999). The geological record of RSL indicators throughout the ISB are located at consistently higher elevations than geophysical models predict (McCabe, 1997).	Different geophysical models, using somewhat different input parameters, all predict similar 'low' RSL positions during ice retreat in the ISB region (Wingfield, 1995; Lambeck, 1996; Lambeck and Purcell, 2001).

Table 12:1 Comparison of principal landform, sedimentary, biofacies and other evidence supporting the glacial marine and terrestrial ice models along the ISB coasts of Wales. Not all lines of evidence are presented in this table. Note that some field evidence can support either model depending on its interpretation

The glacial terrestrial depositional model offers alternative interpretations of this field evidence (Table 12.1). For example, flat-topped landforms are argued to have formed in ice-marginal glacial lakes rather than in full marine seas, with their elevations therefore independent of RSL stage (McCarroll, 2001). Sediment sorting and stratification within glacial landforms is argued to reflect the deformation of subglacial sediment beneath a terrestrial ice-sheet during eustatically-low RSL (Evans and O'Cofaigh, 2003) which is supported by geophysical models of ice loading and unloading history in the ISB (Lambeck and Purcell, 2001). The presence of marine microfaunas within some of these sediments also reflects glacial reworking rather than *in situ* deposition (Austin and McCarroll, 1992).

Evidence supporting the glacial marine model has come mainly from sites on the Irish side of the ISB, whereas sites along the British coast, including Wales, mainly support the terrestrial ice model. This geographical difference between the western and eastern ISB may be an artefact of the sampling strategy and area of focus of key research workers, or it may represent a fundamental difference in ice retreat patterns across the ISB (Knight, 2001). This is a plausible explanation, given major differences in coastline shape and bathymetry between eastern and western ISB coasts, but this hypothesis has not been investigated in detail. Furthermore, there is as yet no clear methodology or diagnostic criteria for distinguishing between glacial marine and terrestrial landforms and sediments. Possible criteria have been outlined by a number of workers (*e.g.* Hart and Roberts, 1994; Carr, 2001; Knight, 2001; McCarroll and Rijsdijk, 2003) but these are based largely on the subsequent interpretation of field data rather than the presence or absence of diagnostic features in the field, and have not been rigorously applied. This means that the Late Devensian glacial environments of the ISB are still not known with certainty.

Dating evidence, illustrating the temporal stages of ice retreat in the ISB, is equivocal and neither proves nor disproves the glacial marine model. Although the dating evidence supports climate forcing of ice activity in the ISB region, it does not identify the pathway through which the climate signal was imprinted on ice source regions (Knight, 2003). For example, increased precipitation from a cooler post-Heinrich event North Atlantic can lead to a positive mass balance and increased activity of terrestrial ice. Alternatively, changes in ocean circulation, sea-surface temperature and RSL can destabilise and drive activity of marine-based ice-margins. Whether an 'atmospheric' or 'oceanic' pathway prevailed in the ISB is unknown, but it is reasonable to consider that both pathways were important in climate forcing of ISB ice activity. Evidence for the impact of ISB ice on coastal margins of eastern Ireland and Wales is briefly described below.

Sediment deposition along the east coast of Ireland

Debate on the glacial marine versus glacial terrestrial origin of Late Devensian sediment successions along the eastern (Irish Sea) coast of Ireland is important because it also has implications for coeval RSL (thus ice loading/unloading history) and controls on ice retreat patterns. Generally, it can be said that coastal sites further north show clear evidence for deposition in a waterlain (interpreted as marine) environment; sites further south, in counties Dublin, Wicklow and Wexford, are more equivocal (McCabe, 1997; McCarroll, 2001). Most debate has therefore centred upon these southern sites, despite most recent work (and dating control) coming from the more northerly sites. This clearly identifies priorities for future field investigations.

Sites in coastal counties Wicklow and Wexford (southeast Ireland) are also complex because there is disagreement on the landward extent of ISB ice and the southerly extent of terrestrial Irish ice, and their phase relationships (Farrington, 1947; Synge, 1977; McCabe, 1987). These areas are further complicated by the presence of glacial diamictons (tills) attributed to a pre-Late Devensian glaciation (Synge, 1964b, 1977) which is of uncertain age (Knight *et al.*, 2004). Deposits of this pre-Late Devensian glaciation are not discussed here.

The Screen Hills complex (location shown on Fig. 12.1), which is not topographically controlled, comprises a large area (150 km^2) of hummocky (sometimes ridged) topography, and is developed in tectonically-deformed, water-sorted gravel to mud beds (Synge, 1977; Thomas and Summers, 1983; Eyles and McCabe, 1989). In detail, sedimentary units are generally laterally continuous over several km distance, and comprise flat-lying sands and gravels with more discontinuous and interbedded mud and diamicton units (Thomas and Summer, 1983). Sand beds may be rippled or trough cross-bedded; mud beds generally contain a mixture of *in situ* and derived cold-water and temperate marine molluscs and foraminifera (Huddart, 1981; Thomas and Kerr, 1987). Tight upright folds and thrusts are common in the sand and mud beds; large-scale south-west-going shear planes extend to the surface through the entire sediment sequence (Thomas and Summers, 1983; Eyles and McCabe, 1989). Possible interpretations of the Screen Hills complex include kettled kame (Synge, 1977), kame-moraine (Thomas and Summers, 1983) and ice-contact braid-delta (Eyles and McCabe, 1989). Deposition in a glaciolacustrine (terrestrial) environment was followed by repeated ice advance-retreat events causing thrusting at the ice-margin (Thomas and Summer, 1983; Evans and O'Cofaigh, 2003), similar to inferred processes at St Bees (England) (Williams *et al.*, 2001) and Dinas Dinlle (Wales) (Harris *et al.*, 1997). Alternatively, sediment deposition took place by delta progradation into the marine environment with thrusts caused by proglacial bulldozing of saturated sediments (Eyles and McCabe, 1989; McCabe, 1997), similar to inferred processes at Bride (Isle of Man) (Eyles and Eyles, 1984). Both models demonstrate onshore flow of ISB ice, and the microfauna within the mud beds (Knocknasillogue Member) largely represent reworked (warm) interglacial rather than (cool) glacial species (Huddart, 1981) and therefore cannot discriminate between the models (Evans and O'Cofaigh, 2003). Furthermore, independent sampling from the Knocknasillogue Member (by Huddart, 1981 and Haynes *et al.*, 1995) reveals rather different species lists and therefore raises issues of field sampling strategy, representativeness of sampled microfauna assemblages, and confidence of microfaunal interpretation.

Sediments at Killiney Bay are dominated by glacial diamictons which are interbedded with discontinuous sand and gravel beds (Hoare, 1975; Rijsdijk *et al.*, 1999). These sediments have been variously interpreted as being of subglacial (van der Meer *et al.*, 1994; Rijsdijk *et al.*, 1999) and glacial marine origin (Eyles and McCabe, 1989). Silt to gravel-infilled clastic dikes are present in the lowermost few metres of the exposure, and are overlain by 5–15 m of undisturbed diamictons (Rijsdijk *et al.*, 1999). Similar structures are recorded just south of Killiney Bay at Greystones (McCabe and O'Cofaigh, 1995) where they are interpreted as formed by soft-sediment deformation (due to loading) with coeval water-escape. Rijsdijk *et al.* (1999) argue that the clastic dikes are hydrofracture infills developed subglacially in overconsolidated till during ice advance. Such overconsolidation might also be attributed to compression at the Irish Sea glacier margin during advance to the site (Hoare, 1975). Mud and diamicton beds in Killiney Bay (at Shanganagh in the central part of the bay) yielded sparse, mixed cool-water and temperate microfauna (Haynes *et al.*, 1995).

Glacial landforms, including end-moraines, in the Wicklow Mountains demonstrate the presence of a terrestrial mountain ice-cap with associated valley glaciers during the Late Devensian glaciation, and possibly during earlier glaciations also (Farrington, 1934; Charlesworth, 1937; Synge, 1973; McCabe, 1987). ^{36}Cl rock-exposure age determinations from the Wicklow Mountains indicate latest ice retreat at 17,100 ± 900 cal. years BP at Lough Nahanagan (Bowen *et al.*, 2002). ISB ice moving onshore onto adjacent coastal lowlands during the Late Devensian may have interacted with mountain ice in the eastern Wicklow Mountains (Farrington, 1934), although an antiphase relationship between mountain and lowland ice is also likely (McCabe, 1987).

These examples from the ISB coast of Ireland demonstrate the dynamic nature of ISB ice interaction with its substrate and surrounding landscape (including topographic control and ice-waterbody feedback), and suggest caution in interpreting from incomplete field evidence. A strict lithostratigraphic approach to the relative-age relationships of sediments in eastern Ireland (*e.g.* Bowen, 1973; Warren, 1985) is unhelpful because it does not consider the effects of time-transgressive ice retreat or lateral-equivalence of different facies. The lithostratigraphic approach also requires sediments to be (often arbitrarily) classified into formations and members based upon generalised characteristics (such as clast lithology), which both hides complexity and cannot consider processes such as sediment reworking or stratigraphic inversion. A glacial systems approach (Eyles and McCabe, 1989) is more useful because it can link subglacial and proglacial evidence together, but this approach is deterministic and encourages interpretations of single sites to be rejected where they do not fit the regional model.

Sediment deposition along the coasts of Wales
A number of sites along the coast of Wales are also pertinent to the glacial marine-terrestrial debate where there has been disagreement over their interpretation. Examples of alternative interpretations of several sites are shown in Table 12.2 and some are discussed below.

At Mwnt (Cardigan) deformed silt-rich glacial diamictons are interbedded with gravel-rich pods, lenses and overfolds. These sediments are contained within a bedrock-bounded basin. Structures observed within gravel beds at Mwnt are interpreted as syndepositional soft-sediment deformation caused by loading (Rijsdijk, 2001) and suggest that the sediments were highly water-saturated. Eyles and McCabe (1989) interpreted this section as a 'valley fill complex' dominated by mass flow and debris flow processes into a waterlain setting, likely glacial marine. Other explanations include deposition into a proglacial lake (McCarroll and Rijsdijk, 2003), which may fit with a regional picture of impounded ice-marginal lakes (Charlesworth, 1929; Edwards, 1996). Mud beds at this site are fossiliferous but contain a mixed marine fauna with strong derived components (Haynes *et al.*, 1995).

Cross-bedded sands and gravels found within isolated hills at Banc-y-Warren (Cardigan) dip steeply towards the southwest and are cut by minor normal faults. Eyles and McCabe (1989) suggest this site is a glacial marine delta with an upper surface level of 140 m OD where sediments prograded southward into a full marine sea. Other interpretations link the sediments at Banc-y-Warren to the development of glacial Lake Teifi located immediately to the south (Helm and Roberts, 1975; Fletcher and Siddle, 1998). Here, deltaic sediments prograded into the ice-marginal glacial lake which was impounded by the ISB glacier. Allen (1982) argued that Banc-y-Warren represents a proglacial outwash plain and that no deep

Site (grid reference)	Glacial marine interpretation	Alternative interpretations
Dinas Dinlle (SH 436564)	Ice-contact glacial marine	Thrust moraine (Harris *et al.*, 1997)
		Drumlin (Hart, 1995)
		Ice-pushed glaciomarine apron (Haynes *et al.*, 1995)
Mwnt (SN 194519)	Valley fill complex	Density-driven deformation of gravels into muds deposited in local proglacial lake (McCarroll and Rijsdijk, 2003)
Banc-y-Warren (SN 204475)	Gilbert-type delta at 140 m OD	Glaciofluvial outwash (Allen, 1982; Fletcher and Siddle, 1998)
		Glaciolacustrine delta (Helm and Roberts, 1975)
Bryncir (SH 480446)	Gilbert-type delta at 84 m OD	Kame terrace (Whittow and Ball, 1970)
		Alluvial fan (Thomas *et al.*, 1998)
Abermawr (SM 883347)	Valley infill complex	Irish Sea Drift overlain by outwash and periglacial slope deposits (John, 1970)
		Subglacially sheared sediments formed during Irish Sea advance and retreat cycle followed by paraglacial resedimentation (McCarroll and Rijsdijk, 2003)

Table 12:2 Alternative interpretations of some Late Devensian glacial marine sites (Eyles and McCabe, 1989) along the ISB coasts of Wales

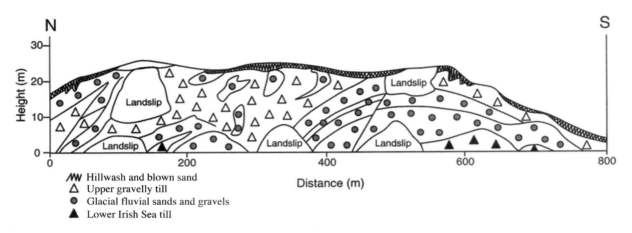

Fig. 12.2 Sketch of the glacial sediments at Dinas Dinlle (northern section) (after Campbell and Bowen, 1989)

glacial lake was present. The presence of varved bottom sets within deeper parts of the Teifi basin (Fletcher and Siddle, 1998) is similar to marine muds found offshore (Edwards, 1996), but this neither proves nor disproves a glacial marine hypothesis.

At Dinas Dinlle (Lleyn Peninsula) numerous workers have described the contorted succession of glacial diamicton (till) units separated by a glacial sand and gravel unit (*e.g.* Reade, 1893; Jehu, 1909; Synge, 1964a; Saunders, 1968; Whittow and Ball, 1970; Hart, 1995; Harris *et al.*, 1997). These sediments form a ridge-like landform that overlies a bedrock basement that rises steeply towards the south (Fig. 12.2). Sediments at Dinas Dinlle have been viewed as part of the Clynog Fawr 'end moraine' (Synge, 1964a; Saunders, 1968), but this ice limit may not have been associated with a major ISB ice advance (McCarroll, 1991). Eyles and McCabe (1989) argued that the sediments at Dinas Dinlle are ice-contact glacial marine with a marine mud drape and influenced by downslope resedimentation processes, presumably into a marine environment. Patterns of sediment thrusting and napping along deep-seated listric shears, however, suggest ice-marginal advance into a thick, loose sediment pile causing an estimated shortening of 54% (Harris *et al.*, 1997). This deformation style suggests a terrestrial ice origin during low RSL (McCarroll and Rijsdijk, 2003), and is supported by the mud beds being barren of marine microfauna (Haynes *et al.*, 1995). The presence of resedimented diamictons down the southern (lee) slope at Dinas Dinlle, however, suggests shearing of cohesive sediment was not the only deformation process. This can be evaluated further at Trwyn Maen Dylan (SH 427524), 4 km south of Dinas Dinlle. At this site, located at the foot of the bedrock upland to the south, there is clearer evidence for waterlain deposition of glacial sediments (Fig. 12.3). Here, a massive grey diamicton with a likely Irish Sea and local Welsh source (Harris *et al.*, 1997) is overlain by a sand and gravel unit that mantles the exposure. These two units are separated by a sharply-erosional planar contact (Fig. 12.3a). On the south side of the exposure sand, gravel and mud beds onlap to the south and are interbedded with gravely diamictons which show shears and soft-sediment deformation (Figs. 12.3b, c). The sand beds (10–30 cm thick) show inverse grading, are undulating, and pinch out or grade laterally into gravel. Massive to poorly-laminated clay layers are generally laterally continuous across the exposure and may grade upwards into silt and fine sand. Loading structures and sediment drapes over isolated, striated clasts are common (Fig. 12.3d). The presence of interbedded sand and mud layers and the presence of loading structures strongly suggests deposition in a waterlain environment, thereby contrasting with terrestrial ice interpretations at Dinas Dinlle. Whether deposition took place in a glacial lake or marine environment is unknown.

Fig. 12.3 Sediments at Trwyn Maen Dylan. a. top left) Sand and gravel unit overlying basal grey glacial diamicton. The field of view is 5 m wide. b, top right) Complexly-interbedded sand, gravel and mud beds. Trowel for scale is 29 cm long. c, lower left) Striated boulders pressed into fine sand and silt bed which shows dish-like soft sediment deformation structures. Trowel for scale is 29 cm long. d, lower right) Deformation of subjacent sediment caused by clast impaction. The clear absence of clast rotation suggests a proglacial waterlain rather than a subglacial origin. Pencil for scale is 15 cm long

Consideration of these three sites, as examples, shows that field evidence is often equivocal and glacial marine and glacial terrestrial depositional settings cannot be clearly distinguished. This is particularly the case along the ISB coast of Wales. Here, sites of sediment deposition are often confined to within small bedrock basins, buttressed against bedrock headlands or in embayments. This means that deposition at any one site would have been dominated by local controls such as basin shape, size and slope angle rather than glacier-wide controls such as RSL or climate forcing. In addition, ice retreat along such a geomorphically diverse coastline is likely to have been more episodic and variable than along the more open Irish ISB coast (Knight, 2003). For these reasons, glacial depositional sites along the ISB coast of Wales tell a more complex story than elsewhere, and are likely to be less easily correlatable. It is also notable that no Welsh glacial site located at the coast has been radiocarbon dated, which makes it difficult to fit the Welsh evidence to the radiocarbon-dated deglacial framework constructed for the rest of the ISB.

Concluding remarks

The ISB, as a complex marine and geological system, illustrates some of the factors affecting Late Devensian glacial processes in a paraglacial shelf setting. As such it can be used as a test-bed for glacier-sediment-waterbody interactions in similar Late Devensian geological settings such as the Kattegat (Baltic Sea), Barents Sea and North Sea. Evidence from ISB coasts in eastern Ireland and Wales highlights a number of issues, including the interpretation of the 'Irish Sea Till'; the relationship of this till to associated sand, gravel and mud beds; the age and significance of mollusc and microfaunas found within these sediments; the overall environment of sediment deposition; and the wider controls on ISB ice retreat and sediment deposition patterns, including RSL changes and climate (Knight, 2001). Disagreement over interpretations of glacial landforms and sediments around ISB coasts does not resolve problems of glacial marine or terrestrial depositional setting, even using such 'diagnostic' indicators as the presence of marine microfauna. Recent arguments for the ISB glacier as an ice-stream (Evans and O'Cofaigh, 2003) merely confirms the dynamic nature of the glacier, not its ultimate controls.

References

CHAPTER 1 Introduction *by* Colin A. Lewis

Andrews, J.T. 1998 'Abrupt changes (Heinrich events) in late Quaternary North Atlantic marine environments: a history and review of data and concepts', *Journal of Quaternary Science* 13, 3-16.

Bartley, D.D. 1960 'Rhosgoch Common, Radnorshire: stratigraphy and pollen analysis', *New Phytologist* 59, 238-262.

Bond, G.C. and Lotti, R. 1995 'Iceberg discharges into the North Atlantic on millennial time scales during the last deglaciation', *Science* 267, 1005-1010.

Bos, J.A.A., Bohncke, S.J.P., Kasse, C. and Vandenberghe, J. 2001 'Vegetation and climate during the Weichselian Early Glacial and Pleniglacial in the Niederlausitz, eastern Germany - macrofossil and pollen evidence,' *Journal of Quaternary Science* 16, 269-289.

Boulton, G.S., Jones, A.S., Clayton, K.M. and Kenning, M.J. 1977 'A British ice-sheet model and patterns of glacial erosion and deposition in Britain', in Shotton, F.W. (Editor) *British Quaternary studies: recent advances*, Clarendon Press, Oxford, 231-246.

Boulton, G.S., Smith, G.D., Jones, A.S. and Newsome, J. 1985 'Glacial geology and glaciology of the last mid-latitude ice sheets', *Journal of the Geological Society of London* 142, 447-474.

Bowen, D.Q. 1973 'The Pleistocene history of Wales and the border', *Geological Journal* 8, 207-224.

Bowen, D.Q. 1974 'The Quaternary of Wales', in Owen, T.R. (Editor), *The Upper Palaeozoic and post-Palaeozoic Rocks of Wales*, University of Wales Press, Cardiff, 373-426.

Bowen, D.Q. 1978 *Quaternary geology*, Pergamon, Oxford.

Bowen, D.Q. 1980 The *Llanelli Landscape: the geology and geomorphology of the country around Llanelli*, Llanelli Borough Council, Llanelli.

Bowen, D.Q. 1981 'The South Wales End-Moraine: fifty years after', in Neale, J. W. and Flenley, J. (Editors) *The Quaternary in Britain*, Oxford, 60-67.

Bowen, D.Q. 1999 'Only four major glaciations in the Brunhes Chron?', *International Journal of Earth Science* 88, 276-284.

Bowen, D.Q. (Editor) 1999 *A revised correlation of Quaternary deposits in the British Isles*, Geological Society Special Report No. 23.

Bowen, D.Q. 2000 'Tracing climate evolution', *Earth Heritage* (January), 8-9.

Bowen, D.Q. 2004 'Ice Ages (Interglacials, Interstadials and Stadials)', in Goudie, A. S. (Editor) *Encyclopedia of Geomorphology*, Routledge, London, 549-554.

Bowen, D.Q. 2005a 'Pleistocene climates', in Fairbridge, R. W. and Gornitz, V. S. (Editors) *The encyclopaedia of Paleoclimate*, van Nostrand, New York.

Bowen, D.Q. 2005b 'The Last Glacial Maximum', in Fairbridge, R. W. and Gornitz, V. S. (Editors) *The encyclopaedia of Paleoclimate*, van Nostrand, New York.

Bowen, D.Q., Phillips, F.M., McCabe, A.M., Knutz, P.C. and Sykes, G.A., 2002 'New data for the Last Glacial Maximum in Great Britain and Ireland', *Quaternary Science Reviews* 21, 89-101.

Broeker, W.S. 1987 'Unknown surprises in the greenhouse', *Nature* 328, 123-126.

Broeker, W.S. 1994 'Massive iceberg discharges as triggers for global climate change', *Nature* 372, 421-424.

Buckland, W. 1841 Quoted from Davies,1969.

Cane, M. and Clement, A.C. 1999 'A role for the Pacific coupled ocean-atmosphere system on Milankovitch and millennial timescales: Part II: Global impacts', in Clark, P.U.,Webb, R.S. and Keigwin, L.D. (Editors) *Mechanisms of Global climate change at millennial time scales*, American Geophysical Union, Washington D.C., 373-384.

Carr, S. 2001 'A glaciological approach for the discrimination of Loch Lomond Stadial glacial landforms in the Brecon Beacons, South Wales', *Proceedings of the Geologists' Association* 112, 253-262.

Charlesworth, J K. 1929 'The South Wales end-moraine', *Quarterly Journal of the Geological Society of London* 85, 335-355.

Clark, P.U., McCabe, A.M., Mix, A.C. and Weaver, A.J. 2004 'Rapid rise of sea level 19,000 years ago and its global implications', *Science* 304, 1141-1144.

Clarke, G.K.C. and Marshall, S. 1999 'A glaciological perspective on Heinrich Events', in Clark, P.U., Webb, R.S. and Keigwin, L.D. (Editors) *Mechanisms of Global climate change at Millenial time scales*, American Geophysical union, Washington D.C., 373-384.

Coope, G.R. 1977 'Fossil coleopteran assemblages as sensitive indicators of climatic changes during the Devensian (Last) cold stage', *Philosophical Transactions of the Royal Society of London* B280, 313-340.

Coope, G.R. and Brophy, J.A. 1972 'Late-glacial environmental changes indicated by a coleopteran succession from North Wales', *Boreas* 1, 97-142.

Coope, G.R. and Sands, C.H.S. 1966 'Insect faunas of the last glaciation from the Tame Valley, Warwickshire', *Proceedings of the Royal Society of London* B165, 389-412.

Coope, G.R., Field, M.H., Gibbard, P.L., Greenwood, M. and Richards, A.E. 2002 'Palaeontology and biostratigraphy of Middle Pleistocene river sediment in the Mathon Member, at Mathon, Herefordshire, England', *Proceedings of the Geologists' Association* 113, 237-258.

Crabtree, K. 1972 'Late-glacial deposits near Capel Curig, Caernarvonshire', *New Phytologist* 71, 1233-1243.

Croll, J. 1875 *Climate and time.*

Dansgaard, W., Johnsen, S.J., Clausen, H.B. and Langway Jr., C.C. 1971 'Climate record revealed by the Camp Century ice core', in Turekian, K.J. (Editor) *The Late Cenozoic Glacial Ages*, Yale University Press, New Haven, 37-56.

Dansgaard, W., Johnsen, S.J., Moller, J. and Langway, C.C.Jr. 1969 'One thousand centuries of climatic record from Camp Century on the Greenland ice sheet', *Science* 166, 377-381.

David, T.E. 1883 'On the evidence of glacial action in south Brecknockshire and east Glamorgan', *Quarterly Journal of the Geological Society of London* 39, 39-54.

Davies, G.L. 1963 'Dr Anthony Farrington and the Geographical Society of Ireland', *Irish Geography* 4, 311-316.

Davies, G.L. 1969 *The Earth in decay*, Macdonald, London.

de la Beche, H.T. 1839 'Report on the geology of Cornwall, Devon and West Somerset', *Memoir of the Geological Survey of Great Britain.*

Dwerryhouse, A.R. and Miller, A.A. 1930 'Glaciation of the Clun Forest, Radnor Forest and some adjoining districts', *Quarterly Journal of the Geological Society of London* 86, 96-129.

Eberl, B. 1930 *Die Eiszeitenfolge im nördlichen Alpenvorlande*, Augsburg.

Ellis-Gruffydd, I.D. 1977 'Late Devensian glaciation in the Upper Usk Basin', *Cambria* 4, 46-55.

Emiliani, C. 1955 'Pleistocene temperatures', *Journal of Geology* 63, 538-575.

EPICA 2004 'Eight glacial cycles from an Antarctic ice core', *Nature* 429, 623-628.

Farrington, A 1942 'The granite drift near Brittas, on the border between County Dublin and County Wicklow', *Proceedings of the Royal Irish Academy* 50B, 279-291.

Farrington, A. 1934 'The glaciation of the Wicklow Mountains', *Proceedings of the Royal Irish Academy* 42B, 173-209.

Farrington, A. 1944 'the glacial drifts of the district around Enniskerry, Co. Wicklow', *Proceedings of the Royal Irish Academy* 50B, 133-157.

Farrington, A. 1949 'The glacial drifts of the Leinster Mountains', *Journal of Glaciology* 1, 220-223.

Farrington, A. 1957 'The Ice Age in the Dublin District', *Institute of Chemists of Ireland Journal* 5, 23-27.

Farrington, A. 1966 'The last glacial episode in the Wicklow mountains', *Irish Naturalists Journal* 15, 226-229.

Foster, H.D. 1970 'Sarn Badrig, a sub-marine moraine in Cardigan Bay, North Wales', *Zeitschrift für Geomorphologie* 14, 475-486.

Geike, A. 1863 Quoted from Davies (1969) op. cit.

George, T.N. 1933 'The glacial deposits of Gower', *Geological Magazine* 70, 208-232.

Godwin, H. 1955 'Vegetational history at Cwm Idwal, a Welsh plant refuge', *Svensk botanisktidsskrift* 49, 35-43.

Godwin, H. and Mitchell, G. F. 1938 'Stratigraphy and development of two raised bogs near Tregaron, Cardiganshire', *New Phytologist* 37, 425-454.

Godwin, H. and Newton, 1938 'The submerged forest at Borth and Ynyslas, Cardiganshire', *New Phytologist* 37, 333-344.

Heinrich, H. 1988 'Origin and consequences of cyclic ice rafting in the northeast Atlantic Ocean during the past 130 000 years', *Quaternary Research* 29, 142-152.

Howard, F.T. 1903/4 'Notes on glacial action in Brecknockshire and adjoining districts', *Transactions of the Cardiff Naturalists Society* 5.

Huijer, B. and Vandenberghe, J. 1998 'Climatic reconstruction of the Weichselian Pleniglacial in northwestern and central Europe', *Journal of Quaternary Science* 13, 391-417.

Hunt, R. 1886 'Buckland, William (1784-1856)', in Stephen, L. (Editor) *Dictionary of National Biography* VII, Smith Elder, London, 206-208.

Imbrie, J. and Imbrie, K.P. 1979 *Ice Ages. Solving the mystery*, Macmillan, London.

Jansson, K.N. and Glasser, N.F. 2004 'Palaeoglaciology of the Welsh sector of the British–Irish Ice Sheet', *Journal of the Geological Society of London* 161, 1-13.

Jessen, K. and Farrington, A. 1938 'The bogs at Ballybetagh, near Dublin, with remarks on Late-Glacial conditions in Ireland', *Proceedings of the Royal Irish Academy* 44B, 205-260.

Jouzel, J., Cattani, O., Dreyfus, G., Falourd, S., Masson-Delmotte, V., Nouet, J., Oerter, H., Johnsen, S., Rarrenin, F. and Stenni, B. 2004 'A high resolution isotopic ice core record covering the last 800,000 years', *Eos, Transactions of the American Geophysical Union* 85, 47.

Lambeck, K. 1993a 'Glacial rebound of the British Isles - I. Preliminary model results', *Geophysical Journal International* 115, 941-959.

Lambeck, K. 1993b 'Glacial rebound of the British Isles - II. A high-resolution, high precision model', *Geophysical Journal International* 115, 960-990.

REFERENCES

Lewis, C.A. 1979 'Periglacial wedge-casts and patterned ground in the midlands of Ireland', *Irish Geography* 12, 10-24.

Lewis, C.A. 1984 'Quaternary studies', *Irish Geography* 17, Supplement: The Geographical Society of Ireland Golden Jubilee 1934-1984, 73-96.

Lowe, J.J. and Walker, M.J.C. 1997 *Reconstructing Quaternary environments*, 2nd edition, Longman, Harlow.

MacAyeal, D.R. 1993 'Binge/purge oscillations of the Laurentide ice sheet as a cause of North Atlantic Heinrich Events', *Paleoceanography* 8, 775-784.

Mallet, R. 1838 Quoted from Davies, G.L. (1969).

Mitchell, G.F. 1956 'Post-boreal pollen diagrams from Irish raised bogs', *Proceedings of the Royal Irish Academy* 52B, 185-251.

Mitchell, G.F. 1960 'The Pleistocene history of the Irish Sea', *Advancement of Science*, 17, 313-325.

Mitchell, G.F. 1971 'Fossil pingos in the south of Ireland', *Nature* 230, 43-44.

Mitchell, G.F., Penny, L.F., Shotton, F.W. and West, R.G. 1973 A correlation of Quaternary deposits in the British Isles. *Geological Society of London*, Special Report No. 4, 99.

Moore, J.J. 1970 'The pollen diagram for Mynydd Illtyd', in Lewis, C.A. (Editor) *The glaciations of Wales and adjoining regions*, Longman, London, 168-173.

Moore, P.D. 1966 *Investigations of peats of central Wales*, unpublished PhD thesis, University of Wales.

Moore, P.D. 1968 'Human influences upon vegetational history in north Cardiganshire', *Nature* 217, 1006-1009.

Moore, P.D. 1970 'Studies in the vegetational history of Mid Wales. II. The late-glacial period in Cardiganshire', *New Phytologist* 69, 363-375.

Moore, P.D. 1972 'Studies in the vegetational history of Mid Wales. III. Early Flandrian pollen data from west Cardiganshire', *New Phytologist* 71, 947-959.

Moore, P.D. 1978 'Studies in the vegetational history of Mid Wales. V. Stratigraphy and pollen analysis of Llyn Mire in the Wye Valley', *New Phytologist* 80, 281-302.

Morgan, A. 1973 'Late Pleistocene environmental changes indicated by fossil insect faunas of the English Midlands', *Boreas* 2, 173-212.

Paterson, W.S.B. 2001 *The physics of glaciers*, 3rd edition, Butterworth Heinemann, Oxford.

Penck, A. and Brückner, E. 1909 *Die Alpen im Eiszeitalter*, Tauchnitz, Leipzig.

Phillips, F.M., Bowen, D.Q. and Elmore, D. 1994 'Surface exposure dating of glacial features in Great Britain using cosmogenic Chlorine-36: preliminary results', *Mineralogical Magazine* 58A, 722-723.

Pissart, A. 1963a 'Les traces de "pingos" du Pays de Galles (Grande-Bretagne) et du Plateau des hautes Fagnes (Belgique)', *Zeitschrift für Géomorphologie* 7, 147-165.

Pissart, A. 1963b 'Des replats de cryoturbation au Pays de Galles', *Biuletyn Peryglacjalny* 12, 119-135.

Pocock, T.I. 1925 'Terraces and drifts of the Welsh Border and their relation to the drift of the English Midlands', *Zeitschrift für Gletscherkunde* 14, 10-38.

Prestwich, J. 1892 'The raised beaches and head or rubble drift of the south of England', *Quarterly Journal of the Geological Society of London* 48, 263-343.

Ramsay, A.C. 1852 Quoted from Davies, G.L. (1969).

Ramsay, A.C. 1860 *The old glaciers of Switzerland and North Wales*, London.

Raymo, M.E. and Ruddiman, W.F. 1992 'Tectonic forcing of late Cenozoic climate', *Nature* 359, 117-122.

Raymo, M.E., Oppo, D.W., Flower, B.P., Hodell, D.A. and McManus, J.F. 2004 'Stability of North Atlantic water masses in face of pronounced climate variability during the Pleistocene', *Paleoceanography* 19,

Robertson, D.W. 1988 *Aspects of the Late glacial and Flandrian environmental history of the Brecon Beacons, Fforest Fawr, Black Mountain and Abergavenny Black Mountains, South Wales (with emphasis on the Late-glacial and early Flandrian periods)*, unpublished PhD thesis, University of Wales.

Ruddiman, W.F. 2000 *Earth's climate: past and future*, W.H. Freeman, New York.

Seddon, B. 1957 'Late-glacial cwm glaciers in Wales', *Journal of Glaciology* 3, 94-99.

Seddon, B. 1962 'Late-glacial deposits at Llyn Dwythwch and Nant Ffrancon, Caernarvonshire', *Philosophical Transactions of the Royal Society of London* B244, 459-481.

Shackleton, N.J. and Opdyke, N.D. 1973 'Oxygen isotope and palaeomagnetic stratigraphy of Equatorial Pacific core V28-238: oxygen isotope temperatures and ice volumes on a 10^5 year and a 10^6 year scale', *Quaternary Research* 3, 39-55.

Shakesby, R.A. and Matthews, J.A. 1993 'Loch Lomond Stadial glacier at Fan Hir, Mynydd Du (Brecon Beacons), South Wales: critical evidence and palaeoclimatic implications', *Geological Journal* 28, 69-79.

Siegert, M.J. 2001 *Ice sheets and Late Quaternary environmental change*, Wiley, Chichester.

Sissons, J.B. 1976 *The geomorphology of the British Isles: Scotland*, Methuen, London.

Spurrell, F.C.J. 1886 'A sketch of the history of the rivers and denudation of West Kent', *Report of the West Kent Natural History Society*.

Trotman, D.M. 1963 *Data for Late Glacial and Post Glacial history in South Wales*, unpublished PhD thesis, University of Wales.

Walker, M.J.C. 1980 'Late-Glacial history of the Brecon Beacons, South Wales', *Nature* 287, 133-135.
Walker, M.J.C. 1982 'The Late-Glacial and Early Flandrian deposits at Traeth Mawr, Brecon Beacons, South Wales', *New Phytologist* 90, 177-194.
Walker, M.J.C., Coope, G.R., Sheldrick, C., Turney, C.S.M., Lowe, J.J., Blockley, S.P.E. and Harkness, D.D. 2003 'Devensian Lateglacial environmental changes in Britain: a multi-proxy record from Llanilid, South Wales, UK', *Quaternary Science Reviews* 22, 475-520.
Washburn, A.L. 1973 *Periglacial processes and environments*, Edward Arnold, London.
Watson, E. 1965a 'Grèzes litées ou éboulis ordonnés tardiglaciaires dans la région d'Aberystwyth, au centre du Pays de Galles', *Bulletin de l'Association de Géographes Français* 338-339, 16-25.
Watson, E. 1965b 'Periglacial structures in the Aberystwyth region of central Wales', *Proceedings of the Geologists' Association* 76, 443-462.
Watson, E. 1966 'Two nivation cirques near Aberystwyth, Wales', *Biuletyn Peryglacjalny* 15, 79-101.
Watson, E. 1969 'The periglacial landscape of the Aberystwyth region', in Bowen, E.G., Carter, H. and Taylor, J.A. (Editors) *Geography at Aberystwyth*, University of Wales Press, Cardiff, 35-49.
Watson, E. 1971 'Remnants of pingos in Wales and the Isle of Man', *Geological Journal* 7, 381-392.
Watson, E. 1981 'Characteristics of ice-wedge casts in west central Wales', *Biuletyn Peryglacjalny* 28, 164-177.
Whittow, J.B. 1984 *The Penguin dictionary of physical geography*, Penguin Books, Harmondsworth.
Wilson, R.C.L., Drury, S.A. and Chapman, J.L. 2000 *The Great Ice Age: climate change and life*, Routledge (Open University), London.
Wright, W.B. 1914 *The Quaternary Ice Age*, Macmillan, London.

CHAPTER 2 Stratigraphy *by* Andrew E. Richards

Andrews, J.T., Gilbertson, D.D. and Hawkins, A.B. 1984 'The Pleistocene succession of the Severn Estuary: a revised model based on amino-acid racemization studies', *Journal of the Geological Society of London* 141, 967-974.
Boulton, G.S. and Worsley, P. 1965 'Late Weichselian glaciation in the Cheshire-Shropshire Basin', *Nature* 207, 704-706.
Bowen, D.Q. 1978 *Quaternary Geology*, Pergamon, Oxford.
Bowen, D.Q. 1994 'Late Cenozoic Wales and Southwest England', *Proceedings of the Ussher Society* 8, 209-213.
Bowen, D.Q. 1999 (Editor) *A Revised Correlation of Quaternary Deposits in the British Isles*, Geological Society of London Special Report 23.
Bowen, D.Q., Phillips F.M., McCabe, A.M., Knutz P.C. and Sykes G.A 2002 'New data for the Last Glacial Maximum in Great Britain and Ireland', *Quaternary Science Reviews* 21, 89-101.
Campbell, S. and Bowen, D.Q. 1989 *Quaternary of Wales*, Geological Conservation Review Series, 2, Nature Conservancy Council, Peterborough.
Evans, D.J.A. and Ó Cofaigh, C. 2003 'Depositional evidence for marginal oscillations of the Irish Sea ice stream in southeast Ireland during the last glaciation', *Boreas* 32, 76-101.
Gilbertson, D.D. and Hawkins, A.B. 1978a 'The Pleistocene succession at Kenn, Somerset', *Bulletin of the Geological Survey of Great Britain*, 66, London.
Gilbertson, D.D. and Hawkins, A.B. 1978b 'The col-gully and glacial deposits at Court Hill, Clevedon, near Bristol, England', *Journal of Glaciology* 20, 173-188.
Green, H.S. 1981 'Pontnewydd Cave in Wales- a new Middle Pleistocene hominid site', *Nature* 294, 707-713.
Green, H.S. 1984 *Pontnewydd Cave. A Lower Palaeolithic hominid site in Wales: The first report*, National Museum of Wales, Cardiff.
Hey, R.W. 1991 'Pre-Anglian glacial deposits in the British Isles', in Ehlers, J., Gibbard, P.L., and Rose, J. (Editors) *Glacial Deposits in Great Britain and Ireland*, Balkema, Rotterdam.
Horton, A. 1974 *The sequence of Pleistocene deposits exposed during the construction of the Birmingham motorways*, Report of the Institute of Geological Sciences, 74/1.
Horton, A. 1989 'Quinton', in Keen, D.H. (Editor) *The Pleistocene of the West Midlands, Field Guide*, Quaternary Research Association, Cambridge, 69-76.
Jones, R and Keen, D.H. 1993 *Pleistocene deposits in the British Isles,* Arnold, London.
Keen, D.H., Coope, G.R., Jones, R.L., Field, M.H., Griffiths, H.I., Lewis, S.G. and Bowen, D.Q. 1997 'Middle Pleistocene deposits at Frog Hall Pit, Streeton-on-Dunsmore, Warwickshire, English Midlands, and their implication for the age of the type Wolstonian', *Journal of Quaternary Science* 12, 183-208.
Kelly, M.R. 1964 'The Middle Pleistocene of North Birmingham', *Philosophical Transactions of the Royal Society of London* B247, 533-592.
Lewis, C.A. 1970 (Editor) *The glaciations of Wales and adjoining regions*, Longman, London.
Maddy, D., Keen, D.H., Bridgland, D.R., and Green, C.P. 1991 'A revised model for the Pleistocene development of the River Avon, Warwickshire', *Journal of the Geological Society of London* 151, 221-233.

Maddy, D. Green, C.P., Lewis, S.G. and Bowen, D.Q. 1995 'Pleistocene geology of the Lower Severn Valley, U.K.', *Quaternary Science Reviews* 14, 209-222.

Maddy, D. 1999 'The English Midlands', in Bowen, D.Q. (Editor) *A revised correlation of Quaternary deposits in the British Isles, Geological Society of London Special Report* 23.

McCabe, A.M. and Clark, P.U. 1998 'Ice sheet variability around the North Atlantic Ocean during the last glaciation', *Nature* 392, 373-377.

Mitchell, G.F., Penny, L.F., Shotton, F.W. and West, R.G. 1973 *A correlation of Quaternary deposits in the British Isles*, Geological Society of London Special Reports 4.

O'Cofaigh C. and Evans, D.J.A. 2001 'Deforming bed conditions associated with a major ice stream of the last British ice sheet', *Geology*, 29, 795-798.

Perrin, R.M.S., Rose, J. and Davies, H. 1979 'The distribution, variation, and origins of pre-Devensian tills in eastern England', *Philosophical Transactions of the Royal Society of London*, B287, 535-570.

Richards, A.E. 2002 'A multi-technique study of the glacial stratigraphy of Co. Clare and Co. Kerry, southwest Ireland', *Journal of Quaternary Science* 17, 261-276.

Richards, A.E., Waller, M.P. and Bloetjes, 0. 2002 'Late Midlandian Cold Stage deposits at Loop Head, Co. Clare', *Irish Journal of Earth Science* 20, 61-76.

Ruddiman, W.F. 1987 'Synthesis of the ocean/ice-sheet record', in Ruddiman, W.F. and Wright, H.E. (Editors) *North America and adjacent oceans during the last deglaciation*, Geological Society of America, Boulder, Colorado.

Salvador, A. (Editor) 1994 *International Stratigraphic Guide*, Geological Society of America (2nd edition), Boulder, Colorado.

Shackelton, N.J., Berger, A. and Peltier, W.R. 1990 'An alternative astronomical calibration of the lower Pleistocene timescale based on ODP site 677', *Transactions of the Royal Society of Edinburgh: Earth Sciences* 81, 251-261.

Shotton, F.W. 1968 'The Pleistocene succession around Brandon, Warwickshire', *Philosophical Transactions of the Royal Society of London*, B254, 387-400.

Sumbler M.A.1983 'A new look at the type Wolstonian glacial deposits of central England', *Proceedings of the Geologists' Association* 94, 23-23 1.

Wills L.J. 1937 *The Pleistocene History of the West Midlands*. Section C – Geology, Annual Report, British Association for the Advancement of Science, 71-94.

Worsley, P. 1970 'The Cheshire-Shropshire Lowlands', in Lewis, C.A. (Editor) *The glaciations of Wales and adjoining regions*, Longman, London, 83-106.

Worsley, P. 1991 'Glacial deposits of the lowlands between the Mersey and Severn River', in Ehlers, J., Gibbard, P.L. and Rose, J. (Editors) *Glacial Deposits in Great Britain and Ireland*, Balkema, Rotterdam, 203-211.

CHAPTER 3 North-west Wales by Danny McCarroll

Addison, K., 1983 *Classic Glacial Landforms of Snowdonia*, The Geographical Association, Classic Landforms Guide, 3, Sheffield, 48.

Addison, K., 1988 *The Ice Age in Cwm Idwal*, Second Edition, K. and M.K. Addison, Shropshire, 16.

Addison, K. and Edge, M.J. 1992 'Early Devensian interstadial and glacigenic sediments in Gwynedd, North Wales', *Geological Journal* 27, 181-190.

Allen, P.M. and Jackson, A.A. 1985 *Geology of the country around Harlech*. Memoir of the British Geological Survey, Sheet 135 with part of 149.

Austin, W.E.N. and McCarroll, D. 1992 'Foraminifera from the Irish Sea glacigenic deposits at Aberdaron, western Lleyn, North Wales: palaeoenvironmental implications', *Journal of Quaternary Science* 7, 311-317.

Ballantyne, C.K 2001 'Cadair Idris: a Late Devensian palaeonunatak'. in Walker, M.J.C. and McCarroll, D. (Editors) *The Quaternary of West Wales: Field Guide*, Quaternary Research Association, London. 126-131.

Ballantyne, C.K., McCarroll, D., Nesje, A., Dahl, S-O, Stone, J.O. and Fifield, L.K. 1998 'The last ice sheet in north-west Scotland: reconstruction and implications', *Quaternary Science Reviews* 17, 1149-1184.

Battiau-Queney, Y. 1984 'The pre-glacial evolution of Wales', *Earth Surface Processes and Landforms* 9, 229-252.

Boulton, G.S. 1977 'A multiple till sequence formed by a Late Devensian Welsh ice-cap: Glanllynnau, Gwynedd', *Cambria* 4, 10-31.

Bowen, D.Q. 1970 'South-east and central South Wales', in Lewis, C.A. (Editor) *The glaciations of Wales and adjoining regions*. Longman, London, 197-227.

Bowen, D.Q. 1973 'The Pleistocene succession of the Irish Sea', *Proceedings of the Geologists' Association* 84, 249-272.

Bowen, D.Q. (Editor) 1999 *A revised correlation of Quaternary deposits in the British Isles*, Geological Society Special Report 23, 174.

Bowen, D.Q., Phillips, F.M., McCabe, A.M., Knutz, P.C. and Sykes, G.A. 2002 'New data for the Last Glacial Maximum in Great Britain and Ireland', *Quaternary Science Reviews* 21, 89-101.

Bowen, D.Q., Rose, J., McCabe, A.M. and Sutherland, D.G. 1986 'Correlation of Quaternary glaciations in England, Ireland, Scotland and Wales', *Quaternary Science Reviews* 5, 299-340.

Brown, E.H. 1960 *The relief and drainage of Wales*, University of Wales Press, Cardiff.
Campbell, S. and Bowen, D. Q., 1989 *Quaternary of Wales*, Geological Conservation Review, Nature Conservancy Council, Peterborough.
Campbell, S. and Thompson, I.C. 1991 'The palaeoenvironmental history of Late Pleistocene deposits at Moel Tryfan, North Wales: evidence from Scanning Electron Microscopy (SEM)', *Proceedings of the Geologist's Association* 102, 123-134.
Campbell, S., Wood, M., Addison, K., Scourse, J.D. and Jones, R.E. 1995 'Notice of raised beach deposits at Llanddona, Anglesey, North Wales', *Quaternary Newsletter* 77, 1-5.
Chambers, F.M., Addison, K., Blackford, J.J. and Edge, M.J. 1995 'Palynology of organic beds below Devensian glacigenic sediments at Oen-y-Bryn, Gwynedd, North Wales', *Journal of Quaternary Science* 10, 157-173.
Coope G.R. 1977 'Fossil coleopteran assemblages as sensitive indicators of climatic changes during the Devensian (Last) cold stage', *Philosophical Transactions of the Royal Society of London* B280, 313-340.
Coope, G.R. and Brophy, J.A. 1972 'Late-glacial environmental changes indicated by a coleopteran succession from North Wales', *Boreas* 1, 97-142.
Crimes, T.P., Chester, D.K. and Thomas, G.S.P. 1992 'Exploration of sand and gravel resources by geomorphological analysis in the glacial sediments of eastern Lleyn Peninsula, Gwynedd, North Wales', *Engineering Geology* 32, 137-156.
Currant, A. and Jacobi, R. 2001 'A formal mammalian biostratigraphy for the Late Pleistocene of Britain', *Quaternary Science Reviews* 20, 1707-1716.
Curry, A.M., Walden, J. and Cheshire, D.A. 2001 'The Nant Ffrancon 'protalus rampart': evidence for late Pleistocene paraglacial landsliding in Snowdonia, Wales', *Proceedings of the Geologists' Association* 112, 317-330.
Embleton, C. 1964 'The deglaciation of Arfon and southern Anglesey and the origin of the Menai Straits', *Proceedings of the Geologists' Association* 75, 407-430.
Eyles, N and McCabe, A.M. 1989 'The Late Devensian (<22,000 BP) Irish Sea Basin: the sedimentary record of a collapsed ice sheet margin', *Quaternary Science Reviews* 8, 307-351.
Foster, H. D. 1970 'Sarn Badrig, a sub-marine moraine in Cardigan Bay, North Wales', *Zeitschrift für Geomorphologie* 14, 475-486.
Gemmell, C., Smart, D. and Sugden, D. 1986 'Striae and former ice flow directions in Snowdonia', North Wales, *Geographical Journal* 152, 19-29.
Gibbons, W. and McCarroll, D. 1999 *Geology of the country around Aberdaron and Bardsey Island,* Memoir of the British Geological Survey. London, HMSO.
Gray, J. M., 1982 'The last glaciers (Loch Lomond Advance) in Snowdonia, North Wales', *Geological Journal* 17, 111-133.
Gray, J.M. and Lowe, J.J. 1982 'Problems in the interpretation of small-scale erosional forms on glaciated bedrock surfaces: examples from Snowdonia, N. Wales', *Proceedings of the Geologists' Association* 93, 403-414.
Green, H.S. 1984 *Pontnewydd Cave. A Lower Palaeolithic hominid site in Wales: The first report*, National Museum of Wales, Cardiff, 227.
Greenly, E., 1919 *The Geology of Anglesey*, Memoirs of the Geological Survey of Great Britain. HMSO, London.
Greenly, E. 1938 'The age of the mountains of Snowdonia', *Quarterly Journal of the Geological Society of London* 94, 117-122.
Harris, C. 1991 'Glacial deposits at Wylfa Head, Anglesey, North Wales: evidence for Late Devensian deposition in a non-marine environment', *Journal of Quaternary Science* 6, 67-77.
Harris, C. and McCarroll, D. 1990 'Glanllynnau. In Addison, K., Edge, M.J. and Watkins, R. (Editors) *The Quaternary of North Wales: Field Guide*. Quaternary Research Association, Coventry, 38-47.
Harris, C., Williams, G., Brabham, P., Eaton, G. and McCarroll, D., 1997 'Glaciotectonized Quaternary sediments at Dinas Dinlle, Gwynedd, North Wales, and their bearing on the style of deglaciation in the eastern Irish Sea', *Quaternary Science Reviews* 16, 109-127.
Harrison, R.K. 1971 'The petrology of the Upper Triassic rocks in the Llanbedr (Mochras Farm) borehole', *Institute of Geological Sciences Report*, 71/18, 37-72.
Hughes, P.D. 2002a 'Nunataks and the surface altitude of the last ice sheet in southern Snowdonia, Wales', *Quaternary Newsletter* 97, 19-25.
Hughes, P.D. 2002b 'Nunataks and the surface altitude of the last ice sheet in southern Snowdonia, Wales: a reply to McCarroll and Ballantyne (2002)', *Quaternary Newsletter* 98, 15-17.
Jehu, T.J. 1909 'The glacial deposits of western Caernarvonshire', *Transactions of the Royal Society of Edinburgh* 47, 17-56.
Jones, O.T. 1952 'The drainage system of Wales and adjoining regions', *Quarterly Journal of the Geological Society of London* 107, 201-225.
Lowe, J.J. and Lowe, S. 2001 'Llyn Gwernan' in Walker, M.J.C. and McCarroll, D. (Editors) *The Quaternary of West Wales: Field Guide*, Quaternary Research Association, London, 132-152.
McCarroll, D. 1991 'Ice directions in western Lleyn and the status of the Gwynedd re-advance of the last Irish Sea glacier', *Geological Journal*, 26, 137-143.

McCarroll, D. 1995 'Geomorphological evidence from the Lleyn Peninsula constraining models of the magnitude and rate of isostatic rebound following deglaciation of the Irish Sea Basin', *Geological Journal* 30, 157-163.

McCarroll, D. 2001 'Deglaciation of the Irish Sea Basin: a critique of the glacimarine model', *Journal of Quaternary Science* 16, 393-404.

McCarroll, D. 2002 'Amino-acid geochronology and the British Pleistocene: secure stratigraphical framework or a case of circular reasoning?', *Journal of Quaternary Science* 17, 647-651.

McCarroll, D. and Ballantyne, C.K. 2000 'The last ice sheet in Snowdonia', *Journal of Quaternary Science* 15, 765-778.

McCarroll, D. and Ballantyne, C.K. 2002 'Nunataks and the surface of the last ice sheet in southern Snowdonia, Wales: a comment on Hughes (2002)', *Quaternary Newsletter* 98, 10-14.

McCarroll, D. and Harris, C. 1992 'The glacigenic deposits of western Lleyn, North Wales: Terrestrial or marine?', *Journal of Quaternary Science* 7, 19-29.

McCarroll, D. and Rijsdijk, K. 2003 'Deformation styles as a key for interpreting glacial depositional environments', *Journal of Quaternary Science* 18, 234-256.

Mitchell, G.F. 1960 'The Pleistocene history of the Irish Sea', *Advancement of Science* 17, 313-325.

Mitchell, G.F. 1972 'The Pleistocene history of the Irish Sea: a second approximation', *Scientific Proceedings of the Royal Dublin Society* Series A 4, 181-199.

Nicholas, T.C. 1915 'The geology of the St Tudwal's Peninsula (Caernarvonshire)', *Quarterly Journal of the Geological Society of London* 71, 83-143.

Rose, J. 'Moelwyn Mawr rock glacier, Blaenau Ffestiniog, Gwynedd', In Walker, M.J.C. and McCarroll, D. (Editors) *The Quaternary of West Wales: Field Guide*. Quaternary Research Association, London, 153-155.

Rowlands, B.M. 1971 'Radiocarbon evidence of the age of an Irish Sea glaciation of the Vale of Clwyd', *Nature* 230, 9-11.

Rowlands, B. M., 1979 'The Arenig Region: a study in the Welsh Pleistocene', *Cambria* 6, 13-31.

Saunders, G.E. 1968 'A fabric analysis of the ground moraine of the Lleyn Peninsula of south-west Caernarvonshire', *Geological Journal* 6, 105-118.

Sharp, M., Dowdeswell, J.A. and Gemmell, J.C 1989 'Reconstructing past glacier dynamics and erosion from glacial geomorphic evidence: Snowdon, North Wales', *Journal of Quaternary Science* 4, 115-130.

Synge, F.M. 1964 'The glacial succession in west Caernarvonshire', *Proceedings of the Geologists' Association* 75, 431-444.

Thomas, G.S.P. 1989 'The Late Devensian glaciation along the western margin of the Cheshire-Shropshire lowland', *Journal of Quaternary Science* 4, 167-181.

Thomas, G.S.P., Chester, D.K. and Crimes, P. 1998 'The late Devensian glaciation of the eastern Lleyn peninsula, North Wales: evidence for terrestrial depositional environments', *Journal of Quaternary Science* 13, 255-270.

Walden, J. and Addison, K. 1995 'Mineral magnetic analysis of a 'weathering' surface within glacigenic sediments at Glanllynnau, North Wales', *Journal of Quaternary Science* 10, 367-378.

Whittow, J.B. and Ball, D.F., 1970 'North-west Wales', in Lewis, C.A. (Editor) *The glaciations of Wales and adjoining areas*, Longman, London, 21-58.

Young, T.P., Gibbons, W. and McCarroll, D. 2002 *Geology of the country around Pwllheli*, Memoir of the British Geological Survey, Sheets 134, 151

CHAPTER 4 North-East Wales by Geoffrey S. P. Thomas

Ball, D.F. 1982 *The sand and gravel resources of the country south of Wrexham*, Mineral Assessment Report, Institute of Geological Sciences 106, 67.

Boulton, G.S. and Worsley, P. 1965 'Late Weichselian glaciation in the Cheshire-Shropshire Basin', *Nature* 207, 704-706.

Bowen, D.Q. 1999 'A revised correlation of Quaternary deposits in the British Isles', *Geological Society Special Report* 23, 174.

Bowen, D.Q., Phillips, F.M., McCabe, A.M., Knutz, P.C., Sykes, G.A., 2002 New data for the Last Glacial Maximum in Great Britain and Ireland, *Quaternary Science Reviews* 21, 89-101.

Brown, E.H. and Cooke, R.U. 1977 'Landforms and related glacial deposits in the Wheeler Valley area, Clwyd', *Cambria* 1, 32-45.

Cambell, S. and Bowen, D.Q. 1989 *Quaternary of Wales*, Geological Conservation Review Series 2, Nature Conservancy Council, Peterborough.

Cannell, B. 1982 *The sand and gravel resources of the country around Shrewsbury, Shropshire:description of 1:25000 sheets SJ41 and SJ5*, Mineral Assessment Report, Institute of Geological Sciences 90, 149.

Cheel, R.J. and Rust, B.R. 1982 'Coarse-grained facies of glaciomarine deposits near Ottawa, Canada', in Davidson-Arnot, R. (Editor) *Research in Glacial, Glaciofluvial and Glaciolacustrine Systems*, GeoBooks, Norwich, 271-292.

Clemmensen, L.B. and Houmark-Nielsen, M. 1981 'Sedimentary features of a Weichselian glacio-lacustrine delta', *Boreas* 10, 22-245.

Derbyshire, E. 1962 'Late-glacial drainage in part of north-east Wales: an alternative hypothesis', *Proceedings of the Geologists' Association* 73, 327-334.

Dunkley, P.N. 1981 *The sand and gravel resources of the country north of Wrexham, Clwyd*, Mineral Assessment Report, Institute of Geological Sciences 61, 90.

Embleton, C. 1957 'Some stages in the drainage evolution of part of north-east Wales', *Transactions of the Institute of British Geographers* 23, 19-35.

Embleton, C. 1961 'The geomorphology of the Vale of Conway, North Wales, with particular reference to deglaciation', *Transactions of the Institute of British Geographers* 29, 47-70.

Embleton, C. 1964a 'Late-glacial drainage in part of north-east Wales', *Proceedings of the Geologists' Association* 67, 393-404.

Embleton, C. 1964b 'Sub-glacial drainage and supposed ice-dammed lakes in north-east Wales', *Proceedings of the Geologists' Association* 73, 31-38.

Embleton, C. 1970 'North-eastern Wales', in Lewis, C.A. (Editor) *The glaciations of Wales and adjoining areas*, Longman, London, 59-82.

Fishwick, A. 1977 'The Conway basin', *Cambria* 1, 56-64.

Francis, E.A. 1978 *Keele Field Handbook*, Quaternary Research Association, Keele, 101.

Garrod, D.A.E. 1926 *The Upper Palaeolithic Age in Britain*, Oxford University Press, Oxford.

Green, H.S. and Walker, E. 1991 *Ice Age Hunters: Neanderthals And Early Modern Hunters In Wales*, National Museum Of Wales, Cardiff.

Hall, H.F. 1870 'On the glacial and post-glacial deposits in the neighbourhood of Llandudno', *Geological Magazine* 7, 509-513.

Livingstone, H.J. 1986 *Quaternary Geomorphology of the part of the Elwy Valley and the Vale of Clwyd, North-east Wales*, unpublished PhD thesis, University of London.

Livingstone, H.J. 1990 'Elwy Valley and Vale of Clwyd', in Addison, K., Edge, M.J. and Watkins, R. (Editors) *The Quaternary of North Wales: Field Guide*, Quaternary Research Association, Coventry, 144-153.

Mackintosh, D. 1874 'Additional remarks on boulders, with particular reference to a group of very large and far-travelled erratics in Llanarmon parish, Denbighshire', *Quarterly Journal of the Geological Society of London* 30, 711-721.

Mackintosh, D. 1879 'The erratic blocks of the West of England and the East of Wales', *Quarterly Journal of the Geological Society of London* 35, 425-452

Peake, D.S. 1961 'Glacial changes in the Alyn River system and their significance in the glaciology of the North Wales border', *Quarterly Journal of the Geological Society of London* 117, 335-366.

Peake, D.S.1981 'The Devensian Glaciation of the North Welsh Border', in Neale, J. and Flenley, J. (Editors) *The Quaternary in Britain*, Pergamon, London, 49-59.

Pedley, H.M., Andrews J., Ordonez, S., del Curad, M.A.G., Gonzales, J-A., Martin, and Taylor, D. 1996 'Does climate control the morphological fabric of freshwater carbonates? A comparative study of Holocene barrage tufas from Spain and Britain', *Palaeogeography, Palaeoclimatology, Palaeoecology* 121, 239-257.

Pocock, R.W. and Wray, D.A. 1925 *Geology of Wem*, Memoir Geological Survey of England and Wales.

Pocock, R.W., Whitehead, T.H., Wedd, C.B. and Robertson, T. 1938 *The Shrewsbury District*, Memoir Geological Survey of England and Wales.

Poole, E.G. and Whiteman, A.J. 1961 'The glacial drifts of the southern part of the Shropshire-Cheshire basin', *Quarterly Journal of the Geological Society of London* 117, 91-130.

Preece, R.C., Turner, C and Green, H.S. 1982 *Field excursion to the tufas of the Wheeler valley and to Pontnewydd and Cefn Caves*, Quaternary Research Association Field Guide, 1-9.

Reade, T.M. 1885 'The drift deposits of Colwyn Bay', *Quarterly Journal of the Geological Society of London* 41, 102-107.

Rowlands, B.M., 1971 'Radiocarbon evidence for the age of an Irish Sea glaciation in the Vale of Clwyd', *Nature* 230, 9-11.

Shaw, J. 1972 'Sedimentation in the ice-contact environment with examples from Shropshire (England)', *Sedimentology* 18, 23-62.

Shaw, J. 1972. 'The Irish Sea glaciation of north Shropshire along the border of north-east Wales', *Field Studies* 3, 603-631.

Strahan, A. 1886 'On the glaciation of south Lancashire, Cheshire and the Welsh Border', *Quarterly Journal of the Geological Society of London* 42, 369-381.

Strahan, A. 1890 *The geology of the neighbourhood of Flint, Mold and Ruthin,* Memoir Geological Survey of Great Britain.

Thomas, G.S.P. 1984 'A Late Devensian glaciolacustrine fan-delta at Rhosesmor, Clwyd, North Wales', *Geological Journal* 19, 125-142.

Thomas, G.S.P. 1985 'The Late Devensian glaciation along the border of north-east Wales'. *Geological Journal* 20, 319-340.

Thomas, G.S.P. 1989 'The Late Devensian glaciation along the western margin of the Cheshire-Shropshire Lowland', *Journal of Quaternary Science* 4, 167-181.

Tooley, M.J. 1974 'Sea-level change during the last 9,000 years in North-West England, in Johnson, R.H. (Editor) *The Geomorphology of North-West England*, Manchester University Press, Manchester, 94-121.

Warren, P.T., Price, D., Nutt, M.J.C. and Smith, E.G. 1984 *Geology of the country around Rhyl and Denbigh*, Memoir British Geological Survey, 245.

Wedd, C.B., Smith, B., and Wills, L.J. 1928 *The geology of the country around Wrexham. Part 2: Coal Measures and newer formations*, Memoir Geological Survey of England and Wales, 237.
Whittow, J.B. and and Ball, D.F. 1970 'North-west Wales', in Lewis, C.A. (Editor) *The glaciations of Wales and adjoining regions*, Longman, London, 21-58.
Willls, L.J. 1924 'The development of the Severn valley in the neighbourhood of Iron-Bridge and Bridgnorth', *Quarterly Journal of the Geological Society of London* 80, 274-314.
Wilson, A.C., Mathers, S.J. and Cannell, B. 1982 *The Middle Sands, a prograding sandur succession; its significance in the glacial evolution of the Wrexham-Shrewsbury region*, Report of the Institute of Geological Sciences 82/1, 30-35.
Worsley, P. 1970 'The Cheshire-Shropshire lowlands', in Lewis, C.A. (Editor) *The glaciations of Wales and adjoining regions*, Longman, London, 83-106.
Worsley, P. 1975 'An appraisal of the Glacial Lake Lapworth concept', in Phillips, A.D.M and Turton, B.J. (Editors) *Environment, Man and Economic Change*, Longman, London, 90-118.
Worsley, P. 1985 'Pleistocene history of the Cheshire-Shropshire plain', in Johnson, R.H. (Editor) *The Geomorphology of North-west England*, Manchester University Press, Manchester, 201-221.

CHAPTER 5 The Cheshire-Shropshire Plain *by* Peter Worsley

Andrew, R. and West, R.G. 1977 'Appendix. Pollen from Four Ashes, Worcs', *Philosophical Transactions of the Royal Society of London*, Series B, 280, 242-246.
Bonny, A.P., Mathers, S.J. and Hayworth, E.Y. 1986 'Interstadial deposits with Chelford affinities from Burland, Cheshire', *Mercian Geologist* 10, 151-160.
Boulton, G.S. and Worsley, P. 1965 'Late Weichselian glaciation of the Cheshire – Shropshire Plain', *Nature* 207, 704-706.
Bowen, D.Q., Phillips, F.M., McCabe, A.M., Knutz, P.C. and Sykes. G.A. 2002 'New data for the Last Glacial Maximum in Great Britain and Ireland', *Quaternary Science Reviews* 21, 89-101.
Coope, G.R. 1959 'A Late Pleistocene insect fauna from Chelford, Cheshire', *Proceedings of the Royal Society of London*, Series B,151, 70-86.
Coope, G.R., Shotton, F.W. and Strachan, I. 1961 'A Late Pleistocene fauna and flora from Upton Warren, Worcestershire', *Philosophical Transactions of the Royal Society of London*, Series B, 244, 379-421.
Earp, J.R. and Taylor, B.J. 1986 *Geology of the country around Chester and Winsford*, Memoir British Geological Survey, 119.
Evans, W.B., Wilson, A.A., Taylor, B.J. and Price, D. 1968 *Geology of the country around Macclesfield, Congleton, Crewe and Middlewich*, Memoir of the Geological Survey, 328.
Glasser, N.F. and Sambrook Smith, G.H. 1999 'Glacial meltwater erosion of the Mid-Cheshire Ridge: implications for ice dynamics during the Late Devensian glaciation of northwest England', *Journal of Quaternary Science* 14, 703-710.
Gresswell, R.K. 1964 'The origin of the Mersey and Dee Estuaries', *Geological Journal* 4, 77-86.
Gurney, S.D. and Worsley, P. 1996 'Relict cryogenic mounds at Owlbury, near Bishop's Castle, Shropshire'. *Mercian Geologist* 14, 14-21.
Hamblin, R.J.O. 1986 'The Pleistocene sequence of the Telford District', *Proceedings of the Geologists' Association* 97, 365-377.
Hamblin, R.J.O. and Coppack, B.C. 1995 *Geology of Telford and the Coalbrookdale Coalfield*, Memoir of the British Geological Survey, 158.
Hollis, J.M. and Reed, A.H. 1981 'The Pleistocene deposits of the Southern Worfe Catchment', *Proceedings of the Geologists' Association* 92, 59-74.
Howell, F.T. 1973 'The sub-drift surface of the Mersey and Weaver catchment and adjacent areas', *Geological Journal* 8, 285-296.
Johnson, R.H, and Musk, L.F.1974 'A comment on Dr F.T. Howell's paper on the sub-drift surface of the Mersey and Weaver catchment and adjacent areas', *Geological Journal* 9, 209-210.
Jones, R.C.B., Tonks, L.H. and Wright, W.B.1938 *Wigan District*. Memoir of the Geological Survey of Great Britain, 244.
Knowles, A. 1985 'The Quaternary history of north Staffordshire', in Johnson, R.H. (Editor) *The geomorphology of North-west England,* Manchester University Press, Manchester, 220-236.
Leviston, D. 2001 'Subglacial meltwater channels at Lymm Dam, Cheshire', *Proceedings of the Geologists' Association* 112, 147-154.
Lewis, C.A. (Editor) 1970 *The glaciations of Wales and adjoining regions*, Longman, London, 378.
Mitchell, G. F., Penny, L.F., Shotton, F.W. and West, R.G. 1973 'A correlation of the Quaternary deposits in the British Isles', *Geological Society of London Special Report* 4, 99.
Morgan, A.1973 'Late Pleistocene environmental changes indicated by fossil insect faunas of the English Midlands', *Boreas* 2, 173-212.
Morgan, A.V. 1973 'The Pleistocene geology of the area north and west of Wolverhampton, Staffordshire, England', *Philosophical Transactions of the Royal Society of London*, Series B, 265, 233-297.

Paul, M.A. 1983 'The supraglacial landsystem', in Eyles, N. (Editor) *Glacial Geology*, Pergamon Press, Oxford, 71-90.

Paul, M.A. and Little, J.A. 1991 'Geotechnical properties of glacial deposits in lowland Britain', in Ehlers, J., Gibbard, P.L. and Rose, J. (Editors) *Glacial deposits of Great Britain and Ireland*, Balkema, Rotterdam, 389-404.

Poole, E.G. and Whiteman, A.J. 1966 *Geology of the country around Nantwich and Whitchurch*, Memoir of the Geological Survey Great Britain, 154.

Rees, J.G. and Wilson, A.A. 1998 *Geology of the country around Stoke-on-Trent*, Memoir of the British Geological Survey, 152.

Rendell, H., Worsley, P., Green, F and Parks, D. 1991 'Thermoluminesence dating of the Chelford Interstadial', *Earth and Planetary Science Letters* 103, 182-189.

Sambrook Smith, G.H. and Glasser, N.F. 1998 'Late Devensian ice sheet characteristics; a palaeohydrological approach', *Geological Journal* 33, 149-158.

Shaw, J. 1972a 'The Irish Sea glaciation of north Shropshire – some environmental reconstruction', *Field Studies* 3, 603-631.

Shaw, J. 1972b 'Sedimentation in the ice-contact environment with examples from Shropshire, England', *Sedimentology* 18, 23-62.

Simpson, I.M. and West, R.G. 1958 'On the stratigraphy and palaeobotany of a Late-Pleistocene organic deposit at Chelford, Cheshire', *New Phytologist* 57, 239-250.

Shotton, F.W. 1967 'Age of the Irish Sea Glaciation of the Midlands', *Nature* 215, 1366.

Shotton, F.W. 1977 *The English Midlands: Guidebook for Excursion A2*, Inqua, Norwich, 51.

Shotton, F.W. and West, R.G. 1969 'Stratigraphical table of the British Quaternary', *Proceedings of the Geological Society of London* 1656, 155-157.

Stevenson, I.P. and Mitchell, G.H. 1955 *Geology of the country between Burton- upon-Trent, Rugeley and Uttoxeter*, Memoir of the Geological Survey Great Britain, 178.

Taylor, B.J., Price, R.H. and Trotter, F.M. 1963 *Geology of the country around Stockport and Knutsford*, Memoir of the Geological Survey Great Britain, 183.

Thomas, G.S.P. 1989 'The Late Devensian glaciation along the western margin of the Cheshire-Shropshire lowland', *Journal of Quaternary Science* 4, 167-181.

Thompson, D.B. and Worsley, P. 1966 'A Late-Pleistocene molluscan fauna from the drifts of the Cheshire Plain', *Geological Journal* 5, 197-207.

Thompson, D.B. and Worsley, P. 1967 'Periods of ventifact formation in the Permo-Triassic and Quaternary of the North East Cheshire Basin', *The Mercian Geologist* 2, 279-297.

Watson, E. 1977 'The periglacial environment of Great Britain during the Devensian', *Philosophical Transactions of the Royal Society of London*, Series B, 280, 183-198.

Wills, L.J. 1924 'The development of the Severn Valley in the neighbourhood of Iron-Bridge and Bridgnorth', *Quarterly Journal of the Geological Society of London* 80, 274-314.

Wills, L.J. 1950 *The palaeogeography of the Midlands*, 2nd edition, Liverpool University Press, Liverpool, 147.

Wilson, A.C., Mathers, S.J. and Cannell, B. 1982 *The Middle Sands, a prograding sandur succession; its significance in the glacial evolution of the Wrexham – Shrewsbury region*, Report Institute of Geological Sciences 82/1, 30-35.

Worsley, P. 1966a 'Some Weichselian fossil frost wedges from north east Cheshire', *The Mercian Geologist* 1, 357-365.

Worsley, P. 1966b 'Fossil frost wedge polygons at Congleton, Cheshire, England', *Geografiska Annaler* 48A, 211-219.

Worsley, P. 1967a 'Problems in naming the Pleistocene deposits of the North-east Cheshire Plain', *The Mercian Geologist* 2, 51-55.

Worsley, P. 1967b *Some aspects of the Quaternary evolution of the Cheshire Plain*, unpublished PhD thesis, University of Manchester, 388.

Worsley, P. 1970 'The Cheshire-Shropshire Lowlands', in Lewis, C.A. (Editor) *The glaciations of Wales and adjoining regions*, Longman, London, 83-106.

Worsley, P. 1975 'An appraisal of the Glacial Lake Lapworth concept', in Phillips, A.D.M. and Turton, B.J. (Editors), *Environment, man and economic change*, Longman, London, 90-118.

Worsley, P. 1976 'Correlation of the Last glaciation glacial maximum and the extra-glacial fluvial terraces in Britain – a case study', in Easterbrook, D.J. and Sibrava, V. (Editors) *Quaternary glaciations of the northern hemisphere*. Bellingham and Prague, IGCP 73/1/24, Report 3, 274-284.

Worsley, P. 1980 'Problems in radiocarbon dating the Chelford Interstadial of England', in Cullingford, R.A., Davidson, D.A. and Lewin, J. (Editors) *Timescales in Geomorphology*, Wiley, Chichester, 289-304.

Worsley, P. 1991 'Glacial deposits of the lowlands between the Mersey and Severn rivers', in Ehlers, J., Gibbard, P.L. and Rose, J. (Editors) *Glacial deposits in Great Britain and Ireland*, Balkema, Rotterdam, 203-211.

Worsley, P. 1992 'A pre-Devensian mammoth from Arclid, Cheshire', *Proceedings of the Geologists' Association* 103, 75-77.

Worsley, P. 2002 'Physical environment', in Phillips, A.D.M. and Phillips, C.P. (Editors) *A new historical atlas of Cheshire*, Cheshire County Council, Chester, 4-7.

Worsley, P., Coope, G.R., Good, T.R., Holyoak, D.T. and Robinson, J.E. 1983 'A Pleistocene succession from beneath Chelford Sands at Oakwood Quarry, Chelford, Cheshire', *Geological Journal*, 18, 307-324.

Wright, W.B. 1914 *The Quaternary Ice Age*, Macmillan, London, 464pp.

CHAPTER 6 The lower Severn valley *by* Darrel Maddy and Simon G. Lewis

Antoine, P. 1994 'The Somme valley terrace system (northern France): a model of river response to Quaternary climatic variations since 800,000 BP', *Terra Nova* 6, 453-464.

Barclay, W.J., Brandon, A., Ellison, R.A. and Moorlock, B.S.P. 1992 'A Middle Pleistocene palaeovalley-fill west of the Malvern Hills', *Journal of the Geological Society of London* 149, 75-92.

Beckensale, R.P. and Richardson, L. 1964 'Recent findings on the physical development of the lower Severn valley', *Geographical Journal* 130, 87-105.

Bowen, D.Q. (Editor) 1999 *A correlation of Quaternary deposits in the British Isles*. Geological Society of London Special Report 23.

Bowen, D.Q., Hughes, S., Sykes, G.A. and Miller, G.H. 1989 'Land-sea correlations in the Pleistocene based on isoleucine epimerization in non-marine molluscs', *Nature* 340, 49-51.

Boulton, W.S. 1917 'Mammalian remains in the glacial gravels at Stourbridge', *Proceedings of the Birmingham Natural History and Philosophical Society* 14 107-112.

Brandon, A. and Hains, B.A., 1981 *Geological notes and local details for 1:10,000 sheets SO43NE, SO43SE, SO53NW, SO54SW (Hereford City)*, Institute of Geological Sciences, Kegworth.

Bridgland, D.R. 1994 *Quaternary of the Thames*. Geological Conservation Review Series No.7. Chapman and Hall, London, 441.

Bridgland, D.R. 2000 'River terrace systems in north-west Europe: an archive of environmental change, uplift and early human occupation', *Quaternary Science Reviews* 19, 1293-1303.

Bridgland, D.R. and Maddy, D. 1995 'River terrace deposits: long Quaternary terrestrial sequences', Abstracts, XIV International Quaternary Association Congress, Berlin, 37.

Bridgland, D.R., Keen, D.H. and Maddy, D. 1986 'A reinvestigation of the Bushley Green Terrace typesite, Hereford and Worcester', *Quaternary Newsletter* 50, 1-6.

Brown, A.G. 1982 *Late Quaternary Palaeohydrology, Palaeoecology and Floodplain development of the Lower River Severn*, unpublished PhD thesis, University of Southampton.

Bull, W. B. 1991 *Geomorphic responses to climate change*, Oxford University Press, Oxford, 329.

Clayton, K.M. 1977 'River Terraces', in Shotton, F.W. (Editor) *British Quaternary Studies: Recent Advances*, Claendon Press, Oxford, 53-168.

Cloetingh, S., Gradstein, F.M., Kooi, H., Grant, A. C. and Kaminski, M. 1990 'Plate reorganisation: a cause of rapid late Neogene subsidence and sedimentation around the North Atlantic', *Journal of the Geological Society of London* 147, 495-506.

Coope, G. R., 1962 'The Pleistocene coleopterous fauna with arctic affinities from Fladbury Worcestershire', *Quarterly Journal of the Geological Society of London* 118, 103-123.

Coope, G.R., Shotton, F.W. and Strachan, I. 1961 'A Late Pleistocene fauna and flora from Upton Warren, Worcestershire', *Philosophical Transactions of the Royal Society* B244, 379-421.

Cope, J.C.W. 1994 'A latest Cretaceous hotspot and the southeasterly tilt of Britain', *Journal of the Geological Society of London* 151, 905-908.

Dawson, M.R. 1988 'Diamict deposits of the pre-Late Devensian Glacial age underlying the Severn Main Terrace at Stourport, Worcestershire: their origins and straigraphic implications', *Proceedings of the Geologists' Association* 99, 125-32.

Dawson, M.R. 1989 'Chelmarsh', in Keen, D.H. (Editor) *The Pleistocene of the West Midlands: Field Guide*, Quaternary Research Association, Cambridge, 80-85.

Dawson, M.R. and Bryant, I.D. 1987 'Three-dimensional facies geometry in Pleistocene outwash sediments, Worcester, U.K', in Ethridge, F.G. (Editor) *Recent developments in fluvial sedimentology*, Society of Economic Paleontologists and Mineralogists Special Publication 39, 191-196.

Fairbridge, R.W. 1961 'Eustatic changes in sea-level', *Physics and Chemistry of the Earth* 4, 99-185.

Goodwin, M.D. 1999. *Evidence for Late Middle Pleistocene glaciation in the British Isles*, unpublished PhD Thesis, Cheltenham and Gloucester College of Higher Education.

Gray, J.W. 1911 'North and Mid Cotteswolds, and the Vale of Moreton during the Glacial Epoch', *Proceedings of the Cotteswold Naturalists Field Club* 17, 257-274.

Gray, J.W. 1912 'The Lower Severn plain during the Glacial Epoch', *Proceedings of the Cotteswold Naturalists Field Club* 17, 365-380.

Gray, J.W. 1914 'The drift deposits of the Malverns and their supposed glacial origin', *Proceedings of the Birmingham Natural History and Philosophical Society* 13, 1-18.

Gray, J.W. 1919 'Notes on the Cotteswold-Malvern region during the Quaternary Period', *Proceedings of the Cotteswold Naturalists Field Club* 20, 99-141.

Harrison, W.J. 1898 'The ancient glaciers of the Midland counties of England', *Proceedings of the Geologists' Association* 15, 400-408

Hey, R.W. 1958 'High-level gravels in and near the Lower Severn Valley', *Geological Magazine* 95, 161-168

Hey, R.W. 1991 'Pleistocene Gravels in the Lower Wye Valley', *Geological Journal* 26, 123-136.

Keen, D. H. and Bridgland, D., 1986 'An interglacial fauna from Avon No 3 terrace at Eckington, Worcestershire', *Proceedings of the Geologists' Association* 97, 303-307.

Lambeck, K. 1993 'Glacial rebound of the British Isles I: Preliminary model results', *Geophysical Journal International* 115, 941-959.

Lambeck, K. 1995 'Late Devensian and Holocene shorelines of the British Isles and North Sea from models of glacio-hydro-isostatic rebound', *Journal of the Geological Society of London* 152, 437-448.

Leopold L.B. and Bull, W.B. 1979 'Base level, aggradation and grade', *Proceedings of the American Philosophical Society* 123, 168-202.

Leeder, M.R. and Stewart, M.D. 1996 'Fluvial incision and sequence stratigraphy: alluvial responses to relative sea-level fall and their detection in the geological record', in Hesselbo, S. P. and Parkinson, D.N. (Editors) *Sequence stratigraphy in British Geology*, Geological Society Special Publication 103, 25-39.

Lucy, W.C. 1872 ' The gravels of the Severn, Avon and Evenlode, and their extension over the Cotteswold Hills', *Proceedings of the Cotteswold Naturalists Field Club* 5, 71-125.

Maddy, D. 1989 *The Middle Pleistocene development of the rivers Severn and Avon*, unpublished PhD thesis, University of London.

Maddy, D. 1997 'Uplift Driven Valley Incision and River Terrace Formation in Southern England', *Journal of Quaternary Science* 12, 539-545.

Maddy, D. 1999a 'Reconstructing the Baginton River Basin and its implications for the early development of the River Thames drainage system', in Andrews, P. and Banham, P. (Editors) *Late Cenozoic Environments and Homonid Evolution: a tribute to Bill Bishop*, Geological Society of London, 169-182.

Maddy, D. 1999b 'English Midlands', in Bowen, D.Q. (Editor) *A correlation of Quaternary deposits in the British Isles*, Geological Society of London Special Report 23, 28-44.

Maddy, D. 2002 'An evaluation of climate, crustal movement and base level controls on the Middle-Late Pleistocene development of the River Severn, UK', *Netherlands Journal of Geosciences* 81, 329-338.

Maddy, D., Keen, D.H., Bridgland, D.R. and Green, C.P. 1991 'A revised model for the Pleistocene development of the River Avon, Warwickshire', *Journal of the Geological Society of London* 148, 473-484.

Maddy, D., Green, C.P., Lewis, S.G. and Bowen, D.Q. 1995 'Pleistocene Geology of the Lower Severn Valley, UK', *Quaternary Science Reviews* 14, 209-222

Maddy, D. and Bridgland, D.R. 2000 'Accelerated uplift resulting from Anglian glacioisostatic rebound in the Middle Thames valley, UK? : Evidence from the river terrace record', *Quaternary Science Reviews* 19, 1581-1588.

Maddy, D., Bridgland, D.R. and Green, C.P. 2000 'Crustal uplift in southern England: Evidence from the river terrace records', *Geomorphology* 33, 167-181.

Maddy, D., Bridgland, D.R. and Westaway, R. 2001 'Uplift-driven valley incision and climate-controlled river terrace development in the Thames Valley, UK', *Quaternary International*, 79, 23-36.

Maw, G. 1864 'On the drifts of the Severn in the neighbourhood of Colebrook Dale and Bridgnorth', *Quarterly Journal of the Geological Society of London* 20, 130.

Miller, A. A., 1935 'The entrenched meanders of the Herefordshire Wye', *Geographical Journal* 85, 160-178.

Morgan, A.V. 1973 'The Pleistocene geology of the area north and west of Wolverhampton', *Philosophical Transactions of the Royal Society of London* B265, 233-297

Murchison, R.I. 1836 'Gravel and Alluvia in Worcestershire and Gloucestershire', *Proceedings of the Geological Society* 2, 230-36.

Murchison, R.I. 1839 *Silurian System*, John Murray, London.

Schumm, S.A. 1993 'River response to baselevel change: implications for sequence stratigraphy', *Journal of Geology* 101, 279-294.

Shackleton, N.J. and Opdyke.N.D. 1973 'Oxygen isotope stratigraphy of equitorial Pacific core V28-238: oxygen isotope temperatures and ice volumes on a 10^5 year and 10^6 year scale', *Quaternary Research* 3, 39-55.

Shackleton, N.J., Berger, A. and Peltier, W.R. 1990 'An Alternative Astronomical Calibration of the Lower Pleistocene Timescale Based on ODP Site 677', *Transactions of the Royal Society of Edinburgh* 81, 252-261.

Shotton, F. W., 1953 'The Pleistocene deposits of the area between Coventry, Rugby and Leamington and their bearing on the topographic development of the Midlands', *Philosophical Transactions of the Royal Society of London* B237, 209-260.

Shotton, F.W. and Coope, G.R. 1983 'Exposures in the Power House Terrace of the River Stour, Wilden, Worcestershire, England', *Proceedings of the Geologists' Association* 94, 33-44.

Shotton, F.W., Keen, D.H., Coope, G.R., Currant, A.P., Gibbard, P.L., Aalto, M., Peglar, S.M., and Robinson, J.E. 1993 'The Middle Pleistocene deposits at Waverley Wood Pit, Warwickshire, England', *Journal of Quaternary Science* 8, 293-325.

Symonds, W.S. 1861 'On the drifts of the Severn, Avon, Wye and Usk', *Proceedings of the Cotteswold Naturalists Field Club*, 3, 31-39

Tomlinson, M. E., 1925 'River terraces of the Lower Valley of the Warwickshire Avon', *Quarterly Journal of the Geological Society of London* 81, 137-170.

Van den Berg, M.W. 1996 *Fluvial sequences of the Maas: a 10Ma record of neotectonics and climate change at various timescales*, unpublished PhD thesis, University of Wageningen.

Veldkamp, A. and Van Dijke, J.J. 2000 'Simulating internal and external controls on fluvial terrace stratigraphy: a qualitative comparison with the Maas record', *Geomorphology* 33, 225-236.

Watts, A.B., Mckerrow, W.S. and Fielding, E. 2000 'Lithospheric flexure, uplift, and landscape evolution in south-central England', *Journal of the Geological Society of London* 157, 1169-1177.

Westaway, R. 2001 'Flow in the lower continental crust as a mechanism for the Quaternary uplift of the Rhenish Massif, north-west Europe', in Maddy, D., Macklin, M., and Woodward, J. (Editors) *River Basin Sediment Systems: Archives of Environmental Change*, Balkema, Rotterdam, 87-168.

Westaway, R., Maddy, D. and Bridgland, D.R. 2002 'Flow in the lower continental crust as a mechanism for the Quaternary uplift of southeast England', *Quaternary Science Reviews* 21, 559-60

Whitehead, P.F., 1989 'Development and deposition of the Avon valley terraces', *The Pleistocene of the West Midlands: Field Guide,* in Keen, D.H (Editor). Cambridge, Quaternary Research Association, 37-41.

Williams, G.J. 1968 'The buried channel and superficial deposits of the lower Usk and their correlation with similar features in the lower Severn', *Proceedings of the Geologists' Association* 79, 325-48.

Wills, L.J. 1924 'The development of the Severn Valley in the neighbourhood of Iron-Bridge and Bridgnorth', *Quarterly Journal of the Geological Society of London* 80, 274-314

Wills, L.J. 1937 *'The Pleistocene History of the West Midlands ,Section C – Geology'*, British Association for the Advancement of Science Annual Report, 71-94.

Wills, L.J. 1938 'The Pleistocene development of the Severn from Bridgnorth to the sea', *Quarterly Journal of the Geological Society of London* 94, 161-242.

Wills, L.J. 1948 *The palaeogeography of the Midlands*, Liverpool University Press, Liverpool.

CHAPTER 7 West Wales *by* James L. Etienne, Michael J. Hambrey, Neil F. Glasser and Krister N. Jansson

Allen, A. 1960 'Seismic refraction investigations of the pre-glacial valley of the River Teifi near Cardigan', *Geological Magazine* 97, 276-282.

Allen, J.R.L. 1982 'Late Pleistocene (Devensian) glaciofluvial outwash at Banc-y-Warren, near Cardigan (west Wales)', *Geological Journal* 17, 31-47.

Ballantyne, C.K. 2002 'Paraglacial geomorphology', *Quaternary Science Reviews* 21, 1935-2017.

Ballantyne, C.K. and Harris, C. 1994 *The periglaciation of Great Britain*, Cambridge University Press, Cambridge, 330.

Blundell, D.J., Griffiths, D.H. and King, R.F. 1969 'Geophysical investigations of buried river valleys around Cardigan Bay', *Geological Journal* 6, 161-180.

Bowen, D.Q. 1973a 'The Pleistocene history of Wales and the borderland', *Geological Journal* 8, 207-224.

Bowen, D.Q. 1973b 'The Pleistocene succession of the Irish Sea', *Proceedings of the Geologists' Association* 84, 249-272.

Bowen, D.Q. 1977 'The coast of Wales', in Kidson C. and Tooley M.J. (Editors) *The Quaternary History of the Irish Sea*, Geological Journal Special Issue 7, Liverpool, 223-226.

Bowen, D.Q. and Gregory, K.J. 1965 'A glacial drainage system near Fishguard, Pembrokeshire', *Proceedings of the Geologists' Association* 74, 275-281.

Bowen, D.Q. and Lear, D.L. 1982 'The Quaternary Geology of the lower Teifi Valley', in M.G. Bassett, (Editor) *Geological Excursions in Dyfed, south-west Wales*, National Museum of Wales, Cardiff, 297-302.

Bowen, D.Q. and Henry, A. (Editors) 1984 *Wales: Gower, Preseli, Fforest Fawr*, Quaternary Research Association Field Guide, Cambridge.

Bowen, D.Q., Phillips, F.M., McCabe, A.M., Knutz, P.C. and Sykes, G.A. 2002 'New data for the Last Glacial Maximum in Great Britain and Ireland', *Quaternary Science Reviews* 21, 89-101.

British Geological Survey. 1989 *Aberystwyth*, Sheet 163. 1:50,000, British Geological Survey, Keyworth, Nottingham.

British Geological Survey. 1994 *Aberaeron*, Sheet 177. 1:50,000, Unpublished, British Geological Survey, Keyworth, Nottingham.

British Geological Survey. 1994 *Llanilar*, Sheet 178. 1:50,000, British Geological Survey, Keyworth, Nottingham.

British Geological Survey. 1995 *Cadair Idris*, Sheet 49. 1:50,000, British Geological Survey, Keyworth, Nottingham.

British Geological Survey. 1997 *Geology of the Afon Teifi Catchment*, Map 1: Drift Deposits. 1:50,000', British Geological Survey, Keyworth, Nottingham.

British Geological Survey. 2003 *Cardigan and Dinas Island*, Sheet 196: Solid and Drift Geology, 1:50,000', British Geological Survey.

Campbell, S. and Bowen, D.Q. (Editors) 1989 *The Quaternary of Wales*, Geological Conservation Review Series 2, Nature Conservancy Council, Peterborough.

Carruthers, R.M., Chacksfield, B.C. and Heaven, R.E. 1997 'A geophysical investigation to determine the thickness and nature of glacial deposits in the buried valleys of Afon Teifi between Cardigan and Llanbydder', in Waters R.A.,. Davies J.R, Wilson D. and Prigmore J.K. (Editors) *A geological background for planning and development in the Afon Teifi Catchment*, British Geological Survey, Technical Report WA/97/35.

Cave, R. and Hains, B.A. 1986 *Geology of the country between Aberystwyth and Machynlleth*, Memoir of the British Geological Survey, Sheet 163 (England & Wales).

Cave, R. and Pratt, W.T. 1995 'Tonfanau', in Dobson M.R. (Editor) *The Aberystwyth District*, Geologists' Association Guide 54, 59-62.

Challinor, J. 1933 'The 'Incised Meanders' near Ponterwyd, Cardiganshire', *Geological Magazine* 70, 90-92.

Charlesworth J.K. 1929 'The South Wales end moraine', *Quarterly Journal of the Geological Society of London* 85, 335-58.

Davies, D.C. 1988 *The amino-stratigraphy of British Pleistocene glacial-deposits*, unpublished PhD thesis, University of Wales.

Davies, J.R. Fletcher, C.J.N., Waters, R.A., Wilson, D. Woodhall, D.G. and Zalasiewicz, J.A. 1997 *Geology of country around Llanilar and Rhayader Memoir for 1:50 000 geological sheets 178 and 179 (England and Wales)*, British Geological Survey.

Davies, J.R., Waters, R.A., Wilby, P.R., Williams, M., and Wilson, D. 2003 *Geology of the Cardigan and Dinas Island district – a brief explanation of the geological map sheet 1:50, 000 Sheet 193 (including part of sheet 210)*, British Geological Survey.

Edwards, E. 1997 *Sedimentology and dating of glacigenic sequences; eastern Irish Sea Basin*, unpublished Ph.D thesis, University of Wales.

Etienne, J.L., Hambrey, M.J., Davies, J.R., Waters, R.A. and Glasser, N.F. 2001 'Poppit Sands', in Walker M.J.C. and. McCarroll (Editors) *The Quaternary of West Wales: Field Guide*, Quaternary Research Association, London, 51-54.

Etienne J.L., Hambrey, M.J., Glasser, N.F., Davies J.R., Waters, R.A., Wilby, P.R., and Wilson, D. 2003 'Some observations on recent ideas about the Teifi valley, West Wales – a reply to D. Lear', *Quaternary Newsletter* 100, 26-31.

Eyles, N. and McCabe, A.M. 1989 'The Late Devensian (<22, 000BP) Irish Sea Basin: the sedimentary record of a collapsed ice sheet margin', *Quaternary Science Reviews* 8, 307-51.

Fletcher, C.J.N. 1994 *The geology of the St. Dogmaels area*, British Geological Survey Technical Report WA/94/63.

Fletcher, C.J.N. and Siddle, H.J. 1998 'Development of glacial Llyn Teifi, west Wales: evidence for lake-level fluctuations at the margins of the Irish Sea ice sheet', *Journal of the Geological Society of London* 155, 389-99.

Foster, H,D. 1970 'Sarn Badrig, a submarine moraine in Cardigan Bay, North Wales', *Zeitschrift für Geomorphologie*, 14, 475-86

Francis, T.J.G. 1964 'A seismic refraction section of the pre-glacial Teifi Valley near Cenarth', *Geological Magazine* 101, 108-12.

Garradrd R.A. and Dobson, M.R. 1974 'The nature and maximum extent of glacial sediments off the West coast of Wales', *Marine Geology* 16, 31-44.

George, T.N. 1932 'The Quaternary beaches of Gower', *Proceedings of the Geologists' Association* 43, 291-324.

Glasser, N.F. and Hampbrey M.J. 2001 'Styles of sedimentation beneath Svalbard valley glaciers under changing dynamic and thermal regimes', *Journal of the Geological Society of London* 158, 697-707.

Glasser, N.F., Etienne, J.L., Hambrey, M.J., Davies, J.R., Waters, R.A. and Wilby, P.R. (In Press) 'Geomorphological and sedimentary evidence for multiple glaciations in the lower Afon Teifi region, Wales, UK', *Boreas*.

Gurney, S.D. 1995 'A reassessment of the relict Pleistocene "pingos" of west Wales: hydraulic pingos or mineral palsas?', *Quaternary Newsletter* 77, 6-16.

Hambrey, M.J., Davies, J.R., Glasser, N.F., Waters, R.A., Dowdeswell, J.A., Wilby, P., Wilson, D. and Etienne, J.L. 2001 'Devensian glacigenic sedimentation and landscape evolution in the Cardigan area of southwest Wales', *Journal of Quaternary Science* 16, 455-82.

Harris, C. 1996 'The origin of the Head deposits at Morfa Bychan, Dyfed, Wales: reply to Sybil Watson's comments', *Quaternary Newsletter* 78, 47-50.

Harris, C. 1998 'The micromorphology of paraglacial and periglacial slope deposits: a case study from Morfa Bychan, west Wales, UK', *Journal of Quaternary Science* 13, 73-84.

Harris, C. 2001a 'Llanon', in Walker M.J.C. and McCarroll, D. (Editors) *The Quaternary of West Wales: Field Guide*, Quaternary Research Association, London, 65-66.

Harris, C. 2001b 'Morfa Bychan', in Walker M.J.C. and McCarroll, D (Editors) *The Quaternary of West Wales: Field Guide*, Quaternary Research Association, London, 62-64.

Helm, D.G. and Roberts, B. 1975 'A re-interpretation of the origin of sands and gravels around Banc-y-Warren, near Cardigan, west Wales', *Geological Journal* 10, 131-146.

REFERENCES

Holmes, G.E., Hopkins, D.M. and Foster, H.L. 1968 'Pingos in central Alaska', *U.S. Geological Survey Bulletin*, 1241-1340.

Hutchinson, J. N. and Millar, D.L. 2001 'The Craig Goch landslide, Meirionydd, mid Wales', in: Walker, M.J.C. and McCarroll, D. (Editors) *The Quaternary of West Wales: Field Guide*, Quaternary Research Association, London, 113-125.

Jansson, K.N. and Glasser N.F. 2004 'Palaeoglaciology of the Welsh sector of the British-Irish Ice Sheet', *Journal of the Geological Society of London* 161, 1-13.

John, B.S. 1970 'Pembrokeshire', in C.A. Lewis, (Editor) *The glaciations of Wales and adjoining regions*, Longman, London, 229-265.

John, B.S. 1973 'Vistulian periglacial phenomena in south-west Wales', *Biuletyn Peryglacjalny* 22, 185-213.

Jones, O.T. 1965 'The glacial and post-glacial history of the lower Teifi Valley', *Journal of the Geological Society of London* 121, 247-81.

Jones, O.T. and Pugh, W.J. 1935 'The geology of the districts around Machynlleth and Aberystwyth', *Proceedings of the Geologists' Association* 46, 247-300.

Keeping, W. 1882 'The glacial geology of central Wales', *Geological Magazine* 19, 251-257.

Lambeck, K. 1993 'Glacial rebound of the British Isles: 2. A high-resolution, high precision model', *Geophysics Journal International* 15, 960-990.

Lambeck, K. 1995 'Late Devensian and Holocene shorelines of the British Isles and North Sea from models of glacio-hydro-isostatic rebound', *Journal of the Geological Society of London* 152, 437-448.

Lambeck, K. 1996 'Glaciation and sea-level change for Ireland and the Irish Sea since Late Devensian / Midlandian time', *Journal of the Geological Society of London* 153, 853-872.

Lambeck, K. and Purcell, A.P. 2001 'Sea-level change in the Irish Sea since the Last Glacial Maximum: constraints from isostatic modelling', *Journal of Quaternary Science* 16, 497-506.

Lear, D.L. 1986 *The Quaternary deposits of the lower Teifi Valley*, unpublished PhD thesis, University of Wales.

Lear, D.L. 2003a 'Some observations on recent ideas about the Teifi Valley, west Wales', *Quaternary Newsletter* 99, 19-24.

Lear, D.L. 2003b 'Some observations on recent ideas about the Teifi Valley, west Wales – a reply to Etienne *et al.*', *Quaternary Newsletter* 100, 32-34.

Mitchell, G.F. 1960 'The Pleistocene history of the Irish Sea', *Advancement of Science* 17, 313-325.

Mitchell, G.F. 1972 'The Pleistocene history of the Irish Sea: a second approximation', *Scientific Proceedings of the Royal Dublin Society* Series A4, 181-199.

Nunn, K.R. and Boztas, M. 1977 'Shallow seismic reflection profiling on land using a controlled source', *Geoexploration* 15, 87-97.

Owen, G. 1997 'Origin of an esker-like ridge – erosion or channel fill? Sedimentology of the Monington 'esker' in southwest Wales', *Quaternary Science Reviews* 16, 675-684.

Péwé, T.L. 1969 'The periglacial environment', in Péwé T.L. (Editor) *The periglacial environment*, McGill-Queens University Press, Montreal, 1-9.

Pratt, W.T., Woodhall, D.G. and Howells, M.F. 1995 *Geology of the country around Cadair Idris*, Memoir British Geological Survey 1:50 000 geological sheet 149 (England and Wales).

Reade, T.M. 1896 'Notes on the drift of the mid-Wales coast', *Quarterly Journal of the Geological Society of London* 53, 341-348.

Rijsdijk, K.F. 2001 'Density-driven deformation structures in glacigenic consolidated diamicts: examples from Traeth y Mwnt, Cardiganshire, Wales, UK', *Journal of Sedimentary Research* 71, 122-135.

Thomas, G.S.P., Summers, A.J. and Dackombe, R.V. 1982 'The Late-Quaternary deposits of the middle Dyfi Valley, Wales', *Geological Journal* 17, 297-309.

Vincent, P.J. 1976 'Some periglacial deposits near Aberystwyth, Wales, as seen with a scanning electron microscope', *Biuletyn Peryglacjalny* 25, 59-76.

Walker, M.J.C. and James, J.H. 2001 'A pollen diagram from a ground ice depression ('pingo U') Cledlyn Valley', in Walker M.J.C.and McCarroll D. (Editors) *The Quaternary of West Wales: Field Guide*, Quaternary Research Association, London, 71-75.

Walker, M.J.C. and McCarroll, D. (Editors) *The Quaternary of West Wales: Field Guide*, Quaternary Research Association, London.

Walker, M.J.C., Coope, G. R., Sheldrick, C., Turney, C. S. M., Lowe, J. J., Blockley S. P. E. and Harkness, D.D. 2003 'Devensian Lateglacial environmental changes in Britain: a multi-proxy environmental record from Llanilid, South Wales, UK', *Quaternary Science Reviews* 22, 475-520.

Waters, R.A. Davies, J.R., Wilson, D. and Prigmore, J.K. 1997 'A geological background for planning and development in the Afon Teifi Catchment', *British Geological Survey, Technical Report* WA/97/35.

Watson, E. 1965 'Periglacial structures in the Aberystwyth region of central Wales', *Proceedings of the Geologists' Association* 76, 443-62.

Watson, E. 1970 'The Cardigan Bay Area', in Lewis C.A. (Editor) *The Glaciations of Wales and Adjoining Regions*, Longman, London, 125-145.

Watson, E. 1971 'Remains of pingos in Wales and the Isle of Man', *Geological Journal* 7, 381-392.
Watson, E. 1976 'Field excursions in the Aberystwyth region', *Biuletyn Peryglacjalny* 26, 79-112.
Watson, E. 1981 'Characteristics of ice-wedge casts in west central Wales', *Biuletyn Peryglacjalny* 28, 163-177.
Watson, E. and Watson, S. 1967 'The periglacial origin of the drifts at Morfa Bychan, near Aberystwyth', *Geological Journal* 5, 419-440.
Watson, E. and Watson, S. 1972 'Investigations of some pingo basins near Aberystwyth, Wales', *Report of the 24th International Geological Congress, Montreal*, Section 12, 212-223.
Watson, S. 1996 'The origin of the Head deposits at Morfa Bychan, Dyfed, Wales', *Quaternary Research Association Newsletter* 78, 45-46.
Wilby, P.R. 1998 'The Quaternary Sequence of the Buried Valley of the Teifi near Cardigan: A sedimentological investigation of three cored boreholes', *British Geological Survey Technical Report* WA/98/33C.
Williams, K.E. 1927 'The glacial drifts of western Cardiganshire', *Geological Magazine* 64, 205-227.
Wood, A. 1959 'The erosional history of the cliffs around Aberystwyth', *Geological Journal* 2, 271-279.
Worsley, P. 1987 'Permafrost stratigraphy in Britain: a first approximation', in Boardman J. (Editor) *Periglacial Processes and Landforms in Britain and Ireland*, Cambridge University Press, Cambridge, 89-99.

CHAPTER 8 The upper Wye and Usk regions *by* Colin A. Lewis and Geoffrey S. P. Thomas

Ballantyne, C.K. and Harris, C. 1994 *The periglaciation of Great Britain*, Cambridge University Press, Cambridge.
Barclay, W.J. 1989 *Geology of the South Wales Coalfield, Part II: the country around Abergavenny*, Memoir of the Geological Survey of Great Britain.
Bartley, D.D. 1960 'Rhosgoch Common, Radnorshire: stratigraphy and pollen analysis', *New Phytologist* 59, 238-262.
British Geological Survey 2002 *Talgarth*. England and Wales Sheet 214, Solid and Drift Geology, 1:50 000.
Bowen, D.Q. 1970 'South east and central south Wales,' in Lewis, C.A. (Editor) *The glaciations of Wales and adjoining regions*, Longman, London, 197-227.
Bowen, D.Q. 1973 'The Pleistocene history of Wales and the Borderland', *Geological Journal* 9, 207-224.
Bowen, D.Q. 1981 'The South Wales end-moraine: fifty years after', in Neale, J. and Flenley, J. (Editors) *The Quaternary of Britain*, Pergamon, Oxford, 60-67,
Brown, E.H. 1960 *The Relief and Drainage of Wales*, University of Wales Press, Cardiff.
Campbell, S. and Bowen, D.Q. 1989 *Quaternary of Wales*, Geological Conservation Review 2, Nature Conservancy Council.
Carr, S. 2001 'A glaciological approach for the discrimination of Loch Lomond Stadial glacial landforms in the Brecon Beacons, South Wales', *Proceedings of the Geologists' Association* 112, 253-262.
Charlesworth, J.K. 1929 'The South Wales end-moraine', *Quarterly Journal of the Geological Society of London* 85, 335-358.
Clarke, B.B. 1936-7 'The post-Cretaceous geomorphology of the Black Mountains', *Proceedings of the Birmingham Natural History and Philosophical Society* 16, 155.
Crimes, T.P., Lucas, G.R., Chester, D.K., Thomas, G.S.P., James, P.A., McCall, G.J., Hunt, N.C., Chapman, A. and Lancaster, K. 1992 *An appraisal of the land based sand and gravel resources of South Wales*. Contract report to the Department of the Environment (Welsh Office), Cardiff, 88.
Dwerryhouse, A.R. and Miller, A.A. 1930 'The glaciation of Clun Forest, Radnor Forest and some adjoining districts', *Quarterly Journal of the Geological Society of London* 86, 96-129.
Ellis-Gruffydd, I.D. 1972 *The glacial geomorphology of the Upper Usk Basin (South Wales) and its right bank tributaries*, unpublished PhD thesis, University of London.
Ellis-Gruffydd, I. D. 1977 'Late Devensian glaciation in the Upper Usk Basin', *Cambria* 4, 46-55.
Fuls, A. 2003 Personal communication.
Gray, J.M. 1982 'The last glaciers (Loch Lomond Advance) in Snowdonia, North Wales', *Geological Journal* 17, 111-133.
Griffiths, J.C. 1940 *The glacial deposits west of the Taff*, unpublished PhD thesis, University of Wales.
Haslett, S.K. 2003 'Early to mid-Holocene (Mesolithic-Neolithic) development of the Olway valley (Central Gwent, UK) and its archaeological potential', *Monmouthshire Antiquary* 19, 3-19.
Howard, F.T. 1903-4 'Notes on glacial action in Brecknockshire and adjoining districts,' *Transactions of the Cardiff Naturalist Society*, 36.
Lewis, C.A. 1970a 'The upper Wye and Usk regions', in Lewis, C.A. (Editor) *The glaciations of Wales and adjoining regions*, Longman, London, 147-173.
Lewis, C.A. 1970b 'The glaciations of the Brecknock Beacons, Wales', *Brycheiniog* 14, 97-120.
Lowe, J.J. and Walker, M.J.C. 1997 *Reconstructing Quaternary environments*, 2nd edition, Longman.
Luckman, B.B. 1970 'The Hereford basin', in Lewis, C.A. (Editor) *The glaciations of Wales and adjoining regions*, Longman, London, 175-196.
M'Caw, L.S. 1936 *The Black Mountains. A physical, agricultural and geographical survey*, unpublished MA thesis, Manchester University.

Moore, J.J. 1970 'The pollen diagram from Mynydd Illtyd', in Lewis, C.A. (Editor) *The glaciations of Wales and adjoining regions*, Longman, London, 168-171.

Moore, P.D. 1978 'Studies in the vegetational history of Mid-Wales. Stratigraphy and pollen analysis of Llyn Mire in the Wye valley,' *New Phytologist* 80, 281-302.

Pissart, A. 1963 'Les traces de "pingos" du Pays de Galles (Grande-Bretagne) et du Plateau des Hautes Fagnes (Belgique)', *Zeitschrift für Geomorphologie* 7, 147-165.

Potts, A.S. 1971 'Fossil cryonival features in central Wales', *Geografiska. Annaler* 53A, 39-51.

Reade, T.M. 1894-5 'The moraine of Llyn Cwm Llwch, on the Beacons of Brecon', *Proceedings of the Liverpool Geological Society* 7, 270-276.

Robertson, D.W. 1988 *Aspects of the Late glacial and Flandrian environmental history of the Brecon Beacons, Fforest Fawr, Black Mountain and Abergavenny Black Mountains, South Wales (with emphasis on the Late-glacial and early Flandrian periods)*, unpublished PhD thesis, University of Wales.

Robertson, T. 1933 *The geology of the South Wales Coalfield. Part V. The country around Merthyr Tydfil*, Memoir of the Geological Survey of Great Britain.

Shakesby, R.A. 1992 *Classic landform guides. No. 13. Classic landforms of the Brecon Beacons*, Geographical Association.

Shakesby, R.A. 2002 *Classic landforms of the Brecon Beacons*, Geographical Association, Sheffield.

Shakesby, R.A. and Matthews, J.A. 1993 'Loch Lomond Stadial glacier at Fan Hir, Mynydd Du (Brecon Beacons), South Wales: critical evidence and palaeoclimatic implications', *Geological Journal* 28, 69-79.

Shakesby, R.A. and Matthews, J.A. 1996 'Glacial activity and paraglacial landsliding activity in the Devensian Lateglacial: evidence from Craig Cerrig-gleisiad and Fan Dringarth, Fforest fawr (Brecon Beacons), South Wales', *Geological Journal* 31, 143-158.

Sissons, J.B. 1980 'The Loch Lomond Advance in the Lake District, northern England', *Transactions Royal Society of Edinburgh: Earth Science* 71, 13-27.

Symonds, W.S. 1872 *Records of the Rocks*, Methuen.

Thomas, G.S.P. 1997 'Geomorphology of the Middle Usk valley', in Lewis, S.G. and Maddy, D. (Editors) *The Quaternary of the South Midlands and the Welsh Marches: Field Guide*, Quaternary Research Association, Cambridge, 49-60,

Thomas, G.S.P., Summers, A.J. and Dackombe, R.V. 1982 'The Late-Pleistocene deposits of the middle Dyfi, Wales', *Geological Journal* 17, 297-309.

Thomas, T.M. 1959 'The geomorphology of Brecknock', *Brycheiniog* 5, 55-156.

Walker, M.J.C. 1980 'Late-Glacial history of the Brecon Beacons, South Wales', *Nature* 287, 133-135.

Walker, M.J.C. 1982 'The Late-Glacial and Early Flandrian deposits at Traeth Mawr, Brecon Beacons, South Wales', *New Phytologist* 90, 177-194.

Williams, G.J. 1968a *Contributions to the Pleistocene geomorphology of the middle and lower Usk*, unpublished PhD thesis, University of Wales.

Williams, G.J. 1968b 'The buried channel and superficial deposits of the lower Usk and their correlation with similar features in the lower Severn', *Proceedings of the Geologists' Association* 79, 325-348.

CHAPTER 9 Herefordshire *by* Andrew E. Richards

Aber, J.S. and Ruszczynska-Szenajch, H. 1997 'Glaciotectonic origin of Elblag Upland, northern Poland, and glacial dynamics in the southern Baltic region', *Sedimentary Geology* 111, 119-134.

Barclay, W.J., Brandon, A. Ellison, R.A. and Moorlock, B.S.P. 1992 'A Middle Pleistocene palaeovalley fill west of the Malvern Hills', *Journal of the Geological Society of London* 149, 75-92.

Bonny, A.P. 1992 'Late Anglian and early Hoxnian pollen spectra from the Cradley Silts, Colwall, Hereford and Worcester', *Journal of the Geological Society of London*, 149, 89-91.

Brandon A. and Hains B.A. 1981 *Geological notes and local details for 1:10,000 sheets SO43NE, SO44SE, SO53NW and SO54SW Hereford City, part of 1:50,000 sheets 198 (Hereford) and 215 (Ross-on-Wye)*, Institute of Geological Sciences, Geological Notes and Local Details Technical Report 3, 51.

Brandon, A. 1989 *Geology of the country between Hereford and Leominster* Memoir British Geological Survey for 1,50,000 geological sheet 198 *(England and Wales)*.

Coope G.R., Field M.H., Gibbard P.L., Greenwood M and Richards A.E. 2002 'Palaeontology and biostratigraphy of Middle Pleistocene river sediment in the Mathon Member, at Mathon, Herefordshire, England', *Proceedings of the Geologists' Association* 113, 237-258.

Croot, D.G., 1987 'Glacio-tectonic structures, a mesoscale model of thin-skinned thrust sheets?', *Journal of Structural Geology* 9, 797-808.

Cross, P. and Hodgson, J.M. 1975 'New evidence for the glacial diversion of the River Teme near Ludlow, Salop', *Proceedings of the Geologists' Association* 3, 313-331.

Dawkins, W. B. 1869 'On the distribution of the British postglacial mammals', *Quarterly Journal of the Geological Society of London* 25, 192-217.

Dwerryhouse, A. R. and Miller, A. A. 1930 'Glaciation of Clun Forest, Radnor Forest and some adjoining districts', *Quarterly Journal of the Geological Society of London* 86, 96-129.

Grindley, H. E. 1905 'Glaciation of the Wye Valley', *Transactions Woolhope Naturalists' Field Club*, 163-164.

Grindley, H. E. 1918 'Superficial deposits of the Middle Wye', *Transactions Woolhope Naturalists' Field Club*, ii-iii.

Grindley, H. E. 1954 'The Wye Glacier, in Herefordshire', *Woolhope Club Centenary Volume*, chapter 3.

Hey, R. W. 1959 'Pleistocene deposits on the west side of the Malvern Hills', *Geological Magazine* 96, 403-17.

Hey, R.W. 1991 'Pleistocene gravels of the lower Wye Valley', *Geological Journal* 26, 123-136.

Jackson A.A. and Sumbler M.G. 1983 *Geological notes and local details for 1:10,000 sheet SO45 (South-west Leominster), Quaternary deposits with special emphasis on potential resources of sand & gravel part of 1:50,000 sheet 198 (Hereford)*, Institute of Geological Sciences.

Luckman, B.H. 1970 'The Hereford Basin', in Lewis, C.A. (Editor) *The glaciations of Wales and adjoining regions*, Longman, London, 175-196.

Maddy, D. 1999 'English Midlands', in Bowen, D.Q. (Editor) *A Revised Correlation of Quaternary Deposits in the British Isles*, Geological Society Special Report 23, 28-44.

Pocock, T.I. 1925 'Terraces and drifts of the Welsh border and their relation to the Drift of the English Midlands', *Zeitschrift Gletscherkunde*, 13, 10-38.

Pocock, T.I. 1940 'Glacial drift and river terraces of the Herefordshire Wye', *Zeitschrift Gletscherkunde* 27, 98-117.

Richards, A.E. 1994 *The Pleistocene stratigraphy of Herefordshire*, unpublished PhD thesis, University of Cambridge.

Richards, A.E. 1997 'Middle Pleistocene glaciation in Herefordshire', in Lewis, S.G. and Maddy, D. (Editors) *The Quaternary of the South Midlands and the Welsh Marches: Field Guide*, Quaternary Research Association, Cambridge.

Richards, A.E. 1998 'Re-evaluation of the Middle Pleistocene Stratigraphy of Herefordshire', *Journal of Quaternary Science* 13, 115-136.

Richards, A.E. 1999 'Middle Pleistocene glaciation in Herefordshire: the sedimentology and structural geology of the Risbury Formation (Older Drift Group)', *Proceedings of the Geologists' Association* 110, 173-192.

Stokes, K. 2003 *The Late Quaternary vegetation history of the southern Welsh Borderland*, unpublished MPhil thesis, Kingston University.

Symonds, W.S. 1872 *Records of the Rocks: Notes on the Geology, Natural History and Antiquities of North and South Wales, Devon and Cornwall*.

Van der Wateren, F.M. 1992 *Structural Geology and Sedimentology of Push Moraines*, unpublished PhD thesis, University of Amsterdam.

Van der Wateren, F.M. 1987 'Structural geology and sedimentation of the Dammer Berge push moraine, Federal Republic of Germany', in Van der Meer, J.J.M. (Editor) *Tills and Glaciotectonics*, Balkema, Rotterdam, 157-182.

Wright, J. 1905 'Note on the grey and finely bedded clay at Bredwardine', *Transactions of the Woolhope Naturalists' Field Club*, 167-8.

CHAPTER 10 South Wales *by* D.Q. Bowen

Aldhouse-Green, S. (Editor) 2000 *Paviland Cave and the Red Lady*, Western Academic and Specialist Press, 314.

Allen, J.R.L. 2000 'Goldcliff Island: geological and sedimentological background', in Bell, M., Caseldine, A. and Neumann, H. (Editors) *Prehistoric Intertidal Archaeology in the Welsh Severn Estuary*, Council for British Archaeology, York, 12-18.

Allen, J.R.L. 2002 'Interglacial high-tide coasts in the Bristol Channel and Severn estuary, southwest Britain: a comparison for the Ipswichian and Holocene', *Journal of Quaternary Science* 17, 69-76.

Allen, J.R.L. in press 'Late Quaternary stratigraphy in the Gwent Levels: the subsurface evidence', *Proceedings Geologists' Association*.

Alley, R.B. 2000 *The Two-Mile Time Machine: ice cores, abrupt climate change, and our future*, Princeton University Press, Princeton and Oxford, 229.

Alley, R.B., Clark, P.U., Keigwin, L.D. and Webb, R.S. 1999 Making sense of millennial-scale climate change, in Clark, P.U, Webb, R.S. and Keigwin, L. (Editors) *Mechanisms of Global Climate Change at Millennial Time Scales*, American Geophysical Union, Washington, D.C., 373-384.

Al-Saadi, R. and Brooks, M. 1973 'A geophysical study of Pleistocene buried valleys in the lower Swansea Valley, Vale of Neath and Swansea Bay', *Proceedings of the Geologists' Association* 84, 135-153.

Anderson, J.G.C. and Owen, T.R. 1979 'The Late Quaternary history of the Neath and Afan valleys, South Wales', *Proceedings of the Geologists' Association* 90, 203-211.

Andrews, J.T. Gilbertson, D.D. and Hawkins, A.B. 1984 'The Pleistocene succession of the Severn Estuary: a revised model based upon amino acid racemization studies', *Journal of the Geological Society of London* 141, 967-974.

Andrews, J.T., Shilts, W.W., and Miller, G.H. 1983 'Multiple deglaciations in the Hudson Bay Lowland, Canada, since deposition of the Missinaibi (last-interglacial?) Formation', *Quaternary Research* 19, 18-37.

REFERENCES

Barclay, W.J., Taylor, K. and Thomas, L.P. 1988 *Geology of the South Wales coalfield, Part 5. The country around Merthyr Tydfil*, British Geological Survey.

Bassinot, F.C., Labeyrie L.D, Vincent E., Quidelleir X., Shackleton N.J. and Lancelot, Y. 1994 'The astronomical theory of climate and the age of the Brunhes-Matuyama magnetic reversal', *Earth and Planetary Science Letters* 126, 91-108.

Belknap, D, F., Kelley, J.T. and Gontz, A.M. 2002 'Evolution of the glaciated shelf and coastline of the northern Gulf of Maine', *Journal of Coastal Research* 36, 37-55.

Bevins, R.E. and Donnelly, R. 1992 'The Storrie Erratic Collection: a reappraisal of the status of the Pencoed 'older drift' and its significance for the Pleistocene of South Wales', *Proceedings of the Geologists' Association* 103, 129-142.

Bowen, D.Q. 1967 'On the supposed ice-dammed lakes of South Wales', *Transactions Cardiff Scientific and Field Naturalists Society* 93, 4-17.

Bowen, D, Q. 1969 'Port Eynon Bay north side: Horton, Western Slade, Eastern Slade', in Bowen, D.Q. (Editor) *Coastal Pleistocene deposits in Wales*, Quaternary Research Association Field Guide, 12-16.

Bowen, D.Q. 1970 'Southeast and central south Wales', in Lewis, C.A, (Editor) *The glaciation of Wales and adjoining regions*, Longman, London, 197-227.

Bowen, D.Q. 1971 'The Quaternary succession of South Gower', in Bassett D.A. and Bassett, M.G, (Editors) *Geological Excursions in South Wales and the Forest of Dean*, National Museum of Wales, Cardiff, 135-142.

Bowen, D.Q. 1973a 'The Pleistocene history of Wales and the borderland', *Geological Journal* 8, 207-224.

Bowen, D.Q. 1973b 'The Pleistocene succession of the Irish Sea', *Proceedings of the Geologists' Association* 84, 249-72.

Bowen, D.Q. 1974 'The Quaternary of Wales', in Owen, T.R. (Editor) *The Upper Palaeozoic and post-Palaeozoic Rocks of Wales*, University of Wales Press, Cardiff, 373-426.

Bowen, D.Q. 1977 'The coast of Wales', in C. Kidson and M.J. Tooley, (Editors) *The Quaternary History of the Irish Sea*, Geological Journal Special Issue. 7, Liverpool, 223-56.

Bowen D.Q. 1981 'The South Wales End-Moraine: fifty years after' in Neale, J.W. and J, Flenley (Editors) *The Quaternary in Britain*, Pergamon, Oxford, 60-67.

Bowen, D.Q. 1994 'Late Cenozoic of Wales and South West England', *Proceedings of the Ussher Society* 8, 209-213.

Bowen, D.Q. 1999a 'Only Four Major Glaciations in the Brunhes Chron?', *International Journal of Earth Science (Geologische Rundschau)* 88, 276-284

Bowen, D.Q. 1999b (Editor) *Revised correlation of Quaternary deposits in the British Isles*, Geological Society of London Special Report 23, 176.

Bowen, D.Q, 2000 'Calibration and correlation with the GRIP and GISP2 Greenland ice cores of radiocarbon ages from Paviland (Goat's Hole), Gower', in Aldhouse-Green, S. (Editor) *Paviland Cave and the Red Lady*, Western Academic and Specialist Press, 61-64.

Bowen, D.Q, 2001 'Revised aminostratigraphy for land-sea correlations from the north-eastern north Atlantic margin', in Goodfriend, G., Collins, M.J., Fogel, M.L., Macko, S.A. and Wehmiller, J.F. (Editors) *Perspectives in Amino Acid and Protein Geochemistry*, New York, 253-262.

Bowen, D.Q, 2003 'Uncertainty in oxygen isotope stage 11 sea level: an estimate 13 ± 2 m above low water from Great Britain', in Droxler, A., Poore, R.Z. and Burkle, L.H. (Editors) *Earth's Climate and Orbital Eccentricity: The Marine Isotope Stage 11*, Geophysical Monograph 137, American Geophysical Union, 131-144.

Bowen, D.Q, in preparation *Pleistocene aminostratigraphy of Great Britain and Ireland*.

Bowen, D.Q. and Gregory, K.J. 1965 'A glacial drainage system near Fishguard, Pembrokeshire', *Proceedings of the Geologists' Association* 74, 275-82.

Bowen D.Q. and Sykes, G.A, 1988 'Correlation of marine events and glaciations on the northeast Atlantic margin', *Philosophical Transactions of the Royal Society* B 318, 619-635.

Bowen, D.Q., Catt, J.A., Jenkins, D.G. and Reid, A.M.B. 1992 *The Paviland End-Moraine*, Gower, South Wales, Unpublished manuscript.

Bowen, D.Q., Hughes, S.A., Sykes, G.A. and Miller, G.H.. 1989 'Land-sea correlations in the Pleistocene based on isoleucine epimerization in non-marine molluscs', *Nature* 340, 49-51.

Bowen, D.Q., McCabe, A.M., Phillips, F.M., Knutz, P.C. and Sykes, G.A. 2002 'New data for the Last Glacial Maximum in Great Britain and Ireland', *Quaternary Science Reviews* 21, 89-101.

Bowen, D.Q., McCabe, A.M., Rose, J, and D.G Sutherland, 1986 'Correlation of Quaternary Glaciations in England, Ireland, Scotland and Wales', *Quaternary Science Reviews* 5, 199-340,

Bowen, D.Q., Sykes, G.A., Reeves, A., Miller, G.H., Andrews, J.T., Brew, J.S. and Hare, P.E. (and contributions from: Wyatt, A.R., Hughes, S., James, H.L.C., Smith, D.B., Mottershead, D.N., Jenkins, D.G., Hollin, J.T., McCabe, A.M. and Harkness, D.D.) 1985 'Amino Acid Geochronology of raised beaches in southwest Britain', *Quaternary Science Reviews* 4, 279-318.

Bradley, I, 1980 *Soils in Dyfed V, Sheet SN24 (Llechryd)*, Soil Survey of England and Wales.

Brown, E.H, 1960a *The Relief and drainage of Wales*, University of Wales Press, Cardiff.

Brown, E.H, 1960b 'The building of southern Britain', *Zeitschrift fur Geomorpholgie* 4, 264-274.
Campbell, J.B. 1977 *The Upper Palaeolithic of Britain: A study of man and nature in the late Ice Age*, Clarendon Press, 2 vols.
Campbell, S. 1984 *The nature and origin of Pleistocene deposits around Cross Hands and west Gower, South Wales*, unpublished PhD thesis, University of Wales.
Campbell, S. and D.Q, Bowen.1989, *Quaternary of Wales*, Nature Conservancy Council, Peterborough, 237.
Case, D.J, 1983 *Quaternary airfall deposits in South Wales: loess and coversands*, unpublished PhD thesis, University of Wales.
Case, D.J. 1977 'Horton', in Bowen, D.Q. (Editor) *Wales and the Cheshire-Shropshire Lowland*, Inqua, Norwich.
Case, D.J. 1984 'Port Eynon Silt (loess)', in Bowen, D.Q. and Henry, A. (Editors) 1984 *Wales: Gower, Preseli, Fforest Fawr*, Quaternary Research Association Field Guide, Cambridge.
Case, D.J. 1993 'Evidence for temperate soil development during the Early Devensian in Gower, South Wales', *Geological Magazine* 130, 113-115.
Charlesworth, J.K. 1929 'The South Wales end moraine', *Quarterly Journal of the Geological Society of London* 85, 335-358.
Clark, P.U., McCabe, A.M., Mix, A.C. and Weaver, A.J. 2004 'Rapid rise of sea level 19,000 years ago and its global implications', *Science* 304, 1141-1144.
Clayden, B. 1977 'Hunts Bay Plateau paleo-argillic brown earth' in Bowen, D.Q, (Editor) *Wales and the Cheshire-Shropshire Lowland,* Inqua, Norwich.
Crampton, C.B, 1966 'Certain effects of glacial events in the Vale of Glamorgan', *Journal of Glaciology* 6, 261-266.
Culver, S.J. and Bull, P.A. 1979 'Late Pleistocene rock basins in South Wales', *Geological Journal* 14, 107-116.
Currant, A, P., Stringer, C, B. and Collcutt, S.N. 1984 'Bacon Hole Cave', in Bowen, D.Q. and Henry A, (Editors) 1984 *Wales: Gower, Preseli and Fforest Fawr*, Quaternary Research Association Field Guide, Cambridge, 38-44.
David, T.E. 1883 'On the evidence of glacial action in south Brecknockshire and east Glamorgan', *Quarterly Journal of the Geological Society of London* 39, 39-54.
Davies, D.C. 1988 *The aminostratigraphy of British Pleistocene glacial deposits,* unpublished PhD thesis University of Wales.
Davies, J.R., Waters, R.A., Wilby, P.R., Wilson, D. and Fletcher, C.J.N. 2003 *Cardigan and Dinas Island. British Geological Survey 1:50,000 Series England and Wales Sheet 193*.
Davies, K.A. 1983 'Amino acid analysis of Pleistocene marine molluscs from the Gower Peninsula', *Nature* 302, 137-138.
Davies, R.J. 1986 *Weathering characteristics of erratics from tills in Gower*, unpublished BSc dissertation, University of Wales, Aberystwyth.
Dixon, E.E.L. 1921 *The Geology of the South Wales Coalfield, Part XIII, The Country around Pembroke and Tenby*, Memoirs of the Geological Survey of Great Britain.
Driscoll, E.M. 1953 *Some aspects of the geomorphology of the Vale of Glamorgan*, unpublished MSc thesis London University.
Driscoll, E.M. 1958 'The denudation chronology of the Vale of Glamorgan', *Transactions of the Institute of British Geographers* 25, 45-57.
Edwards, E. 1997 *Sedimentology and age of glacigenic sequences: east Irish Sea Basin*, unpublished PhD thesis, University of Wales.
Emiliani, C. 1955 'Pleistocene temperatures', *Journal of Geology* 63, 538-578.
Evans, J.G., French, C. and Leighton, D. 1978 'Habitat change in two Late-glacial and Post-glacial sites in southern Britain: the molluscan evidence', in Limbrey, S and Evans, J.G. (Editors) *The effect of man on the landscape of the Lowland Zone*, Research Report 21, Council for British Archaeology, London, 63-75.
Eyles, N. and McCabe, A.M. 1989 'The Late Devensian (<22,000 BP) Irish Sea Basin: the sedimentary record of a collapsed ice sheet margin', *Quaternary Science Reviews* 8, 307-351.
Fletcher, C.J.N. and Siddle, H.J. 1998 'The development of glacial Llyn Teifi, west Wales: evidence for lake-level fluctuations at the margin of the Irish Sea ice sheet', *Journal of the Geological Society of London* 155, 389-400.
George, G.T. 1982 *Sedimentary features of a Pleistocene kame delta sequence at Mullock Bridge*, Dyfed, S335 Summer School, The Open University, 1-12.
George, T.N. 1932 'The Quaternary beaches of Gower', *Proceedings of the Geologists' Association* 43, 291-323.
George, T.N. 1933 'The glacial deposits of Gower', *Geological Magazine* 70, 208-232.
George, T.N. 1938 'Shoreline evolution in the Swansea district', *Proceedings Swansea Scientific and Field Naturalists` Society 2, 23-48.*
George, T.N. 1942 'The development of the Towy and upper Usk drainage pattern', *Quarterly Journal of the Geological Society of London* 98, 89-137.
Gilbertson, D.D. and Hawkins, A.B. 1978 'The Pleistocene Succession at Kenn, Somerset', *Bulletin of the Geological Survey of Great Britain* 66, 44.
Green, C.P. 1973 'Pleistocene river gravels and the Stonehenge problem', *Nature* 243, 214-216.
Griffiths, H.I. 1995 *The application of freshwater ostracods to the study of late Quaternary palaeoenvironments in north-western Europe*, unpublished PhD thesis, University of Wales.

References

Griffiths, J.C. 1940 *The glacial deposits west of the Taff*, unpublished PhD thesis, London University.

Harris, C. and Donnelly, R, 1991 'The glacial deposits of South Wales', in Ehlers, J., Gibbard, P.L. and Rose, J. (Editors) *Glacial Deposits in Great Britain and Ireland*, Balkema, Rotterdam, 279–290,

Harris, C. and Wright, M.D. 1980 'Some last glaciation drift deposits near Pontypridd, South Wales', *Geological Journal*, 1, 7-20.

Henderson, G. and Slowey, N.C. 2000 'Evidence from U-Th dating against Northern Hemisphere forcing of the penultimate deglaciation', *Nature* 404, 61-66.

Henry, A. 1983 *Lithostratigraphy and Biostratigraphy of the Pleistocene of Gower*, unpublished PhD thesis, University of Wales.

Hollingworth, S.E. 1931 'The glaciation of western Edenside and adjoining areas and the drumlins of Edenside and the Solway Firth', *Quarterly Journal of the Geological Society of London* 87, 281-359.

Hughes, S. 1987 *Aminostratigraphy of British Pleistocene non-marine mollusca*, unpublished PhD thesis, University of Wales.

Hunt, C.O. 1999 'Somerset and south Gloucestershire', in Bowen, D.Q. (Editor) *Revised correlation of Quaternary deposits in the British Isles*, Geological Society of London Special Report 23, 75-78.

Jenkins, D.G., Beckinsale, R.D., Bowen, D.Q., Evans, J.A., George, G.T., Harris, N.B.W. and Meighan, I.G. 1985 'The origin of granite erratics in the Pleistocene Patella beach, Gower, South Wales', *Geological Magazine* 122, 297-302,

John, B.S. 1970 'Pembrokeshire', in Lewis, C.A. (Editor) *The Glaciations of Wales and adjoining regions*, Longman, London.

Jones, O.T. 1942 'The buried channel of the Tawe valley near Ynystawe, Glamorganshire', *Quarterly Journal of the Geological Society of London* 98, 61-88.

Jones, O.T. 1965 'The glacial and postglacial history of the Lower Teifi Valley', *Quarterly Journal of the Geological Society of London* 121, 247-281.

Kellaway, G.A. 1971 'Glaciation and the stones of Stonehenge', *Nature* 232, 30-35.

Knight, J. 2001 'Glaciomarine deposition around the Irish Sea basin: some problems and solutions', *Journal of Quaternary Science* 16, 405-418.

Knutz P.C. 2000 *Late Pleistocene Glacial Fluctuations and Palaeogeography on the Continental Margin of North-West Britain*, unpublished PhD thesis, University of Wales.

Knutz, P, C., Hall, I., Zahn, R., Rasmussen, T.L., Kuijpers, A., Moros, M. and Shackleton, N.J. 2002 'Multidecadal ocean variability and NW European ice sheet surges during the last deglaciation', *Geochemistry, Geophysics, Geosystems* 3, 1077

Lambeck, K. and Purcel, A.P, 2001 'Sea level change in the Irish Sea since the last glacial maximum; constraints from isostatic modelling', *Journal of Quaternary Science*, 16, 497-506.

Lear, D.J. 1985 *The Quaternary Geology of the Lower Teifi Valley*, unpublished PhD thesis, University of Wales.

Maddy, D. 2002 'An evaluation of climate, crustal movement and base level controls on the Middle-Late Pleistocene development of the River Severn, UK', *Netherlands Journal of Geosciences* 81, 329-338,

Maddy, D., Lewis, S.G., Green, C.P. and Bowen, D.Q. 1995 'Pleistocene Geology of the Lower Severn Valley, U.K', *Quaternary Science Reviews* 14, 209-222.

McCabe, A.M., Haynes, J.R. and McMillan, N.F. 1986 'Late-Pleistocene tidewater glaciers and glaciomarine sequences from north County Mayo, Republic of Ireland', *Journal of Quaternary Science* 1, 73-84.

McCabe, A.M. and Clark, P.U.1998 'Ice sheet variability around the north Atlantic Ocean during the last glaciation', *Nature* 392, 373-377.

McCabe, A.M. and Clark, P.U. 2003 'Deglacial chronology from County Donegal: implications for deglaciation of the British-Irish ice sheet', *Journal of the Geological Society of London* 160, 847-855.

McCabe, A.M., Clark, P.U. and Clark, J. (in press) 'AMS 14C dating of deglacial events in the Irish Sea Basin and other sectors of the British-Irish ice sheet', *Quaternary Science Reviews*.

McKenny-Hughes, T. 1887 'On the drifts of the Vale of Clwyd and their relation to the caves and cave deposits', *Quarterly Journal of the Geological Society of London* 43, 73-120.

Miller, G, H. 1985 'Aminostratigraphy of Baffin Island shell-bearing deposits', in Andrews, J.T. (Editor), *Quaternary Environments: Eastern Canadian Arctic, Baffin Bay and Western Greenland*, London, Allen and Unwin, 394-427.

Mitrovica, J.X. 2003 'Recent controversies in predicting post-glacial sea-level change', *Quaternary Science Reviews* 22, 127-133.

Mitrovica, J.X., Tamisiea, M.E., David, J.L. and Milne, G.A. 2001 'Recent mass balance of polar ice sheets inferred for pattern of global sea-level change', *Nature* 409, 1026-1029.

Mix, A.C., Pisias, N.G., Rugh, W., Morey, A. and Hagelberg, T.K. 1995 'Benthic foramininfera stable isotope record from Site 849 (0-5 Ma): local and global climate changes', *Proceedings Ocean Drilling Program, Scientific Results* 138, 371-412.

Muhs, D., Simmons, K.R. and Steinke, B. 2002 'Timing and warmth of the Last Interglacial period: new U-series evidence from Hawaii and Bermuda and a new fossil compilation from North America', *Quaternary Science Reviews* 21,1355-1384.

Owen, T.R. 1973 *Geology explained in South Wales*, David and Charles, Newton Abbot.

Phillips, F.M., Bowen, D.Q. and Elmore, D. 1994 'Surface exposure dating of glacial features in Great Britain using cosmogenic Chlorine-36: preliminary results', *Mineralogical Magazine* 58A, 722-723.

Price, A. 1976 *Quaternary deposits of the Middle Teifi Valley, Dyfed, between the Llanllwni and Allt-y-Cafan gorges*, unpublished MSc thesis, University of Wales.

Pringle, J. and George, T.N. 1937 *British Regional Geology South Wales*, London, HMSO.

Raymo, M. 1997 'The timing of major climate terminations', *Paleoceanography* 12, 577-585.

Reid, A. 1985 *A resistivity survey of glacial deposits on the Gower Peninsula*, unpublished MSc thesis, Birmingham University.

Roy, M., Clark, P.U., Barendregt, R.W., Glasmann, J.R. and Enkin, R.J. 2004 'Glacial stratigraphy and paleomagnetism of late Cenozoic deposits of the north central United States', *Geological Society of America Bulletin* 116, 3-41,

Schellman, G. and Radtke, U. 2004 'A revised morpho- and chronostratigraphy of the Late and Middle Pleistocene coral reef terraces on Southern Barbados (West Indies)', *Earth Science Reviews* 64, 157-187.

Shotton, F.W. 1967 'The problems and contributions of methods of absolute dating within the Pleistocene Period', *Quarterly Journal of the Geological Society of London* 122, 357-383.

Sissons, J.B. 1974 'The Quaternary in Scotland: a review', *Scottish Journal of Geology* 10, 311-337.

Smith, R, Bowen, D.Q., Cope, J.W.C. and Reid, A. 2002 'An arietitid ammonite from Gower: its palaeogeographical and geomorphological significance', *Proceedings Geologists' Association* 113, 217-222.

Southgate, G.A. 1985 Thermo-luminescene dating of beach and dune sands – potential of single grain measurements, *Nuclear Tracks and Radiation Measurements*, 10, 743-747.

Sparks, B.W. 1964 'The distribution of non-marine Mollusca in the Last Interglacial in south-east England' *Proceedings of the Malacological Society of London* 36, 7-25.

Squirrell, H. and Downing, R.A. 1969 *The Geology of the South Wales Coalfield, Part I, The Country around Newport*, Memoirs of the Geological Survey of Great Britain.

Stephens, N. and McCabe, A.M. 1977 'Late-Pleistocene ice movements and patterns of Late- and Post-Glacial shorelines on the coast of Ulster', in C. Kidson and M.J. Tooley (Editors) *The Quaternary History of the Irish Sea*, Geological Journal Special Issue 7, Liverpool, 223-56.

Stevenson, A.C. and Moore, P.D. 1982 'Pollen analysis of an interglacial deposit at West Angle, Dyfed, Wales', *New Phytologist* 90, 327-337.

Strahan, A. 1907 *The Geology of the South Wales Coalfield, Part VIII, The Country around Swansea*, Memoirs of the Geological Survey of Great Britain.

Strahan, A. 1907 *The Geology of the South Wales Coalfield, Part IX, The Country around West Gower and Pembrey*, Memoirs of the Geological Survey of Great Britain.

Strahan, A. and Cantrill, T.C. 1904 *The Geology of the South Wales Coalfield, Part VI, The Country around Bridgend*, Memoirs of the Geological Survey of Great Britain.

Strahan, A., Cantrill, T.C., Dixon, E.E.L. and Thomas, H.H. 1907 *The Geology of the South Wales Coalfield, Part VII, The Country around Ammanford*, Memoirs of the Geological Survey of Great Britain.

Strahan, A., Cantrill, T.C., Dixon, E.E.L. and Thomas, H.H. 1909 *The Geology of the South Wales Coalfield, Part X, The Country around Carmarthen*, Memoirs of the Geological Survey of Great Britain.

Strahan, A., Cantrill, T.C., Dixon, E.E.L., Thomas, H.H. and Jones, O.T. 1914 *The Geology of the South Wales Coalfield, Part XI, The Country around Haverfordwest*, Memoirs of the Geological Survey of Great Britain.

Sutcliffe, A.J., Currant, A.P. 1984 'Minchin Hole Cave', in Bowen, D.Q. and Henry, A. (Editors) *Wales: Gower, Preseli and Fforest Fawr*, Quaternary Research Association Field Guide, Cambridge, 33-37.

Sutcliffe, A.J., Currant, A.P. and Stringer, C.B. 1987 'Evidence of sea-level change from coastal caves with raised beaches, terrestrial faunas and dated stalagmites', *Progress in Oceanography* 18, 243-271.

Trotman, D. M. 1963 *Data for Late Glacial and Post Glacial history in South Wales*, unpublished PhD thesis, University of Wales.

van Vliet-Lanöe, B., Laurent, M., Bahain, J.L., Balescu, S., Falgures, C., Field, M. and Keen, D.H. 2000 'Middle Pleistocene raised beach anomalies in the English Channel: regional and global stratigraphic implications', *Journal of Geodynamics* 29, 15-41.

Walker, M.J.C., Coope, G.R., Sheldrick,C., Turney, C.S.M., Lowe, J.J., Blockley, S.P.E. and Harkness, D.D. 2003 'Devensian Lateglacial environmental changes in Britain: a multi-proxy environmental record from Llanilid, South Wales, UK', *Quaternary Science Reviews* 22, 475-520.

Walsh, P., Boulter, M. and Morawiecka, I. 1999 'Chattian and Miocene elements in the modern landscape of western Britain and Ireland', in Smith B.J., Whalley, W.B.and Warke, P.A. (Editors) *Uplift, erosion and stability / perspectives on long-term landscape development*, Geological Society Special Publication 162, 45-64.

Waters, R.A. and Lawrence, D.J.D. 1987 *The Geology of the South Wales Coalfield, Part III, The Country around Cardiff*, British Geological Survey.

Wehmiller, J.F., York, L.L. and Bart, M.L. 1995 'Amino acid racemization geochronology of reworked Quaternary mollusks on US Atlantic coast beaches: implications for chronostratigraphy, taphonomy, and coastal sediment transport', *Marine Geology* 124, 303-337,

Wills, L.J, 1938 'The Pleistocene development of the Severn between Bridgnorth and the sea', *Quarterly Journal of the Geological Society of London* 94, 161-242.

Wilson, D., Davies, J.R., Fletcher, C.J.N. and Smith, M. 1990 *The Geology of the South Wales Coalfield, Part VI, The Country around Bridgend*, British Geological Survey.

Winograd, I.J., Landwehr, J.M., Ludwig, K.R., Coplen, T.B. and Riggs, A.C. 1997 'Duration and structure of the past four interglacials', *Quaternary Research* 48, 141-154.

Wood, A. 1974 'Submerged platform of marine abrasion around the coasts of south-western Britain', *Nature* 252, 563.

Woodland, A.W. and Evans, W.B. 1964 *The Geology of the South Wales Coalfield, Part IV, The Country around Pontypridd and Maesteg*, Memoirs of the Geological Survey of Great Britain.

Zeigler., K.E., Schwartz, J.P., Droxler, A.W., Shearer, M.C. and Peterson, L. 2003 'Caribbean carbonate crash in Pedro Channel at subthermoclinal depth: a case of basin to shelf carbonate fractionation?', in Droxler, A., Poore, R.Z. and Burkle, L.H, (Editors) *Earth's Climate and Orbital Eccentricity: The Marine Isotope Stage 11*, Geophysical Monograph 137, American Geophysical Union, 181-204.

CHAPTER 11 South-west England *by* Stephen Harrison and David Keen

Andrews, J.T., Gilbertson, D.D. and Hawkins, A.B. 1984 'The Pleistocene succession of the Severn estuary: a revised model based on amino-acid racemization studies', *Journal of the Geological Society* 141, 967-974.

Arkell, W.J. 1943 'The Pleistocene rocks at Trebetherick Point, Cornwall: their interpretation and correlation', *Proceedings of the Geologists' Association* 54, 141-170.

Ballantyne, C.K. and Harris, C. 1994 *The Periglaciation of Great Britain*, Cambridge University Press, Cambridge.

Bates, M.R., Keen, D.H. and Lautridou, J-P. 2003 'Pleistocene marine and periglacial deposits of the English Channel', *Journal of Quaternary Science* 18, 319-337.

Bowen D.Q. 1999 (Editor) *A revised correlation of Quaternary deposits in the British Isles*, Geological Society of London Special Report 23.

Bowen, D.Q. 1994 'Late Cenozoic Wales and South-West England', *Proceedings of the Ussher Society* 8, 209-213.

Bowen, D.Q., Sykes, G.A., Reeves, A. Miller, G.H., Andrews, J.T., Brew, J.S. and Hare, P.E. 1985 'Amino acid geochronology of raised beaches in south-west Britain', *Quaternary Science Reviews* 4, 279-318.

Campbell, S., Hunt, C.O., Scourse, J.D. and Keen, D.H. 1998 *Quaternary of South-West England*, Chapman and Hall, London.

Charman, D.J., Newnham, R.M. and Croot, D.G. (Editors) 1996 *The Quaternary of Devon and East Cornwall: Field Guide* Quaternary Research Association, London.

Croot, D.G., Gilbert, A., Griffiths, J. and van der Meer, J.J. 1996 'The character, age and depositional environments of the Fremington Clay Series, North Devon', in Charman, D.J., Newnham, R.M. and Croot, D.G. (Editors) *The Quaternary of Devon and East Cornwall: Field Guide*, Quaternary Research Association, London.

Dewey, H. 1910 'Notes on some igneous rocks from North Devon', *Proceedings of the Geologists' Association* 21, 429-434.

Eyles, N. and McCabe, A.M. 1991 'Glacio-marine deposits of the Irish Sea basin: the role of glacio-isostatic disequilibrium', in Ehlers, J., Gibbard, P.L. and Rose, J. (Editors) *Glacial deposits in Britain and Ireland,* Balkema, Rotterdam.

Flett, J. S. and Hill, J.B. 1912 *Geology of the Lizard and Meneage*, Memoir of the Geological Survey of Great Britain.

Gilbertson, D.D. and Hawkins, A.B. 1978a 'The col-gully and glacial deposits at Court Hill, Clevedon, near Bristol, England', *Journal of Glaciology* 20, 173-188.

Gilbertson, D.D. and Hawkins, A.B. 1978b *The Pleistocene succession at Kenn, Somerset*, Bulletin of the Geological Survey of Great Britain, No 66.

Gilbertson, D.D. and Sims, P.C. 1974 'Some Pleistocene deposits and landforms at Ivybridge, Devon', *Proceedings of the Geologists' Association* 85, 65-77.

Green, C.P., Branch, N.P., Coope, G.R., Field, M.H., Keen, D.H., Wells, J.M., Schwenninger, J-L., Preece, R.C., Schreve, D.C., Canti, M.G. and Gleed-Owen, C.P. in press, 'Marine Isotope Stage 9 environments of fluvial deposits at Hackney, north London UK', *Quaternary Science Reviews*.

Harrison, S., Anderson, E. and Passmore, D.G. 1998 'A small glacial cirque basin on Exmoor, Somerset', *Proceedings of the Geologists' Association* 109, 149-158.

Harrison, S. 2001 'Speculations on the glaciation of Dartmoor', *Quaternary Newsletter* 93, 15-26.

Harrison, S., Anderson, E. and Passmore, D.G. 2001a 'Further glacial tills on Exmoor, southwest England: implications for small ice cap and valley glaciation', *Proceedings of the Geologists' Association* 112, 1-5.

Harrison, S., Anderson, E. and Passmore, D.G. 2001b 'Further glacial tills on Exmoor, southwest England: implications for small ice cap and valley glaciation: reply to H. Prudden', *Proceedings of the Geologists' Association* 112, 286-287.

Keen, D.H. 2001 'Towards a late Middle Pleistocene non-marine molluscan biostratigraphy for the British Isles', *Quaternary Science Reviews* 20, 1657-1665.

Kellaway, G.A., Redding, J.H., Shepard-Thorn, E.R. and Destombes, J.P. 1975 'The Quaternary history of the English Channel', *Philosophical Transactions of the Royal Society* A279, 189-218.

Kidson, C. and Wood, R. 1974 'The Pleistocene stratigraphy of Barnstaple Bay', *Proceedings of the Geologists' Association* 85, 223-237.

Kleman, J. 1994 'Preservation of landforms under ice sheets and ice caps', *Geomorphology* 9, 19-32.

Kleman, J. and Borgström, I. 1994 'Glacial land forms indicative of a partly frozen bed', *Journal of Glaciology* 40, 255-264.

Maw, G. 1864 'On a supposed deposit of boulder-clay in North Devon', *Quarterly Journal of the Geological Society of London* 20, 445-451.

Mitchell, G.F. and Orme, A.R. 1967 'The Pleistocene deposits of the Isles of Scilly', *Quarterly Journal of the Geological Society of London* 123, 59-92.

Mitchell, G.F. 1968 'Glacial gravel on Lundy Island', *Transactions of the Royal Geological Society of Cornwall* 20, 65-68.

Mottershead, D.N. 1977 'The Quaternary evolution of the South Coast of England', in Kidson, C. and Tooley, M.J. (Editors) *The Quaternary history of the Irish Sea*, Geological Journal Special Issue No 7, Liverpool, 299-320.

Ormerod, G.W. 1869 'On some of the results arising from the bedding, joints and spheroidal structure of the granite on the eastern side of Dartmoor', *Quarterly Journal of the Geological Society of London* 23, 418-429.

Pickard, R. 1943 'Glaciation on Dartmoor', *Report and Transactions of the Devonshire Association for the Advancement of Science, Literature and Art* 78, 207-228.

Pillar, J.E. 1917 'Evidences of glaciation in the West', *Transactions of the Plymouth Institution* 16, 179-187.

Preece, R.C. and Parfitt, S. 2000 'The Cromer Forest Bed Formation: new thoughts on an old problem', in Lewis, S.G., Whiteman, C.A. and Preece, R.C. (Editors) *The Quaternary of Norfolk and Suffolk: Field Guide*, Quaternary Research Association, London, 1-28.

Scourse, J.D. and Furze, M.F.A. 1999 *The Quaternary of West Cornwall: Field Guide*, Quaternary Research Association, London.

Scourse, J.D. 1991 'Late Pleistocene stratigraphy and palaeobotany of the Isles of Scilly', *Philosophical Transactions of the Royal Society* B334, 405-448.

Scourse, J.D. 1996 'Late Pleistocene stratigraphy of north and west Cornwall', *Transactions of the Royal Geological Society of Cornwall* 22, 2-56.

Somervail, A. 1897 'On the absence of small lakes, or tarns, from the area of Dartmoor', *Report and Transactions of the Devonshire Association for the Advancement of Science, Literature and Art* 29, 386-389.

Stephens, N. 1970 'The West Country and southern Ireland', in Lewis, C.A. (Editor) *The glaciations of Wales and adjoining regions*, Longmans, London.

Ussher, W.A.E. 1878 'The chronological value of the Pleistocene deposits of Devon', *Quarterly Journal of the Geological Society of London* 34, 449-458.

Ussher, W.A.E. 1879 *The post-Tertiary geology of Cornwall*, Stephen Austin, Hertford.

Van Vliet-Lanoë, B., Laurent, M., Bahain, J.L., Balescu, S., Falguères, C., Field, M.H., Hallégouët, B. and Keen, D.H. 2000 'Middle Pleistocene raised beach anomalies in the English Channel: regional and global stratigraphic implications', *Journal of Geodynamics* 29, 15-41.

CHAPTER 12 The Irish Sea Basin *by* Jasper Knight

Allen, J.R.L. 1982 'Late Pleistocene (Devensian) glaciofluvial outwash at Banc-y-Warren, near Cardigan (west Wales)', *Geological Journal* 17, 31-47.

Andrews, J.T. and Barber, D.C. 2002 'Dansgaard-Oeschger events: is there a signal off the Hudson Strait ice stream?', *Quaternary Science Reviews* 21, 443-454.

Austin, W.E.N. and McCarroll, D. 1992 'Foraminifera from the Irish Sea glacigenic deposits at Aberdaron, western Lleyn, North Wales: palaeoenvironmental implications', *Journal of Quaternary Science* 7, 311-317.

Austin, W.E.N. and Scourse, J.D. 1997 'Evolution of seasonal stratification in the Celtic Sea during the Holocene', *Proceedings of the Royal Irish Academy* 154, 249-256.

Barne, J.H., Robson, C.F., Kaznowska, S.S., Doody, J.P., Davidson, N.C. and Buck, A.L. (Editors) 1997 *Coasts and Seas of the United Kingdom. Region 17, Northern Ireland*, Joint Nature Conservancy Council, Peterborough.

Bowen, D.Q. 1973 'The Pleistocene succession of the Irish Sea', *Proceedings of the Geologists' Association* 84, 249-272.

Bowen, D.Q. (Editor) 1999 *A revised correlation of Quaternary deposits in the British Isles*, Geological Society of London Special Report 23.

Bowen, D.Q., Phillips, F.M., McCabe, A.M., Knutz, P.C. and Sykes, G.A. 2002 'New data for the Last Glacial Maximum in Great Britain and Ireland', *Quaternary Science Reviews* 21, 89-101.

Bowen, D.Q., Rose, J., McCabe, A.M. and Sutherland, D.G. 1986 'Correlation of Quaternary glaciations in England, Ireland, Scotland and Wales', *Quaternary Science Reviews* 5, 299-340.

REFERENCES

Brown, J., Carillo, L., Fernand, L., Horsburgh, K.J., Hill, A.E., Young, E.F. and Medler, K.J. 2003 'Observations of the physical structure and seasonal jet-like circulation of the Celtic Sea and St. George's Channel of the Irish Sea', *Continental Shelf Research* 23, 533-561.

Cameron, T.D.J. and Holmes, R. 1999 'The continental shelf', in D.Q. Bowen (Editor) *A revised correlation of Quaternary deposits in the British Isles*, Geological Society of London Special Report 23, 125-139.

Campbell, S. and Bowen, D.Q. 1989 *Quaternary of Wales*, Geological Conservation Review. Nature Conservancy Council, Peterborough.

Carr S. 2001 'Micromorphological criteria for discriminating subglacial and glacimarine sediments: evidence from a contemporary tidewater glacier, Spitsbergen', *Quaternary International* 86, 71-79.

Carter, R.W.G. 1988 *Coastal Environments*, Academic Press, London.

Chapman, M.R., Shackleton, N.J. and Duplessy, J.-C. 2000 'Sea surface temperature variability during the last glacial-interglacial cycle: assessing the magnitude and pattern of climate change in the North Atlantic', *Palaeogeography, Palaeoclimatology and Palaeoecology* 157, 1-25.

Charlesworth, J.K. 1929 'The south Wales end-moraine', *Quarterly Journal of the Geological Society of London* 85, 335-358.

Charlesworth, J.K. 1937 'A map of the glacier-lakes and the local glaciers of the Wicklow Hills', *Proceedings of the Royal Irish Academy* 44B, 29-36.

Charlesworth, J.K. 1939 'Some observations on the glaciation of north-east Ireland', *Proceedings of the Royal Irish Academy* 45B, 255-292.

Charlesworth, J.K. 1957 *The Quaternary Era*, Edward Arnold, London, 2 vols.

Edwards, M.E. 1996 *Sedimentology and dating of glacigenic sequences: eastern Irish Sea Basin*, unpublished PhD thesis, University of Wales.

Edwards, K.J. and Warren, W.P. (Editors) 1985 *The Quaternary History of Ireland,* Academic Press, London.

Evans, D.J.A. and Ó Cofaigh, C. 2003 'Depositional evidence for marginal oscillations of the Irish Sea ice stream in southeast Ireland during the last glaciation', *Boreas* 32, 76-101.

Eyles, C.H. and Eyles, N. 1984 'Glaciomarine sediments of the Isle of Man as a key to late Pleistocene stratigraphic investigations in the Irish Sea Basin', *Geology* 12, 359-364.

Eyles, N. and McCabe, A.M. 1989 'The Late Devensian (<22,000 BP) Irish Sea Basin: the sedimentary record of a collapsed ice sheet margin', *Quaternary Science Reviews* 8, 307-351.

Farrington, A. 1934 'The glaciation of the Wicklow Mountains', *Proceedings of the Royal Irish Academy* 42B, 173-209.

Farrington, A. 1947 'Unglaciated areas in southern Ireland', *Irish Geography* 1, 89-97.

Fletcher, C.J.N. and Siddle, H.J. 1998 'Development of glacial Llyn Teifi, west Wales: evidence for lake-level fluctuations at the margins of the Irish Sea ice sheet', *Journal of the Geological Society of London* 155, 389-99.

Harris, C., Williams, G., Brabham, P., Eaton G. and McCarroll, D. 1997 'Glaciotectonized Quaternary sediments at Dinas Dinlle, Gwynedd, north Wales, and their bearing on the style of deglaciation in the eastern Irish Sea', *Quaternary Science Reviews* 16, 109-127.

Hart, J. 1995 'Drumlin formation in southern Anglesey and Arvon, northwest Wales', *Quaternary Science Reviews* 10, 3-14.

Hart, J.K and Roberts, D.H. 1994 'Criteria to distinguish between subglacial deformation and glaciomarine sedimentation, I. Deformation styles and sedimentology', *Sedimentary Geology* 91, 191-213.

Haynes, J.R., McCabe, A.M. and Eyles, N. 1995 'Microfaunas from Late Devensian glaciomarine deposits in the Irish Sea Basin', *Irish Journal of Earth Sciences* 14, 81-103.

Helm, D.G. and Roberts, B. 1975 'A re-interpretation of the origin of sands and gravels around Banc-y-warren, near Cardigan, west Wales', *Geological Journal* 10, 131-146.

Hoare, P.G. 1975 'The pattern of glaciation of County Dublin', *Proceedings of the Royal Irish Academy* 75B, 207-224.

Huddart, D. 1981 'Pleistocene foraminifera from south-east Ireland – some problems of interpretation', *Quaternary Newsletter* 33, 28-41.

Jackson, D.I., Jackson, A.A., Evans, D., Wingfield, R.T.R., Barnes, R.P. and Arthur, M.J. 1995 *United Kingdom Offshore Regional Report: the Geology of the Irish Sea*, British Geological Survey, Nottingham.

Jehu, T.J. 1909 'The glacial deposits of western Carnarvonshire', *Transactions of the Royal Society of Edinburgh* 47, 17-56.

John, B.S. 1970 'Pembrokeshire', in C.A. Lewis (Editor) *The Glaciations of Wales and Adjoining Regions*, Longman, London, 229-265.

Johnson, R.H. (Editor) 1985 *The Geomorphology of North-west England*, Manchester University Press, Manchester.

Kidson, C. and Tookley M.J. (Editors) 1977 *The Quaternary History of the Irish Sea*, Geological Journal Special Issue Number 7, Liverpool.

Knight, J. 1999 'Geological evidence for neotectonic activity during deglaciation of the southern Sperrin Mountains, Northern Ireland', *Journal of Quaternary Science* 14, 45-57.

Knight, J. 2001 'Glaciomarine deposition around the Irish Sea basin: some problems and solutions', *Journal of Quaternary Science* 16, 405-418.

Knight, J. 2003 'Evaluating controls on ice dynamics in the north-east Atlantic using an event stratigraphy approach', *Quaternary International* 99-100, 45-57.

Knight, J., Coxon, P. and McCabe, A.M. 2004 'Pleistocene glaciations in Ireland', in Ehlers J. and Gibbard, P.L. (Editors) *Quaternary Glaciations - Extent and Chronology: Part 1: Europe*, Elsevier, Amsterdam.

Knight, P.J. and Howarth, M.J. 1999 'The flow through the north channel of the Irish Sea', *Continental Shelf Research* 19, 693-716.

Lambeck, K. 1996 'Glaciation and sea-level change for Ireland and the Irish Sea since Late Devensian/Midlandian time', *Journal of the Geological Society of London* 153, 853-872.

Lambeck, K. and Purcell, A.P. 2001 'Sea-level change in the Irish Sea since the Last Glacial Maximum: constraints from isostatic modelling', *Journal of Quaternary Science* 16, 497-506.

Lewis, C.A. (Editor) 1970 *The Glaciations of Wales and Adjoining Regions*, Longman, London.

Lloyd, J.M., Shennan, I., Kirby, J.R. and Rutherford, M.M. 1999 'Holocene relative sea-level changes in the inner Solway Firth', *Quaternary International* 60, 83-105.

Maingarm, S., Izatt, C., Whittington, R.J. and Fitches, W.R. 1999 'Tectonic evolution of the southern – central Irish Sea Basin', *Journal of Petroleum Geology* 22, 287-304.

McCabe, A.M. 1987 'Quaternary deposits and glacial stratigraphy in Ireland', *Quaternary Science Reviews* 6, 259-299.

McCabe, A.M. 1996 'Dating and rhythmicity from the last deglacial cycle in the British Isles', *Journal of the Geological Society of London* 153, 499-502.

McCabe, A.M. 1997 'Geological constraints on geophysical models of relative sea-level change during deglaciation of the western Irish Sea Basin', *Journal of the Geological Society of London* 154, 601-604.

McCabe, A.M. and Clarke, P.U. 1998 'Ice-sheet variability around the North Atlantic Ocean during the last deglaciation', *Nature* 392, 373-377.

McCabe, A.M. and Clarke, P.U. 2003 'Deglacial chronology from County Donegal, Ireland: implications for deglaciation of the British-Irish ice sheet', *Journal of the Geological Society of London* 160, 847-855.

McCabe, A.M., Dardis, G.F. and Hanvey, P.M. 1984 'Sedimentology of a Late Pleistocene submarine-moraine complex, County Down, Northern Ireland', *Journal of Sedimentary Petrology* 54, 716-730.

McCabe, A.M. and Eyles, N. 1988 'Sedimentology of an ice-contact glaciomarine delta, Carey Valley, Northern Ireland', *Sedimentary Geology* 59, 1-14.

McCabe, A.M. and Haynes, J.R. 1996 'A Late Pleistocene intertidal boulder pavement from an isostatically emergent coast, Dundalk Bay, eastern Ireland', *Earth Surface Processes and Landforms* 21, 555-572.

McCabe, A.M., Knight, J. and McCarron, S.G. 1998 'Evidence for Heinrich event 1 in the British Isles', *Journal of Quaternary Science* 13, 549-568.

McCabe, A.M. and Ó Cofaigh, C. 1995 'Late Pleistocene morainal bank facies at Greystones, eastern Ireland: an example of sedimentation during ice marginal re-equilibriation in an isostatically depressed basin', *Sedimentology* 42, 647-663.

McCabe, A.M. and Ó Cofaigh, C. 1996 'Upper Pleistocene facies sequences and relative sea-level trends along the south coast of Ireland', *Journal of Sedimentary Research* 66, 376-390.

McCarroll, D. 1991 'Ice directions in western Lleyn and the status of the Gwynedd readvance of the last Irish sea glacier', *Geological Journal* 26, 137-143.

McCarroll, D. 1995 'Geomorphological evidence from the Lleyn Peninsula constraining models of the magnitude and rate of isostatic rebound during deglaciation of the Irish Sea Basin', *Geological Journal* 30, 157-163.

McCarroll, D. 2001 'Deglaciation of the Irish Sea Basin: a critique of the glaciomarine hypothesis', *Journal of Quaternary Science* 16, 393-404.

McCarroll, D., Knight, J. and Rijsdijk, K. (Editors) 2001 'The glaciation of the Irish Sea Basin', *Journal of Quaternary Science* 16, 391-506.

McCarroll, D. and Rijsdijk, K.F. 2003 'Deformation styles as a key for interpreting glacial depositional environments', *Journal of Quaternary Science* 18, 473-489.

Mitchell, G.F. 1960 'The Pleistocene history of the Irish Sea', *Advancement of Science* 17, 313-325.

Mitchell, G.F. 1963 'Moraine ridges across the floor of the Irish Sea', *Irish Geography* 4, 335-344.

Mitchell, G.F. 1972 'The Pleistocene history of the Irish Sea: second approximation', *Scientific Proceedings of the Royal Dublin Society* 4, 181-199.

Mitchell, G.F., Penny, L.F., Shotton, F.W. and West, R.G. 1973 *A correlation of Quaternary deposits in the British Isles*, Geological Society of London Special Publication 4.

Needham, T. and Morgan, R. 1997 'The east Irish Sea and adjacent basins: new faults or old?', *Journal of the Geological Society of London* 154, 145-150.

O'Cofaigh, C. and Evans, D.J.A. 2001 'Sedimentary evidence for deforming bed conditions associated with a grounded Irish Sea glacier, southern Ireland', *Journal of Quaternary Science* 16, 435-454.

Pantin, H.M. and Evans, C.D.R. 1984 'The Quaternary history of the central and southwestern Celtic Sea', *Marine Geology* 57, 259-293.

Ramster, J.W. and Hill, H.W. 1969 'Current system in the northern Irish Sea', *Nature* 224, 59-61.

Reade, T.M. 1893 'The drift beds of the Moel Tryfaen area of the north Wales coast', *Proceedings of the Liverpool Geological Society* 7, 36-79.

Richter, T.O., Lassen, S., Van Weering, T.C.E. and De Haas, H. 2001 'Magnetic susceptibility patterns and provenance of ice-rafted material at Feni Drift, Rockall Trough: implications for the history of the British-Irish ice sheet', *Marine Geology* 173, 37-54.

Rijsdick, K.F. 2001 'Density-driven deformation structures in glacigenic consolidated diamicts: examples from Traeth y Mwnt, Cardiganshire, Wales, U.K.', *Journal of Sedimentary Research* 71, 122-135.

Rijsdick, K.F., Owen, G., Warren, W.P., McCarroll, D. and Van Der Meer, J.J.M. 1999 'Clastic dykes in over-consolidated tills: evidence for subglacial hydrofracturing at Killiney Bay, eastern Ireland', *Sedimentary Geology* 129, 111-126.

Saunders, G.E. 1968 'Glaciation of possible Scottish re-advance age in north west Wales', *Nature* 218, 76-78.

Scourse, J.D., Austin W.E.N., Batemen, R.M., Catt, J.A., Evans, C.D.R., Robinson, J.E. and Young, J.R. 1990 'Sedimentology and micropalaeontology of glacimarine sediments from the central and southwestern Celtic Sea', in Dowdeswell J.A. and Scourse J.D. (Editors) *Glacimarine Environments: Processes and Sediments*, Geological Society Special Publication 53, 329-347.

Scourse, J.D. and Furze, M.F.A. 2001 'A critical review of the glaciomarine model for Irish sea deglaciation: evidence from southern Britain, the Celtic shelf and adjacent continental slope', *Journal of Quaternary Science* 16, 419-434.

Scourse, J.D., Hall, I.R., McCave, I.N., Young, J.R. and Sugdon, C. 2000. 'The origin of Heinrich layers: evidence from the H2 for European precursor events', *Earth and Planetary Science Letters* 182, 187-195.

Shakesby, R.A., Austin, W.E.N. and McCarroll, D. 2000 'Foraminifera from the glacigenic deposits at Broughton Bay, South Wales: evidence for glacimarine or terrestrial ice-sheet deglaciation of the Irish Sea Basin?', *Proceedings of the Geologists' Association* 111, 147-152.

Stephens, N. and McCabe, A.M. 1977 'Late-Pleistocene ice movements and patterns of Late- and Post-Glacial shorelines on the coast of Ulster, Ireland', in Kidson, C. and Tooley (Editors) *The Quaternary History of the Irish Sea*, Geological Journal Special Issue 7, Liverpool.

Stride, A.H.(Editor) 1982 *Offshore Tidal Sands – processes and deposits*, Chapman and Hall, London.

Synge, F.M. 1964a 'The glacial succession in west Caernarvonshire', *Proceedings of the Geologists' Association* 75, 431-444.

Synge, F.M. 1964b 'Some problems concerned with the glacial succession in south-east Ireland', *Irish Geography* 5, 73-82.

Synge, F.M. 1973 'The glaciation of south Wicklow and the adjoining parts of the neighbouring counties', *Irish Geography* 6, 561-569.

Synge, F.M. 1977 'The coasts of Leinster (Ireland)', in Kidson C. and Tooley M.J. (Editors) *The Quaternary History of the Irish Sea*, Geological Journal Special Issue 7, Liverpool, 199-222.

Syvitski, J.P.M. 1991 'Towards an understanding of sediment deposition on glaciated continental shelves', *Continental Shelf Research* 11, 897-937.

Tappin, D.R., Chadwick, R.A., Jackson, A.A., Wingfield, R.T.R. and Smith, N.J.P. 1994 *United Kingdom offshore regional report: the geology of Cardigan Bay and the Bristol Channel*, HMSO, London.

Thomas, G.S.P., Chester, D.K. and Crimes, P. 1998 'The late Devensian glaciation of the eastern Llyn Peninsula, North Wales: evidence for terrestrial depositional environments', *Journal of Quaternary Science* 13, 255-270.

Thomas, G.S.P. and Kerr, P. 1987 'The stratigraphy, sedimentology and palaeontology of the Pleistocene Knocknasillage Member, Co. Wexford, Ireland', *Geological Journal* 22, 67-82.

Thomas, G.S.P. and Summers, A.J. 1983 'The Quaternary stratigraphy between Blackwater Harbour and Tinnaberna, County Wexford', *Journal of Earth Sciences Royal Dublin Society* 5, 121-134.

Van Der Meer, J.J.M., Verbers, A.L.L.M. and Warren, W.P. 1994 'The micromorphological character of the Ballycroneen Formation (Irish Sea Till): a first assessment', in Warren W.P. and Croot D.G. (Editors) *The Formation and Deformation of Glacial Deposits*, Balkema, Rotterdam, 39-49.

Warren, W.P. 1985 'Stratigraphy', in Edwards K.J. and Warren W.P. (Editors) *The Quaternary History of Ireland*, Academic Press, London, 39-65.

Whittow, J.B. and Ball, D.F. 1970 'North-west Wales', in Lewis C.A. (Editor) *The Glaciations of Wales and Adjoining Regions*, Longman, London, 21-58.

Williams, G.D., Brabham, P.J., Eaton, G.P. and Harris, C. 2001 'Late Devensian glaciotectonic deformation at St Bees, Cumbria: a critical wedge model', *Journal of the Geological Society of London* 158, 125-135.

Wingfield, R.T.R. 1987 'Giant sand waves and relict periglacial features on the sea bed west of Anglesey', *Proceedings of the Geologists' Association* 98, 400-404.

Wingfield, R.T.R. 1989 'Glacial incisions indicating Middle and Upper Pleistocene ice limits off Britain', *Terra Nova* 1, 538-548.

Wingfield, R.T.R. 1995 'A model of sea-levels in the Irish and Celtic seas during the end-Pleistocene to Holocene transition', in Preece, R.C. (Editor) *Island Britain: a Quaternary perspective*, Geological Society of London Special Publication 96, 209-242.

Wright, W.B. 1937 *The Quaternary Ice Age*, 2nd edition, Macmillan, London.
Yokoyama, Y., Lambeck, K., De Deckker, P., JohnsonN, P. and Fifield, K. 2000 'Timing of the last glacial maximum from observed sea-level minima', *Nature* 406, 713-716.

INDEX

The index is divided into two parts: A: place-names; B: topics. Readers are advised to consult both parts and to cross reference in order to obtain the best results. Part A gives the major attributes of each place and includes a key to those attributes. The same location may have a number of different attributes. The same name may refer to two or more different locations, so the index should be used with care.

A: Place names

a = absolute date; b = beach; bf = blockfield; bh = borehole; bi = biogenics, coleoptera, palynology, etc.; c = cirque; d = drumlin; de = delta; e = esker; er = erratics; gl = glaciofluvial deposits; gt = glacial trough; f = alluvial fan; h = head; i = involutions; k = kame; ke = kettle; l = proglacial lake; m = moraine (end, terminal, ice-marginal, etc.); mc = meltwater channel (including fan-delta 'feeders', ice-marginal drainage and proglacial lake overflows); n = nivation cirque; o = outwash terrace/sandur; p = pingo remnant; pa = paraglacial; pt = protalus rampart; r = remains of archaeological/faunal interest (including human bones); s = section; st = striation; t = terrace; w = ice/sand wedge.

Aberaeron 86-7, 96-9
Aberarth (pa) 99
Aberdaron (h) 29, 35-7, 178, 182
Aber-Dulas-uchaf (s) 107
Aberdyfi 86
Aberedw (er) 101-2, 104-6
Aberlliw (m) 120, 123
Abergavenny (k) 108-11, 113-9
Abergwesyn (gt) 106
Abermawr (a) 145, 156-7, 185
Abersenni uchaf (m) 120, 123
Aberysgir (m) 120, 123
Aberystwyth 86-8, 91, 98-100
Adzor Bank (s) 132, 135-6
Aeron 86
Afan valley (er) 155
Afon Dugoed (m) 91
Afon Leri (gt) 88
Afon Soch (mc) 36
Afon Terrig (de) 48
Ailsa Craig (er) 179
Ailstone (bi) 79
Allt-y-Goed (h) 95
Alyn (mc) 42, 44-52, 58
Ammanford 153
Anglesey 3-4, 27-9, 33, 35-6, 96, 178-180, 182
Antarctic 4, 12, 15, 72
Aperley (t) 77
Aqualate Mere (ke) 68
Aran Fawddwy (c) 30, 32, 88
Arclid (bi) 59, 63, 65, 70
Arenig/Fawr (er) 30, 32, 41, 43
Arfon (s) 27-8, 33-4
Arrow (k) 101, 138-140, 142
Arthur's Stone (a) 10, 154
Atlantic Ocean 12-5
Australia 12

Avon valley (t) 20, 73-7, 79-80
Aymestrey (de) 140-1, 143

Bach Howey 102, 106, 124
Bacon Hole (a) 151, 153, 159-60, 163-4
Baginton River 74
Bala/Lake 30, 45
Ballybetagh (bi) 7
Baltic Sea 9
Banc-y-Warren (de) 92, 100, 161, 185
Bangor, Lake 54
Bardsey Island (st) 36
Barents Sea 188
Bar Hill (m) 59, 65
Barmouth 30, 32
Barnstaple Bay (er) 165-8, 175
Barra Fan (bh) 14, 160
Battlefield (bh) 55
Bay of Biscay 13
Bay View (bh) 149
Beddgelert 1
Berwyns 30, 32, 41
Bettws (k), 116
Bewdley (t) 79
Bex 1
Bickerton Hills (mc) 71
Birmingham 74-5, 78
Bishops Castle (p) 72
Bishopstone (ke) 138
Black Mountain 120, 126-7
Black Mountains 101-3, 107-8, 116, 118-9, 126-7
Blackwardine (de) 133
Blaenau-isaf (k) 120, 123
Blorenge (c) 113, 116
Boatside Farm (m) 104
Bodenham (t) 134
Bodfari (m) 45-7

Borras (bh) 50-1, 54
Borth 86
Boxbush (ke) 105
Brannam's Pit (st) 167-8
Brecon 120, 122-3, 126
Brecon Beacons (c) 6, 8, 11, 85, 87, 101-2, 108, 118, 120-2, 126-8
Bredwardine (m) 102-3, 142
Breidden 56
Bride (m) 178, 182, 184
Bridgnorth (mc) 59, 70, 73, 75, 79-80
Bridge Sollers (ke) 138-9, 141
Bristol (er) 168-9
Bristol Channel 59, 69, 81-2, 159, 166-7, 178-9
Brittany (er) 170
Broadhaven (b) 149, 152
Brobury Scar (m) 103, 139
Bromyard 129-30, 132-3
Bronllys (mc) 104, 122
Broomfield (m) 55-6
Broughton Bay (a) 145, 154, 178, 182
Brymbo (mc) 50
Bryn Arw (ke) 113, 116
Bryncir (k) 178, 185
Brynich (s) 120
Bryniau Gleision (bi) 113, 124-5
Buckland Hill 113-4
Builth Road (k) 107
Builth Wells 101-2, 104-7, 110, 153
Burghill (l) 132, 135-7
Burlton Court (s) 135, 138
Bushley Green (t) 76, 78
Butterslade (b) 152, 159
Bwlch 101, 109-10, 113-4
Bwlch yr Efengyl 102, 116
Byton (gl) 140

Cadair Idris (c) 30, 32, 39-40, 85, 87
Caerfagu (e) 107
Caergwyle (mc) 49-52
Caernarfon 2, 29, 34
Cambrian Mountains 85, 87-8, 100
Camel estuary 165-6
Canada, 160
Cannock Chase (er) 61
Cantref (mc) 120
Cape of Good Hope 12
Capel Curig 31-2
Capel-y-ffin 116, 127
Cardiff 6, 147, 155
Cardigan 85-8, 92, 156

Cardigan Bay 30, 86, 92, 96-7
Careg Yspar (i) 99
Carey (de) 181
Caribbean 14
Carmarthen 153
Carmarthen Bay 145-9
Carneddau 30-2, 40
Carnedd Dafydd 2, 31
Carnedd Llywelyn 2, 30-1
Carn Meini (er) 147
Castell y Gwynt (bf) 32
Castle Mound (d) 106
Caswell Bay (bi) 164
Cat Hole (r) 153
Cefn Bryn 10
Cefn Caves (r) 28, 43
Celtic Sea 13, 178, 180, 182
Cennarth (er) 156
Cerrig Gleisiad, see Cwm Cerrig Gleisiad
Cheadle 60
Chelford (bi) 59, 61-5, 70, 72
Cheshire 59 ff (Chapter 5)
Chester 42
Churchill (er) 78
Church Stretton (er) 133
Cilieni (m) 107, 120, 123-4
Cilmery (d) 106-7
Cilonw 103
Cippyn (mc) 92-3, 95, 154
Claerwen (w) 106, 108
Clarach (h) 99
Cledlyn valley (p) 98, 100
Clee Hill (er) 133
Clifford (o) 104
Clwyd, Vale of (m) 2, 28, 41-2, 45-7
Clwydian Hills 30, 41, 43-7
Clydach 113-5
Clync (m) 145, 161
Clynnog Fawr (m) 36, 186
Clyn-yr-Ynys (s) 100
Clyro (m) 102-5, 138, 142
Clytha Castle 117-8
Collington (mc) 133
Colorado Plateau 12
Colwyn Bay (er) 41, 45
Congleton (w) 59, 71-2
Conway 42, 58
Conwy valley (see also Vale of Conwy) (mc) 30, 32-3, 42-5
Corris 87
Cors Geirch (t/de) 36, 38

Index

Cors Geuallt (bi), 8
Cothi valley (l) 153
Cotswolds 74, 76
Court Hill (er) 166, 169
Coygan Cave (r) 153
Cradley Brook (l) 129-32, 144
Cradoc (k) 120, 122-3
Crai 120, 123, 127
Craig Maesglase (gt) 88
Craig y Fro (a) 120, 126-7
Craig-y-llyn 155, 162
Craig y March (st) 89-90
Creigiau (e) 155
Crewe (bh) 59, 64-5
Criccieth (ke) 38-9
Crickhowell (k) 102, 109-11, 113-4, 118
Cross Hands 153
Crossways (s) 133
Croyde (er) 167
Crymych (l) 154
Crugiau Cemaes (er) 156
Cwm Cerrig-gleisiad (c) 120, 126-7
Cwm Cewydd (m) 91
Cwm Claisfer 113-4
Cwm Coed-y-Cerrig 108, 116
Cwm Crew (c) 120, 125
Cwm Cynwyn (c) 120, 126
Cwm Du (n) 97-8
Cwm Gwerin (c) 10, 30, 32, 162
Cwm Idwal (c) 10, 30, 32, 162
Cwm Llwch (c) 120, 125
Cwm Oergwm (c) 120, 125-6
Cwm Saerbren 155, 162
Cwm Wysg-ganol (m) 120, 123
Cwm Wysg-uchaf (m) 120, 123
Cwm Ystwyth 88
Cwmyoy 116
Cwrt Newydd (p) 97-8
Cynon 155
Cynrig (mc) 120
Cynwil Elfed (l) 154

Dartmoor 166, 172, 174-5
Davis Strait 13
Dderw (ke) 104-5
Dean Brook 68
Dee River/valley 30, 32, 41-2, 44, 47, 50-2, 59, 64, 68-9, 74
Denbigh 41-2
Denbigh Moors 30, 43, 45
Derndale Hill (m) 138

Digedi 103
Dilwyn (ke) 138
Dimlington (a) 24, 65
Dinas Dinlle (s) 2, 34-5, 178, 182, 184-6
Dinmore Hill (gl) 136-7, 141
Dolauarth (pa) 99
Doldowlod 107
Dolgellau 30, 32
Dorrington (m) 55-7
Downton (mc) 143
Drefach (h), 97, 99
Druidston Haven (b) 149, 156
Dublin 7, 178, 183
Duhonw (m) 106
Dulas (k), 102, 107-8, 121-2
Dyfi 85-8, 97
Dyfi Hills 86-7, 91
Dyffryn Crawnon 124

Eastern Slade (b) 148-9, 152, 159
Ebbw Valley 155
Edwinsford (Rhydodyn) Bridge (l) 153
Elan (m) 106-7
Elidir Fawr (bf) 31-2
Ellesmere (m) 55, 58-9, 65
Elwy River/valley 28, 41-2, 45
English Channel (er) 166, 170
Eppynt 106-8, 122, 124, 153
Erddig (de) 51
Erwbeili (de) 107
Erwood 102, 106
Estavarney (o) 117-8
Ewenni (l) 145, 155
Exmoor 166, 172-6

Fan Fawr (c) 120, 125
Fan Gihyrych (c) 120, 127
Fan Hir (m), 120, 125-7
Farm Wood quarry, Chelford (bi) 63
Felindre 104-5
Felin-newydd (ke) 102, 104, 122
Fenni-Fach (t) 120, 123
Ffordd Las (l) 104-5
Fforest Fawr 6, 108, 120-8, 155
Ffrith Gorge (mc) 50
Firth of Clyde (er) 179
Fishguard (mc) 92, 156, 161
Fladbury (a/bi) 79
Forest of Dean 83
Four Ashes (a/bi) 59, 61-3, 70, 72, 79, 160
Four Crosses (o) 56-7

Franklands Gate (mc) 132, 134-5
Freiburg 1
Fremington Quay (st) 166-8
Freshwater West Bay (b) 149
Frome 130

Gaer (m) 104
Gafenni River 113, 115-6
Gallfaenan caves (r) 43
Garnedd Goch 30
Geneva, Lake 2
Gilfach-yr-Halen (s) 94-5, 99
Gilman Point (s) 149, 160
Gilwern (k) 109-10, 113-5
Gladestry Brook (er) 101
Glais (m) 145
Glandwr (Landore) (m) 145
Glangwye (m) 106
Glan-Irfon (d) 106
Glanllyn (gt) 107
Glanllynnau (ke) 27, 35, 38-9
Glasbury (k) 102-5
Glaslyn (c) 30
Gloppa Hill 41
Gloucester (t) 77, 79
Glyder Fach (bf) 31-2
Glyder Fawr 30
Glyderau massif 30, 32, 40
Glyn Tarell 126
Gnosall (mc) 68
Goat's Hole Cave (a) 153
Goldcliff (b) 149, 152
Golden Valley 101
Gospel Pass (see Bwlch yr Efengyl) 102, 116
Gott (k) 116
Gower 7, 10, 15, 29, 147-54, 160
Graig Ddu (pa) 99
Graig Fawr (gt) 155, 162
Great Orme (er) 41
Greenland 4, 12-5, 170, 176
Gresford (de) 51-3
Greystones (s) 184
Gribin Fawr (c) 91
Groesffordd (mc) 120
Grwyne Fawr (m) 102, 108, 113-4, 116
Grwyne Fechan (m) 102, 113-4, 116
Gulf of Mexico 14
Günz 4
Gwbert Caravan Park 95
Gwernyfed Park (mc) 104-5
Gwersyllt Park (c) 50

Halkyn Mountain (mc) 41, 45
Hangman's Cross (bh), 148
Harlech 28
Hay-on-Wye (m) 101-5, 116, 124, 130, 136-8, 142
Hendre (mc) 44-8
Heol Senni (k) 120, 123
Heol y gaer (mc) 104-5
Hereford 102-3, 107, 132, 134, 137-8, 141-2
Hereford Basin 130, 136-7, 140-3
Hereford Racecourse (m) 138
Herefordshire 129 ff (Chapter 9)
Hermitage (st) 116
Hills (bh) 148-9
Himalayas 12
Hirwaun (d) 145, 155
Holly Brook valley (mc) 131-2, 134
Holywell (t) 45
Honddu 102, 113, 116, 118
Hope 50-1
Hope Mountain 41, 48-50
Hope Quarry (o) 49-50
Horton (b) 148, 152, 156, 159-60
Hoyles Mouth (r) 153
Hudson Strait 13
Humber Brook/valley (l) 131-2, 134
Huntington (er) 101
Hunts Bay (b) 148, 150, 153, 160
Hyatt Sarnesfield (m) 103
Hydfer 123

Iceland 12, 176
Iller 4
Indian Ocean 12-3
Innsbruck 5
Ireland 85, 188
Irfon, River/Vale 106-7, 124
Irish Sea 11, 15, 25, 33-6, 38, 41-53, 55-9, 65-7, 85, 88, 91, 100, 147-8, 155-6, 159, 175, 177 ff, 185
Irish Sea Basin 177 ff
Ironbridge Gorge (mc) 58-9, 68-70
Isle of Man 178
Ivybridge 166, 174

Japan 12

Kattegat (Baltic Sea) 188
Kemys Commander (o) 117-8
Kenn (er) 147, 166, 169
Killiney Bay (s) 178, 184
Kington (k/ke) 7 103, 130, 136-7, 139-40, 142
Kyre Brook (mc) 132-3

INDEX

Lake Buildwas (l) 58
Lake District (er) 3, 41, 44, 65, 79, 85, 124, 127, 179
Lampeter (de) 88, 91
Land's End 170-1
Langland Bay (b) 152, 155-6
Laugharne 153
Laurentide (ice-sheet) 13
Lech 4
Leominster 129-30, 132
Leys 160
Lightmoor Channel (mc) 69
Lilleshall 68
Little Hoyle (r) 153
Little Lodge (mc) 104-5
Little Mill (o) 117-8
Liverpool Bay 41, 68
Llanarthne (m) 145
Llanbedr (m) 116
Llanbella (er) 101-3, 106
Llanberis (gt) 1, 30-2
Llanbradach (m) 145
Llanddona (b) 29
Llandeilo 145, 153
Llandudno 42
Llandudno Junction (er) 41
Llandudoch (bh) 93-4, 154
Llandyfaelog (Gwendraeth Fach) (m) 145
Llandysul (ke) 153-4
Llanellen 117-8
Llanfaredd (m) 106
Llanfechan House (k) 106
Llanfihangel Crucorney (m) 102, 109, 113, 116
Llanfoist (k) 113, 116
Llanfrynach 120
Llangadog (er) 147, 153
Llangattock 113
Llangollen 30, 32
Llangurig 9, 108
Llangynidr (m) 109-11, 113-4
Llangynog (er) 147, 153
Llanidloes 108
Llanilar 87
Llanigon (mc) 104-5
Llanilid (a/ke) 8, 99, 155, 160
Llanllwni (ke) 154
Llan-non (i) 96-9
Llanover (m) 111-2, 117-8
Llanpumpsaint (p) 162
Llanrhystud (h) 96-8
Llantilio Crossenny (l) 109, 119
Llantrithyd (er) 147

Llantood (Llantwyd) (mc) 154
Llanwnnen (de) 154
Llanwrthwl (gt) 107,
Llanwrtyd Wells 106-7
Llanybydder (de) 154
Llan-y-pwll (de) 54
Llawr dre-fawr (k) 107
Llay (bh) 51
Llechryd (l) 154
Lleine (h) 96
Lleyn 2-3, 7, 27, 32-3, 35-8, 85, 178-9, 182, 186
Llowes (er) 104
Llwynau Bach (mc) 104-5
Llwyn-gwyn (k) 116
Llwynpiod Farm (bh) 94
Llydaw (c) 30-1
Llynfi 101-2, 104-5, 113, 121-2, 155
Llyn Gwernan (a/l) 39
Llyn Llydaw (st) 40
Llyn Llygad (c) 88
Llyn Mire (bi) 124
Llyn Tegid 30
Llyn y Fan Fach (c) 145, 162
Llyswen (ke) 102, 104-6
Lodon Valley 131
Loggerheads (l) 48
Long Hole Cave (bi) 153
Longmynd 56
Lough Nahanagan (a/c) 184
Lourtier 1
Lower Broadheath (t) 77
Lower House Farm (m) 104
Lucton (d) 139-40, 142
Lugg, River/valley (t) 129-37, 139-42, 144
Lundy (er) 165-7, 175-6

Machynlleth 85-8, 91
Mad Brook 68
Madley (ke) 138
Maendy Pool (er) 147
Maes Carnog (m) 120, 123
Maesmynan (s) 47
Maesteg 155
Mallwyd (ke) 91
Malvern Hills 75, 129-30
Manorbier (b) 149, 156
Marford Quarry (de) 52-4
Marros (b) 149, 156
Marton Pool (bh) 56
Mathon (bi) 8, 74, 76, 129-31
Mathon Pit (l) 132

Mathry Road (w) 100
Melverly Lake (l) 58
Menai Straits 3
Menascin (mc) 120
Mendips 169
Merbach Hill 137-8
Merioneth 32
Mersey, River 59, 64
Merthyr Tydfil 126, 155
Migneint 30
Mlford Haven (b) 149, 156
Minchin Hole Cave (b) 150-2, 159-60, 163-4
Mindel 4
Minera (de) 50-2
Minllyn (ke) 91
Mochras (bh) 28
Moel Hebog 30
Moel Siabod 30-1
Moel Tryfan (a) 2-4, 34-5
Moelwyn Mawr (pt) 30, 40
Mold (o) 44, 47, 49-50
Monington (o) 96
Montford Bridge (bh) 55
Morfa Bychan (pa) 99
Morfa Conwy (er) 43
Morfa Mawr (h) 97
Morte Point 116
Mortimer's Cross (d) 137, 139-40, 142-3
Moss Pool (ke) 68
Mousecroft Lane Quarry (o) 67
Moylegrove (l) 92
Mullock Bridge (de) 145, 156
Mull of Kintyre 178-9
Mumbles 160
Musselwick (b) 156
Mwnt (s) see Traeth y Mwnt
Myarth (k) 113-4
Mynydd Du 11, 126, 145, 153
Mynydd Eglwysilian 155
Mynydd Illtyd (a, bi) 8 120, 124
Mynydd Llangattock 113-4
Mynydd Llangynidr 113-4
Mynydd Preseli (er) 85, 87, 147-8, 156
Mynydd Sylen 153
Mynydd-y-Bettws 13
Mynydd-y-Glyn 155

Nanhoron (mc) 36
Nannerch (ke,t) 46-7
Nant Cae-garw (p) 108
Nant Cymrun (m) 107
Nant Ffrancon (gt) 30-1, 40

Nant Llaniestyn (mc) 36
Nantmel 107
Nant Olwy (mc) 119
Nant Saethon (mc) 36
Nant-y-benglog (gt) 30
Nant y gwryd (gt) 30-2
Nant y Bwch 116
Nant-y-moch (gt) 88-90, 98
Nant yr arian (gt) 88
Nant Ysgallen (mc) 104-5
Nash Point (h) 153
Neath 145, 155, 161
Nelson (e) 155
Netherlands 9
Nevern (l) 92
Newborough (er) 33
Newbridge on Usk (t) 118-9
Newbridge-on-Wye 107
Newcastle Emlyn 153
Newchurch (m) 102, 106
New Inn (er) 147
Newport (Gwent) 111, 119, 155
Newport (Shropshire) 59
New Quay (h) 86, 97
Newton (er) 147, 160
Newton Fram (de) 132-3
New Zealand 160
North Atlantic Ocean 10, 12-3, 21, 24, 183
North Channel 178-9
North Sea 4, 188
Norton Canon (m) 103
Norton Court (mc) 134
Norwegian Sea 13

Oaker Wood (ke) 139
Ogwen Valley (gt) 30-2
Oldwalls (m) 145
Ogwr 155
Okehampton Common 174
Old Weir (s) 139
Onny 129
Orkney 4
Oricton (k, ke, m) 7, 103, 136-7, 139-40, 142
Oswestry 41-2, 58
Owlbury (p) 59, 72
Oxford 1
Oxwich Bay 149

Pacific Ocean 12-3
Padswood (ke) 44, 49
Painscastle 106
Pant y dwr (l) 154

INDEX

Pant-y-rhyg (mc) 153
Pant Ysgallog (mc) 120, 123
Park Pool (ke) 68
Parson's Bridge (mc) 88
Paviland (m) 145, 148-9
Pembroke 145 ff, 178-9
Pembrokeshire 9, 145 ff
Pedran (l) 154
Pencarreg (de) 154
Pencoed (er) 147, 155
Penmincae (k) 107
Pennines 59
Penpare (de) 92, 100
Penoyre 122
Penperlleni (m) 117-8
Penrhyncoch (gt) 88
Pentraeth (er) 33
Pentre (er) 147
Pentrecwrt (l) 154
Pentre'r-felin (m) 123
Pen-y-bryn (bi) 29, 94
Pen-y-genffordd 101
Pen yr Ole Wen 30-1
Pershore (t) 78
Plas Power 52
Plymouth Sound 174
Point of Ayre 44
Pont-aber-glass-llyn 1
Pontanwn (Gwendraeth Fach) (m) 145
Pontarsais (p) 162
Ponterwyd (mc) 88
Pontesbury 56
Pont-gareg (mc) 154
Ponthenri (Gwendraeth Fawr) (m) 145
Pontnewydd (Gwendraeth Fawr) (m) 145
Pontnewydd Cave (r) 25, 28, 41-4
Pont-Tyweli (m) 145, 153
Pontypool (o) 109-10, 117-8
Pontypridd (er) 155
Pont yr Angel (mc) 104-5
Poppit (b) 86, 95, 149
Port Eynon 148
Porthamel (er) 122
Porthcawl (er) 155, 160
Porth Ceiriad (h) 29
Porth Clais (b) 149
Porthleven (er) 166, 170
Porthmadog 30
Porth Neigwl (mc) 29, 35-6, 38
Porth Oer (b) 29
Porth Ysgo (h) 29
Portway (l) 132, 135-6

Prawle Point (er) 170
Preston gubbels (bh) 55
Pumlumon 6, 85-8, 91, 97-8, 101, 153
Pumpsaint (p) 162
Punchbowl (c) 172-3
Pwll Du/Head (s) 152, 159, 164
Pwllheli (mc) 35-6
Pwyll-glas (m) 45

Radyr (m) 145
Raglan 109, 119
Ragwen Point (b) 149
Red Wharf Bay (b) 29
Rest Bay 160
Rhayader (m) 107-8
Rheidol (gt) 86, 88-9
Rhiangoll (o) 101-2, 114
Rhinog (st) 30, 32, 40
Rhinog Fawr (st) 32
Rhiw-gam (c) 91
Rhondda Fawr 155
Rhosesmor (de) 47-8
Rhos goch (bi) 102, 106, 124
Rhosili (h) 148, 152, 162
Rhuddlan (de) 154
Rhyd-Owen-isaf (m) 120, 123
Rhydymwyn (Lake) (l) 47-8
Rhyl 28
Rhymni valley 155
Ridgeway (o) 76
Rissbury Bridge (mc) 134
Roch-Trefgarne 156
Ruabon Mountain 41
Rugeley (m) 60
Ruthin (d) 42, 45

St Athan (er) 147
St Bees 184
St Bride's Bay 145, 156
St Clears 145, 153
St Dogmaels (s) 93
St George's Channel, 178, 180
St Martin's (er) 171-2
St Maughans Plateau 129, 131-3
St Tudwal's Peninsula (mc) 32, 35-6, 38
Salhouse-Machynys (m) 145
Salisbury Plain 87
Salwarpe River (t) 69-70, 76, 79
Sandiway (a) 59, 65
Saredon Brook (bi) 61
Sarnau (m) 145
Sarn Badrig (m) 29, 32, 96

223

Sarn Galed (mc) 47
Sarn Gynfelyn (m) 29, 96
Sarn y Bwlch (m) 29, 96-7
Saunton (er) 167
Scandinavia 4, 7, 9, 160
Scilly Islands (er) 165, 170-2, 175-6, 180
Scotland 4, 12, 14, 16, 41, 79, 85, 92, 124, 156, 167, 179
Screen Hills (k,ke) 178, 184
Seaisdon-Stourbridge channel 69
Selsey (er) 170
Senni 120, 123-4
Sennybridge 120, 123
Severn, River/valley 25, 41-2, 55-9, 64, 68-70, 73-84, 129, 158-60
Shanganagh (bi) 184
Sheephouse (mc) 104-5
Shelton (bh) 55
Shetland 4
Shobdon (ke,s) 137, 139
Shoot Hill (bh) 55
Shawardine (m) 55-7
Shrewsbury 32, 41-2, 45, 55-9, 64, 66-7
Shropshire 59 ff (Chapter 5)
Sidestrand, Norfolk (bi) 169
Singret (de) 50-4
Slipper Stones (m) 174
Snowdonia 2, 7-8, 29-6, 40-1, 43, 85, 88, 127
Somerset 147
Somerset Levels (er) 165, 168-9, 175
Sourton Common (m) 174
South Hams 174
South Tawton Common (m) 174
Spitsbergen 13, 91
Stafford (bi) 70
Stansbach (a,ke) 139, 141
Start Point (er) 170
Staunton-on-Wye (m) 102-3, 137-9, 142
Stiperstones 56
Stoke Lacy (m) 132-3
Stoke Prior (mc) 132-3, 138
Stoke Prior School (de) 133
Stonehenge (er) 87, 147
Stour, River (t) 76-8
Stourbridge (t) 76, 78-9
Stourport (t) 79-80
Stretford (e) 140
Stretford Brook (l) 131-2, 134, 140
Stretton Sugwas (m) 102-3, 138-9, 143
Sugar Loaf 107, 113, 115
Sutton St Nicholas (mc) 132, 134-5
Svalbard see Spitsbergen

Swanlake (b) 149
Swansea 155
Swansea Bay 7, 145, 147, 155
Sydney (borehole) (bh) 64

Taff 155
Talbot Green (m) 145, 155
Talgarth 102, 104-5, 122
Talybont (m) 88, 108, 110, 113, 120
Tal-y-llyn (gt) 30, 85-6, 89, 91, 98, 101
Talley (m) 153
Tarrenhendre (c) 85, 88
Tarren yr Esgob (m) 116, 127
Taw estuary (er) 166-7
Taw valley (m) 174
Tawe 155
Teifi Pools (gt) 85-6, 88, 90
Teifi, River/valley (l,m) 85-8, 91-2, 96, 98, 100, 147, 153-4, 156, 161, 182, 186
Teme, River/valley (mc) 129-30, 143-4
Tenby (r) 145, 153
Tewkesbury (t) 76, 79
Thames, River/valley (t) 74, 81-3, 165
The Bryn (t) 111-2, 117
The Swamp (ke) 114-5
The Vauld (mc) 132, 134
Tillington (m) 138, 141
Timberline Wood (m) 138
Tippets Brook (mc) 140
Tirdonkin (Pontlassau) (m) 145
Tiruched (mc) 104-5
Tir-y-capel (m), 120, 123
Tonfannau (s) 96-7, 99-100,
Tonna (Aberdulais) (m) 145
Towy 153
Towyn (s) 96
Traeth Mawr (a,bi) 120, 124, 126
Traeth-y-mwnt (s) 94-5, 99, 185
Trebetherick (er) 165-7
Trefnant (m) 45
Tregaron (ke,m) 7, 88, 91, 98
Tregoyd 104-5
Tregoyd Mill (m) 104-5
Tregoyd Wood (mc), 105
Tregunter (m) 102, 104, 121-2
Trellys (a) 161
Tremadog Bay 32
Tremeirchion Caves (r) 24, 28, 44
Trent, River 59, 68, 74
Tre'r Domen (m) 120, 123
Treweren (m) 120, 123

Tretower (o) 113-4
Troed yr harn (k) 120-1
Trothy, River (l) 112, 119
Trwyn Maen Dylan (s) 186-7
Tumers Boat (ke) 138-9, 142,
Twyn y Beddau (mc) 101-3
Tyle Brychgoed-Bailea (m) 120, 123
Tyle-crwn (mc) 102, 121
Tynllwyn (mc) 120
Tywi, River 101, 106-7

Upper Lyde Pit (o) 132, 135
Upton Warren (a,bi) 70, 79
Usk 102, 109-11, 117-9
Usk, River/valley 7, 101-2, 108-20, 122-6, 128

Vale of Clwyd see Clwyd
Vale of Conwy 41
Vale of Ewyas 116, 118, 127
Vale of Ffestiniog 32, 38
Vale of Glamorgan 6, 155
Valois 1
Vennwood (mc) 132, 134
Vosges Mountains 4
Vrnwy (d) 41-2, 45, 56-7

Waen Oer 85
Waen Rydd 107
Walford Heath (bh) 55
Wallog (h) 99
Waun Gron (Pontarddulais) (m) 145
Waverley Wood Farm (bi) 76
Weaver, River 59, 69
Weeping Cross (ke) 67
Wellington Brook (t) 141
Welshpool (m) 42, 45, 56-7
Wenlock Edge 68
Wern Iago (mc) 104-5
West Angle Bay (b) 147, 149, 157
Western Slade (bh) 148-9, 152, 159-60
Westhope Hill 136
West Okement River (m) 174-5
Wexford (er) 178-9, 183
Wheeler, River/valley (ke) 41-2, 44-7, 58
Whitchurch (m) 51, 58-9, 65, 70
White Island (er) 171-2
Whitfield (m) 136-7
Whitford-Pembrey (m) 145
Wicklow 183
Wicklow Head 178-9
Wicklow Mountains (m), 8, 178, 184

Wigmore Basin (l) 140, 143
Windmill Hill (s) 132-3
Winsford Hill (c) 172-3
Wolston (s) 20
Wolverhampton (m) 25, 58-60
Wood Lane Quarry (m,s) 65
Woolridge (t) 76
Worcester (t) 74, 76, 80, 129
Worfe, River (o) 68-9
Worm's Head (b) 152
Wormsley Hill (k,ke) 137-8, 140-3
Wrekin 68
Wrexham (de) 42, 44-5, 50-5, 58-9, 65-6
Würm 4-5
Wye, River/valley (m,t) 6, 73, 75, 77-80, 83, 101-7, 109, 111, 121-2, 125, 127-30, 134, 136-9, 141-2, 144
Wylfa Head (er,s) 33

Yr Aryg (bf) 32
Yr Allt 121
Yr Eifl 35

B: Topics

Ailsa Craig microgranite 179
Allerod 8, 10, 14, 162
Alluvial fan 44, 46-7, 49, 110-1, 113-6
Alpine glaciations/model 4-7
Aminostratigraphy/dates 10, 35, 77, 79, 150-3, 156-7, 159, 161, 166-7, 169
Anglian Stage 19-20, 25, 69, 76, 103, 129-31, 138, 158, 166, 168
Astronomical forcing 10, 12, 17-8
Aurignacion artifacts 44
Axial tilt 5, 18

Bed, definition of 23
Biostratigraphy 21
Blockfield 30, 32
Bolling-Allerod 8, 10, 14, 162
Bond cycles 15
Boyn Hill Terrace 165
Bushley Green Terrace 77, 81

Catastrophism 2
Chattermarks 90
Chelford Interstadial 63-4, 70, 72, 153
Chelford Sands 63-6, 70-2
Chronology, glacial 3
Chronostratigraphy 21
Cirque 30-2, 39-40, 44, 88, 91, 111, 116, 118, 120, 124-8, 145, 162, 172-3

Cold-based ice 4, 11, 91
Cold/Warm Stages, classification of 20
Coleoptera 8, 39, 62-3, 79
Cosmogenic nuclide surface exposure/dates 10, 15, 32, 78, 147, 154, 160, 162, 184
Cromerian 20, 25, 76, 169
Crustal instability 28, 81-3, 159-60
Cryogenic mound, relict 72, 98
Cryoturbation 100

Dansgaard-Oeschger cycles 1, 181
Delta, marine, 27, 36, 38, 181-2
Delta 45, 47-55, 58, 92, 140, 143, 154, 184
Devensian, Early, Middle, Late 38, 60
Devensian Stage, stratotype 61
Dimlington Stadial 24, 65, 72, 98, 100, 129
Donau glaciation 6
Dropstones, 47, 54-5
Drumlin 33, 38, 41-2, 45, 57, 91, 102, 106-7, 137, 139, 140, 142-3, 155, 181-2

Eccentricity 12, 18
Elephant 43
Eister glaciation/stage 60
Equilibrium Line Altitude (ELA) 11
Erratics 1, 3, 6, 33, 36, 41, 43-4, 65, 78-9, 91-2, 101-7, 110-1, 133, 136, 147-9, 153, 155, 159, 165-8, 172
Esker 44-5, 47, 50, 60, 68, 107, 124, 140, 142, 155

Formation, definition of 23
Formations, see also Member, Bed
    Formation:
        Avon Valley 73, 75, 77-80
        Baginton 76
        Barnstaple Bay 24
        Brecknock 24, 146, 148-9, 153, 155-8, 160
        Caerwys 43-4
        Cardigan Bay 180
        Cradley Valley 131
        Croyde Bay 24
        Deganwy 43
        Elenid 24, 146-9, 153-8, 160-1
        Eryri 24
        Glamorgan 24, 146-7, 155, 157-8, 160
        Gwynllwg 43-4
        Hereford 24, 75, 131, 136
        Hiraethog 43
        Humber 130-1, 133
        Kenfig 43-4
        Kenn 24-5, 157-8
        Llanddewi 24-5, 146-9, 153, 157-60

Llandudno 41, 43
Lugg Valley 131
Mathon 73, 75-6, 129-31
Meirion 24, 43-4, 52
Nurseries 24-5, 73, 76, 80
Oakwood, 24-5
Penfro 24-6, 146-7, 149, 157-8
Pennard 146, 148-9, 151, 155-6, 158-60
Pontnewydd 24-5, 41, 43
Ridgacre 24-5, 73, 75, 77-8, 80, 159
Risbury 24-5, 75, 130-2, 134-6, 138
Ruabon 43-4, 52
St Asaph 24, 43-4, 146, 149, 153, 155-8, 160-1
St Martins 24, 180
St Marys 24
Seisdon 24-5
Severn Valley 73, 75, 77
Shrewsbury 24, 65, 67
Starpit 131
Stockport 24, 43-4, 52, 61-3, 66-7, 70-1, 73, 75, 77, 79-80
Tregaron 43-4
Tywi 43-4
Western Irish Sea 180
Wolston 24-5, 73, 75-7, 80, 158
Wye Valley 73, 75, 77, 131
Yew Tree 169
Member: 43, 75-6:
    Bacon Hole 151
    Bretford 75, 80
    Bullingham 75, 79
    Bushley Green 75, 77
    Coddington 131
    Colwall 131
    Cradley Silt 131
    Cropthorne 75, 78
    Dee 43-4, 52
    Eckington 75, 79
    Farm Wood 63-4
    Franklands Gate 131, 134
    Ham Green 131
    Hampton 75, 79
    Holme Lacy 75, 78
    Holt Heath 75, 79, 81, 83
    Horton 148-9, 153, 162
    Hunts Bay 154-5
    Kidderminster Station 75, 78-9, 83
    Kingsfield 131
    Knocknasillogue 184
    Kyre Brook 131
    Limbury 131

Llai 43-4, 52
Marden 131
Moreton on Lugg 131
Newton Farm 131
Pershore 75, 78
Portway 131
Power House 75, 80
Ruabon 43-4, 52
Scilly 180
Singret 43-4, 52
Slade 148, 152-3
Spring Hill 75, 77, 83
Stoke Prior 131, 133
Stoke Lacy 131, 133
Strensham 75
Sutton Walls 131
Thrussington 76
Thurmaston 76
Trewyddel 156
Wasperton 75, 79
Wheeler 43-4
Whitehouse 131-2
Woolridge 75-7, 83
Worcester 75, 79, 81, 83
Bed:
  Bacon Hole 2: 164
  Bacon Hole 4-7: 148
  Minchin Hole 1: 148, 159, 163
  Minchin Hole 2: 148, 163
  Minchin Hole 3: 148, 159, 163
  Minchin Hole 4-6: 148, 163
Freshwater pulses 13-4

Gelifluction 32
Gibbsite 32
Glacial protection 4
Glacial theory, rise of 1-4
Glacial trough, 28, 30-1, 88-9, 91, 106-7, 116
Glacial trough's end 88-9, 106-7
Global Conveyor 12-4
Global systems 12
Greenhouse gasses 15
Grezes litees 9
Günz glaciation 5-6, 60
Grooves 32
Gwynedd re-advance 38

Hay-Clyro moraine 102-5, 137-8, 142
Head 8, 29, 43, 92, 95, 97-9, 148-9, 160, 162, 165, 167
Heinrich Events 13-5, 19, 21, 24-6, 33, 158, 160-1, 170, 181, 183

Hell Bay Gravel 172
Hippopatamus 43, 79
Holocene 20, 110-2, 115, 118-9, 124, 126, 153, 165
Holt Heath Terrace 77
Hottinger Breecia 5
Hoxnian 20, 25, 131, 165-7
Human remains 28, 41, 153

Ice-berg, 2-4, 12-4, 133, 167, 170, 176, 181
Ice core 12, 15
Ice-scoured surface 32
Ice-sheet, models of 11
Ice wedge cast (see also sand wedge), 9, 38, 71-2, 99-100, 108, 127, 162
Interstadial 15
Involution 97-100
Ipswichian 20, 25, 29, 43, 60-2, 95, 98, 164, 168
Irish Sea Drift 180
Isostacy 11, 161, 170, 181-2

Kame/moraine/terrace 60, 66, 91, 102-7, 110-8, 120-4, 138, 142, 154, 184
Kettle holes/moraine 7, 34, 38-9, 44, 46-7, 57, 60, 66-8, 91, 102-6, 112-9, 122, 124, 137-9, 141, 144, 153-5, 184
Kidderminster Terrace 77-8
Kington-Orleton moraine 7, 137 ff

Landslide, 40, 91, 93, 127, 155
Last Glacial Maximum (LGM) 15, 19, 24-5, 29, 33, 63, 70-2, 87, 157, 160, 162
Laurentide ice-sheet, 13
Lithology 8
Lithostratigraphy (see also Formations, with Members, Beds) 6, 8, 21-6, 43, 75, 130-1, 146
Llansantffraid Interglacial 98
Loch Lomond Glaciation/Stadial 16, 27, 39-40, 72, 126-7
Loess, 148-9, 162
Lowestoft Formation 20

Magnetic polarity 11, 18
Main Terrace 69-70, 79
Mammalian fauna 159
Mammoth 28, 43, 44
Medial moraine, 32, 96
Meltwater channel 36, 38, 44-5, 47, 68-71, 86, 88, 92-3, 102-5, 118-22, 130, 133-5, 153-6
Member, definition of 23
Milankovitch cycles 6, 10, 17-8, 21, 26, 73, 81, 83
Millenial variations 12

Mindel glaciation 3-6, 60
Morphostratigraphy 5
Mutual climatic range, 8

Newer Drift 6-7, 59-60, 63, 65, 153
Nivation cirque 9, 98
Numerical dating techniques 10
Nunatak 30, 32, 40

Obliquity 12
Ocean circulation 12-3
Older Drift, 6, 60
Old Man Sandloess 172
Orbital forcing/variations 5-6, 10, 12-3, 17-8, 21
Outwash terraces, 4-5
Oxygen isotope, 10-1, 16-8, 24-5, 28-9, 41, 43, 58, 64, 73-83, 103, 129, 145, 157-61, 166-7, 169-70

Palaeoglaciology 11-2, 91, 126-7
Palynology 7-8, 124-7, 149
Paraglacial, 36, 99
Patella beach 95, 148ff, 163. See Minchin Hole Bed 3
Perched boulders 32
Periglacial environments/landscape 9, 32, 98-100, 108, 162, 172
Permafrost 9, 71-2, 100, 108
Piedmont glacier 124, 130, 136
Pingo/remnant 9, 97-9, 108, 127, 162
Plate tectonics, role of 12
Polythermal glacier 91, 100, 175
Pothloo Breccia 172
Power House Terrace 77
Pre-Anglian glacial limit 19, 26
Precession 12, 18
Pro-glacial lake 38, 47, 51-8, 69, 92-5, 105, 107, 118-9, 132-4, 136, 155, 161, 182-2. See also: Lake Aeron 92-3; Bromyard 133; Erwbeile 107; Ewenni 155; Humber, 133-4; Llanllwni 154, 161; Melverley 57-8; Rhydodyn 153; Rhydymwyn 47-8; Teifi 92-5, 153-6, 161, 185; Wigmore 143
Pro-talus rampart 40, 120, 126

Radiocarbon dating 9
Radiocarbon dates 24, 28, 34, 39, 44, 63, 65, 70, 79-80, 124-5, 127, 141, 145, 149, 153-7, 160-1, 172, 180
Raised beach 10, 39, 150 ff, 165, 167-8
Red Lady of Paviland 153
Riss glaciation 5-6, 60
River terraces 69-70, 76-7, 111-2, 115, 134, 141, 160
Roches moutonnée 4, 30, 90, 106
Rock basins/overdeepened 111 ff, 155

Saale glaciation/Stage 60
Sandur 44, 46-57, 67, 110-1, 113-9, 135-6
Sand wedges 9, 71-2, 78
Schotter see outwash terraces
Scilly Till 172
Sea level, rises/falls of 2-3, 18, 38, 81, 92, 95, 159-60, 162, 181
Sediment-landform assemblages 109 ff
Shells, amino-acid ratios 3
    dating of 2, 34-5
Shelly drift 2-4, 34, 43
Solar radiation, 5
Solifluction 9, 98, 108, 127, 162
Sorted circles 32
Sorted stripes 32, 97-8
Southern Uplands erratics 65
South Wales end-moraine 7, 156
Spring Hill Terrace 77-8
Stadial 15
Stratotypes 19-20
Striations 1-4, 6, 29, 32, 36, 40-1, 89-91, 116, 167-8, 186-7
Surging 14, 21

Tertiary erosion 28
Thermohaline circulation 12-3
Thermokarst 99
Till facies, 22-3
Tor 30, 32, 172
Trafalgar Square 20
Tregarthen Gravel 172
Trimline 30, 32

Uffington Terrace 70
Upton Warren Interstadial 62-3, 69-70, 72, 79, 153,
U-series dates 153, 159

Warm/Cold Stages classification of 20
Weichselian 9, 60
Wet based ice 91
Windermere Interstadial 39
Wolstonian 20, 26, 166, 169-70
Wolvercote Interglacial 166
Wolverhampton Line 25
Woolly rhinoceros 43,
Woolridge Terrace 76-7
Worcester Terrace 70, 77, 79-80
Würm glaciation 5-6, 60

Younger Dryas 10, 14, 16, 87, 98-100, 126-7, 162